Julius Weisbach, Gustav Herrmann

The Mechanics of Hoisting Machinery

Including Accumulators, Excavators, and Pile-Drivers

Julius Weisbach, Gustav Herrmann

The Mechanics of Hoisting Machinery
Including Accumulators, Excavators, and Pile-Drivers

ISBN/EAN: 9783744643580

Printed in Europe, USA, Canada, Australia, Japan

Cover: Foto ©berggeist007 / pixelio.de

More available books at **www.hansebooks.com**

THE MECHANICS

OF

HOISTING MACHINERY

INCLUDING

ACCUMULATORS, EXCAVATORS, AND PILE-DRIVERS

TEXT-BOOK FOR TECHNICAL SCHOOLS AND A GUIDE
FOR PRACTICAL ENGINEERS

BY

DR. JULIUS WEISBACH

AND

PROFESSOR GUSTAV HERRMANN

AUTHORISED TRANSLATION FROM THE SECOND GERMAN EDITION

BY

KARL P. DAHLSTROM, M.E.,

LATE INSTRUCTOR OF MECHANICAL ENGINEERING AT THE LEHIGH UNIVERSITY

WITH 177 ILLUSTRATIONS

New York

THE MACMILLAN COMPANY

LONDON: MACMILLAN & CO., Ltd.

1907

TRANSLATOR'S PREFACE

THE translation herewith presented to the engineering public
has been made from Professor Herrmann's revised edition of
Weisbach's great work on Engineering Mechanics. Of this
work several volumes are already familiar to English readers
through the translations completed successively by Messrs. Coxe,
Du Bois, and Klein, and treating respectively of Theoretical
Mechanics, Steam-engines and Hydraulics, and Machinery of
Transmission. The present section, however, has never heretofore appeared in English print, although its great value has
been recognised by all the above able translators, and by
institutions of learning all over the world. As the need of
suitable text-books for the more advanced courses in the
Mechanics of Machinery has long been felt at our technical
schools, the translator was induced to undertake the work of
editing the volume on Hoisting Machinery, in order to make a
beginning towards alleviating this need.

References in the text to previous volumes of Weisbach's
Mechanics, allude to the English translations unless otherwise
specified. The metric and English measurements are used, the
latter being enclosed in brackets.

The translator is indebted to Professor J. F. Klein of the
Lehigh University for much valuable aid in the preparation
of the work.

October 1893.

CONTENTS

PAGE

Introduction 1

CHAPTER I

Levers and Jacks 6

CHAPTER II

Tackle and Differential Blocks 38

CHAPTER III

Windlasses, Winches, and Lifts 74

CHAPTER IV

Hydraulic Hoists, Accumulators, and Pneumatic Hoists . 114

CHAPTER V

Hoisting Machinery for Mines 160

CHAPTER VI

Cranes and Sheers 228

CHAPTER VII

Excavators and Dredges 286

CHAPTER VIII

Pile-Drivers 307

INTRODUCTION

§ 1. THE object of all hoisting machinery is to *raise* and *lower* masses. Such apparatus is extensively used in extracting mineral products, in raising and distributing building materials, and in granaries, warehouses, machine-shops, and factories.

In all hoisting arrangements the motive power is expended in two ways: *first*, in performing *useful* work, namely, the product Qh of the weight Q of the load and the height h through which its centre of gravity is lifted ; and, *second*, in overcoming *wasteful* resistances. It is usually unnecessary to take into account the energy stored up in the lifted body by virtue of its velocity, since the arrangement is generally such that the velocity of the load when it reaches its destination is equal to zero.

When a hoisting apparatus is intended for intermittent service only, and absorbs but a small amount of power, it is usually operated by hand, as is the case with the various forms of jacks, hand-cranes, etc.

On the other hand, when the machine is to be in continual use, some other source of energy, chiefly steam power, is employed, which is the case in hoisting machinery for mines and nearly all large works of engineering of the present day.

With reference to economy of power, that hoist is generally considered the most efficient in which the ratio of hurtful to useful resistances is least. If no wasteful resistances were present all hoisting machines would be equally efficient as regards expenditure of energy, for according to the principle of virtual velocities we should have for every construction

$$Qh = Ps,$$

where s denotes the distance, in the direction of motion,

through which the point of application of the effort P has been moved while the weight Q has been lifted through a height h. Therefore, in the absence of friction, the theoretical effort, which in the following will be denoted by P_0, would be

$$P_0 = Q \frac{h}{s}.$$

Now let Ww denote the total work performed in overcoming the prejudicial resistances, while the weight Q is being raised or lowered through a height h—that is, let Ww represent the sum of the products obtained by multiplying each prejudicial resistance W into the distance w through which it has been overcome, then the expression for the work performed in raising the weight is

$$Qh + Ww = Ps,$$

or

$$P = \frac{Qh + Ww}{s}.$$

From this follows that, under all circumstances, the actual force P is greater than the theoretical force P_0, as long as the resistances W act in the same direction as the load Q, or as long as the force P acts to raise the load. This constitutes the *forward* motion as distinguished from the *backward* or *reverse* motion, which results when the weight Q is lowered; here the load Q is the cause of the motion, and P is to be considered as the resistance which acts to prevent acceleration.

Let (P) denote the force required to prevent acceleration in the latter case, and let (W)w denote the work performed in overcoming the wasteful resistances; then, for the reverse motion, the prejudicial resistances W are acting in the same direction as (P), and $(P)s + (W)w = Qh$; solving this equation we find

$$(P) = \frac{Qh - (W)w}{s},$$

a result which shows that (P) is less than the theoretical force P_0.

It is customary in hoisting as well as in other machines to designate the ratio

$$\eta = \frac{P_0}{P} = \frac{Qh}{Qh + Ww}$$

of the effort when the hurtful resistances are neglected to the effort actually exerted by the term *efficiency*. This ratio, which according to the above is always less than unity, represents that part or percentage of the effort P which is employed in performing the useful work. Similarly we speak of the efficiency (η) of the hoisting machines for the reverse motion, understanding by this the ratio of the actual effort (P) required when the load Q is being lowered, to the effort P_0 required when hurtful resistances are neglected, and then we have

$$\eta = \frac{(P)}{P_0} = \frac{Qh - (W)w}{Qh}.$$

This value also is always less than unity, and even becomes negative when $(W)w > Qh$. For the limiting case $(W)w = Qh$, we have (η) and consequently (P) equal to zero; in other words, this means that the forces of the machine are balanced without the additional effort (P). Therefore a negative value of (η), for which $(P) = (\eta)P_0$ is also negative, shows that during the lowering of the load Q an additional force (P) is to be applied, which will act in the same sense as Q.

A negative sign (η) may therefore be taken as an indication that the machine is capable of holding the load suspended without running backward when the application of motive power ceases, a property which under certain conditions belongs to the worm-wheel gearing. The efficiency η for the forward motion is of course always positive.

The introduction and use of this fraction to express the efficiency is a great convenience in practical calculations, for even in the most complicated machine the theoretical force

$$P_0 = Q\frac{h}{s}$$

can always be determined from the relations between the distances h and s, and thus the knowledge of the efficiency η immediately gives the actual effort required

$$P = \frac{P_0}{\eta}.$$

But the value of η can easily be computed, when we know the values of the efficiencies of the separate pieces and mechanisms of which the machine consists. In symbols let $\eta_1, \eta_2, \eta_3 \ldots \eta_n$ denote the efficiencies of the several parts of the train, then the efficiency of the whole machine is $\eta = \eta_1\,\eta_2\,\eta_3 \ldots \eta_n$.

Since the simple mechanisms of which all hoisting machines consist can be reduced to a very limited number of classes, as will be seen in the following, it is easily understood that a knowledge of the mean value of η for these simple mechanisms will in most cases lead to results sufficiently exact for practical purposes. As we proceed this will become more evident.

A general remark may here be made, however, in regard to the above mentioned self-locking hoisting apparatus, whose efficiency (η) in the reverse motion was found to be negative, namely, that their efficiency in the forward motion always is comparatively small. The truth of this statement will be evident from the following reasoning.

Assuming the limiting case $(\eta) = 0$, in which the machine is still self-locking, we shall have

$$Qh = (W)w.$$

For the forward motion we have the general expression

$$\eta = \frac{Qh}{Qh + Ww}.$$

Under the supposition that both values W and (W) are equal, and therefore that $Qh = Ww$, we have

$$\eta = \frac{Qh}{Qh + Ww} = \frac{Qh}{Qh + Qh} = \tfrac{1}{2}.$$

In this case, accordingly, we obtain the result that *the efficiency of a hoisting machine which automatically prevents the load from " running down," does not exceed 50 per cent under the most favourable circumstances*, and that it must be even smaller in all cases for which (η) is negative, that is to say $(W)w > Qh$.

As a matter of fact, however, the work performed in overcoming the wasteful resistances has a value Ww for the forward motion which is different from the value $(W)w$ for the reverse motion, inasmuch as the wasteful resistances are dependent upon the forces in action, namely, W upon P and Q, and (W) upon Q and (P). In general we can assume that W is larger than (W), because P always exceeds the value of (P), although in a few exceptional cases the resistance W may be even less than (W). Therefore, although the result obtained above is not strictly general, but holds under the supposition that the wasteful resistances do not consume more work during the reverse, than during the forward motion, we may, nevertheless, assume that in all cases the efficiency of hoisting mechanisms which automatically hold the load suspended without " running down " is small, and therefore their employment is, from economical reasons, not recommended in cases where great expenditure of power is required.

On the other hand, where they are not to be in continued operation, such machines are very useful, owing to the convenience with which they may be worked, and because there is no danger of their accidentally " running down."

NOTE.—Since the relation found above for the efficiency of a machine composed of several mechanisms, also holds good when it runs backward, we find, retaining the same notation, that

$$(\eta) = (\eta_1)\,(\eta_2)\,(\eta_3)\,\ldots\,(\eta_n).$$

From this equation we see that (η) cannot be negative, unless some one of the factors in the right hand member has the negative sign, and we conclude that a machine is capable of supporting the load automatically whenever any one of its mechanisms has this feature. It is hardly necessary to state that we are not to infer a positive value for (η), when two of the factors of the right hand member are negative, as the first of the mechanisms which have this self-locking feature will prevent the load from running down ; as regards the remaining mechanism, we can no longer speak of a *reverse*, only a forward motion in one direction or the other.

CHAPTER I

LEVERS AND JACKS

§ 2. **The Lever** is frequently used for lifting heavy loads by the application of a small effort. The height to which a load can be lifted by one sweep of the lever is usually very slight,

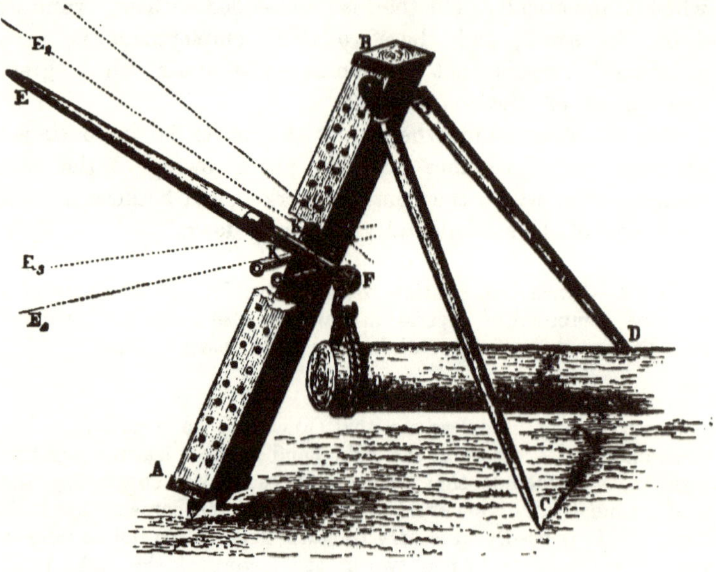

Fig. 1.

a few centimetres (one inch) being the average; therefore, in order to obtain a greater lift, it is necessary to raise the fulcrum of the lever gradually, while the load is being supported in some suitable manner, and then repeat the swinging motion. Various arrangements of *lever-jacks* have been constructed on this principle.

Fig. 1 represents a *German lever-jack*. ABCD is a tripod whose front leg AB is made with a slot allowing the lever EF to pass through.

This leg AB is provided with two rows of holes for the iron bolts K and L, which serve as fulcra for the lever. In order to lift the end Q of a log, for instance, on to a waggon, the end E of the longer arm is depressed to E_1 and the bolt L subsequently moved to L_1, then E is raised from E_1 to E_2,

Fig. 2.

and the bolt K inserted at K_1; afterwards E is again depressed from E_2 to E_3, and L_1 moved to L_2, etc.

Thus, by repeatedly forcing the lever EF up and down and alternately moving the bolts K and L, both lever and load are finally brought to the desired height.

In what is termed the *French* lever-jack, illustrated in Fig. 2, the bolts or pins are shifted automatically while the lever is being moved up and down. This is accomplished by suspending the lever EF through the links KM and LN on pins M and N, which are connected by the spring B. During the reciprocating motion of the lever the pins advance successively from one tooth to another on the toothed post AC.

A disadvantage in common to the two styles of jacks just described arises from the fact that after every lifting movement the load has to be lowered a certain distance during the return of the lever. Designating the angle of sweep of the lever by a, and the distances of the points of application of the load from the bolts K and L by $a = KF$ and $a_1 = LF$, we find that the lift for every forward sweep is $2a \sin \dfrac{a}{2}$, and that the load is lowered $2a_1 \sin \dfrac{a}{2}$ for every return movement—that is to say, the total lift is only

$$h = 2(a - a_1) \sin \frac{a}{2} = 2KL \sin \frac{a}{2} \,.$$

This height h is to be taken as the distance between teeth, or centres of holes in the same row of the post. Neglecting the wasteful resistances of pin friction, the useful work performed by either of these two jacks is found to be

$$\frac{a - a_1}{a} A = \frac{KL}{KF} A,$$

which is only a fraction of the total work A expended at the

Fig. 3.

lever handle E, and it becomes a smaller quantity in the same ratio as the distance between the bolts K and L is reduced.

The *Swedish* lever-jack, Fig. 3, is not subject to this disadvantage. In this apparatus each of the four uprights is provided with a row of holes for the pins K and L, and it is evident that the load, which rests at the middle of the lever

EE, and in the figure is represented by the beam DC, used for uprooting the stub S, can be raised by reciprocating either end of the lever.

This construction is frequently used, in modified form as in Fig. 4, in hoisting gears for operating lock gates. The lever EE is then movable about a pivot C, fixed in the post GG, each side of which it operates alternately on the bolts K and L, which are inserted in the slotted bar AB. This bar is guided in its vertical movement by the pivot and also by the

Fig. 4.

central portion FF of the lever EE, which is likewise slotted in order to prevent side movement.

The manner in which the reciprocating motion of a lever may be utilized, with the aid of a brake, for raising a load, may be learned from vol. iii. 1, § 172, of Weisbach's *Mechanics*.

Denoting the lever arm CK of the load by a, and that of the effort by b, we find the theoretical effort required for lifting the load Q from

$$P_0 = Q\frac{a}{b}.$$

If we now assume the radius of the journal C to be r, and that of the pin K r_1, and let ϕ represent the coefficient of journal friction, we get, after the lever has been swung an angle a, the following equation :

$$Pba = Qaa + \phi(Q + P)ra + \phi Qr_1 a,$$

when the pressure on the journal is expressed by $P + Q$, which gives the required lever effort

$$P = Q\frac{a + \phi(r + r_1)}{b - \phi r},$$

and accordingly the efficiency

$$\eta = \frac{P_0}{P} = \frac{1 - \phi\frac{r}{b}}{1 + \phi\frac{r + r_1}{a}}.$$

For the return motion, or when the load is being lowered, the effort (P) is obtained

$$(P) = Q\frac{a - \phi(r + r_1)}{b + \phi r},$$

and the efficiency

$$(\eta) = \frac{(P)}{P_0} = \frac{1 - \phi\frac{r + r_1}{a}}{1 + \phi\frac{r}{b}}.$$

To determine P by graphical methods, describe in Fig. 5, with C and K as centres, and with ϕr and ϕr_1 as radii, the corresponding friction circles, and take the directions of the reaction Z and the load Q tangential to these circles at c_1 and k_1 for the forward motion, and at c_2 and k_2 for the backward motion. If now EF, which is drawn parallel to these tangents, be made equal to Q, the length $k_1 l_1$, determined by drawing Fc_1, will represent the force P; and the length $k_2 l_2$, determined by producing Fc_2, will give (P).

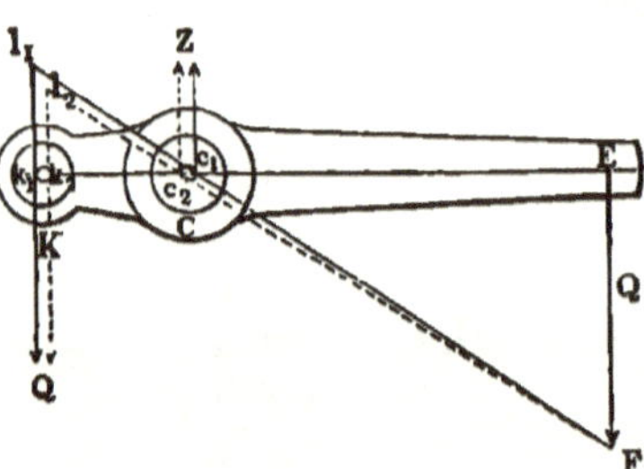

Fig. 5.

In hoisting machines the lever arm a of the load is much smaller than the arm b of the effort. The size of the journal is fixed by the principles of the strength of materials, and it is best to use steel in order to reduce the diameter and diminish the friction as much as possible. As an example, let us assume the very unfavourable case with respect to efficiency, that $r + r_1 = \dfrac{a}{2}$, and let the coefficient $\phi = 0{\cdot}08$; then, after making the supposition that $r = r_1$, we obtain the following table for different ratios $\dfrac{a}{b}$ of the lever arms.

TABLE OF THE EFFICIENCY OF LEVERS.

$$\left(r = r_1 = \frac{a}{4} \right)$$

$\dfrac{a}{b}$	$\frac{1}{2}$	$\frac{1}{3}$	$\frac{1}{4}$	$\frac{1}{8}$	$\frac{1}{16}$
$\eta =$	0·952	0·955	0·957	0·959	0·960

The small difference in the values of η will allow us to assume as a mean $\eta = 0{\cdot}96$, as in most cases $r + r_1$ is smaller than $\dfrac{a}{2}$.

§ 3. **Gear Wheels.**—The oscillating motion of the lever is subject to many inconveniences, and for this reason most hoisting machines are driven by a rotating shaft. To this end let us imagine the two lever arms of the jack to be replaced by two wheels AC and BC (Fig. 6) having the radii r and R respectively, and fixed to a shaft C. The smaller of the two is a

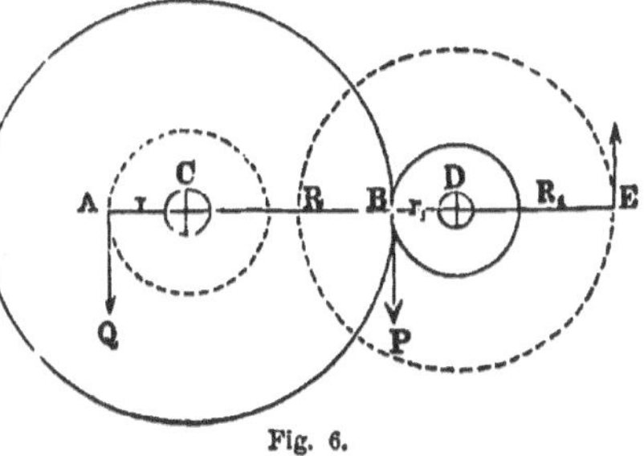

Fig. 6.

spur-wheel gearing with a rack which sustains the load Q.

If now a force P be allowed to act constantly at the circumference of the large wheel, the load Q may be raised without interrupting its motion. When the action of the wheel is to be greatly increased, however, its diameter would have to be made so large as to render this means of increasing the power inconvenient and difficult. In such cases we can make use of the following arrangement. Instead of allowing the driving force to act directly on CB, this wheel is provided with teeth and made to gear with a small pinion of radius $DB = r_1$, which is fixed to the shaft D. The latter may be driven by a crank DE, or another wheel of radius $DE = R_1$. A machine containing one such pair of wheels, as CB and DB, is said to be single-geared.

The action of this simple mechanism is to reduce the motion in the ratio of r_1 to R, for during one turn of the crank E, the point of application of the force moves through a distance $2\pi R_1$, while the shaft C is making only a fractional part $\dfrac{r_1}{R}$ of a revolution, and the load Q is lifted through a distance $2\pi r \dfrac{r_1}{R}$ only. In proportion as the velocity diminishes an increased load is practicable, for which we have the equation

$$Q 2\pi r \frac{r_1}{R} = P 2\pi R_1 ,$$

which gives

$$Q = P\frac{R}{r}\frac{R_1}{r_1} \quad \text{or} \quad P = Q\frac{r}{R}\frac{r_1}{R_1}$$

when all the wasteful resistances are neglected.

If the value of P proves inconveniently large, E may also be made into a spur wheel and allowed to gear into a pinion on a second shaft, which is acted upon by the driving force, and so on indefinitely. Thus, we distinguish windlasses by saying that they are *single, double,* or *treble-geared;* cases where more than three pairs of gears are used are to be counted as exceptions. While we may thus arbitrarily increase the *power* of the windlass, it is of course impossible to increase the *work done* during one revolution of the crank; on the contrary, with each additional pair of gears other wasteful resistances are introduced, which consume work and correspondingly reduce the efficiency of the whole machine.

Owing to the frequency with which cog-wheel gearing

occurs in hoisting apparatus, we deem it necessary to investigate more thoroughly the wasteful resistances which are occasioned by these mechanisms.

In Fig. 7 let the driver having the radius $CA = r$ gear with the larger wheel MA on the shaft M ; let Q be the resistance at A, acting in the direction of the tangent of the pitch circles and opposing the rotation of the shaft M ; then, in the absence of wasteful resistances, the force P required at the end of the lever arm $CB = R$ is given by $P_0 = Q\dfrac{r}{R}$.

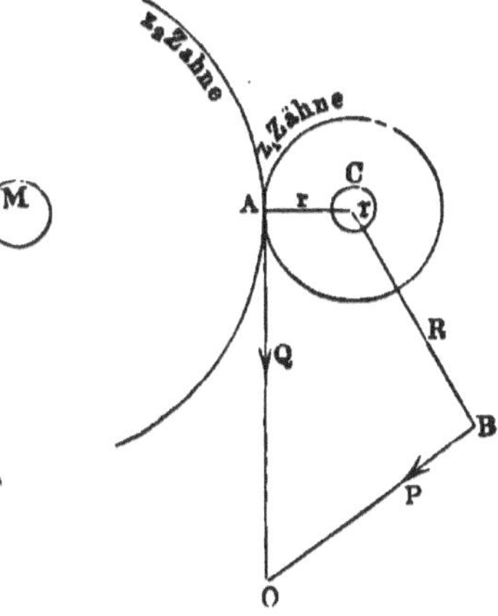

Fig. 7.

But wasteful resistances arise whenever we have relative motion between machine parts, hence in the present case they occur between the teeth at A and between the journal C and its bearing.

The friction between the teeth, according to iii. 1, § 79, is

$$Z = \pi\phi Q\left(\frac{1}{z_1} + \frac{1}{z_2}\right) = \tfrac{1}{3}Q\left(\frac{1}{z_1} + \frac{1}{z_2}\right),$$

when the slight deviation in the direction of the pressure between the teeth from the tangent AO to the pitch circles is neglected, and z_1 and z_2 denote the number of teeth in the wheels CA and MA. Putting

$$\phi\pi\left(\frac{1}{z_1} + \frac{1}{z_2}\right) = \zeta,$$

we find the force which is required at A to overcome the resistance Q of the shaft M to rotation, to be $P_1 = (1 + \zeta)Q$. Hence the *efficiency* of the pair of toothed wheels is

$$\eta = \frac{P_0}{P_1} = \frac{1}{1 + \zeta}.$$

The value of ζ increases as the number of teeth diminishes, and for the gears of windlasses, where z_2 is always considerably larger than z_1, it is materially affected by the number of teeth z_1 of the small wheel or pinion. In most cases this number ranges from 7 to 12; it seldom exceeds 20, and it is only in the simplest arrangements, waggon-jacks for example, that the number of teeth in the pinion is less than 7. Let us denote the velocity ratio $\dfrac{z_1}{z_2}$ of the wheels by ν, then we may write

$$\zeta = \tfrac{1}{3}\left(\frac{1}{z_1} + \frac{1}{z_2}\right) = \frac{0\cdot33}{z_1}(1 + \nu);$$

hence

$$\eta = \frac{1}{1 + \dfrac{0\cdot33}{z_1}(1 + \nu)} = \frac{z_1}{z_1 + 0\cdot33(1 + \nu)}.$$

This value of η can always be easily computed, but to obtain a rapid estimate it will be convenient to use the table below. This table gives the value of the efficiency

$$\eta = \frac{1}{1 + \zeta} = \frac{z_1}{z_1 + 0\cdot33(1 + \nu)},$$

for the number of teeth, $z_1 = 5,\ 6,\ 7,\ 8,\ 10,\ 12,\ 15,\ 20$, in the pinion, and for the velocity ratios, $\nu = 1,\ 0\cdot75,\ 0\cdot5,\ 0\cdot4,\ 0\cdot3,\ 0\cdot2,\ 0\cdot1$, and $\nu = 0$ for the rack.

TABLE OF THE EFFICIENCY OF TEETH.

$$\eta = \frac{z_1}{z_1 + 0\cdot33(1 + \nu)}.$$

$\nu = \dfrac{z_1}{z_2} =$	1	0·75	0·5	0·4	0·3	0·2	0·1	Rack.
$z_1 = 5$	0·883	0·897	0·909	0·916	0·921	0·927	0·932	0·938
6	0·901	0·912	0·923	0·929	0·933	0·938	0·943	0·948
7	0·914	0·923	0·934	0·938	0·942	0·946	0·951	0·955
8	0·924	0·932	0·941	0·946	0·949	0·953	0·957	0·960
10	0·938	0·945	0·952	0·956	0·959	0·962	0·965	0·968
12	0·948	0·954	0·960	0·963	0·965	0·968	0·971	0·973
15	0·957	0·963	0·968	0·970	0·972	0·975	0·977	0·978
20	0·968	0·972	0·975	0·978	0·979	0·981	0·983	0·983

(Number of teeth in the pinion.)

Froin this table we see that for the proportions most frequently occurring in hoisting gear, namely $z_1 = 8$ to 10, and $\nu = \frac{1}{3}$ to $\frac{1}{5}$, we may assume as a mean value

$$\eta = 0\cdot95 \text{ to } 0\cdot96.$$

In bevel gearing we may take the same value for the frictional resistances as for spur wheels having the same number of teeth, for the difference between the expressions

$$\frac{1}{z_1} + \frac{1}{z_2} \text{ and } \sqrt{\frac{1}{z_1{}^2} + \frac{1}{z_2{}^2}}$$

is unimportant in most cases.

To find what influence the friction of the journal C has on the efficiency, let r denote the radius of this journal, and Z the pressure on the bearing. Then the force P, acting with a leverage CB = R opposed to motion of the wheel MA, is found from the equation

$$PR = P_1 r + \phi Zr,$$

where P_1 designates the pressure in the circumference of the pinion of radius r.

If P_1 and P were in the same plane, the pressure Z would be given, as in the case of a bell-crank, by

$$Z = \sqrt{P^2 + 2PP_1 \cos a + P_1{}^2},$$

where a is the angle AOB between the directions of the forces. But the value of P deduced from this equation, even leaving out of consideration the inconvenient form of the expression thus found, would but imperfectly represent the actual circumstances of the case, and the result would be only approximately correct for the few exceptional cases in which the wheel AC and the crank or wheel BC are placed close together. As a rule it is customary to arrange the wheels AC and BC near the bearings of the shaft C. Therefore the supposition that the pressure in the bearings $Z = P + P_1$ would in most cases give a closer approximation to the truth; for we may conceive the pressure P as being taken up by one bearing and P_1 by the other. If there is any objection to this assumption, we should find that a calculation involving the determination of the reaction of each bearing would be a very lengthy one. Such a calculation would be of no practical importance, however, as in all probability there is greater error in accepting the coefficients of friction which are determined empirically than in neglecting the error which arises from the assumption $Z = P + P_1$. Moreover, we may add that when a crank CB is fixed to the shaft, the direction of the force P will continually change, causing the angle a to assume all values between $0°$ and $360°$, and hence the determination of P, referred to above as giving a more exact value, would only hold for a definite position of the crank.

Under the above supposition, therefore, we obtain from

$$PR = P_1 r + \phi(P + P_1)\mathfrak{r} :$$

$$P = P_1 \frac{r}{R} \frac{1 + \phi \frac{\mathfrak{r}}{r}}{1 - \phi \frac{\mathfrak{r}}{R}},$$

and since, in the absence of friction,

$$P_0 = P_1 \frac{r}{R},$$

we have, for the efficiency of the pinion shaft C,

$$\eta = \frac{P_0}{P} = \frac{1 - \phi \frac{\mathfrak{r}}{R}}{1 + \phi \frac{\mathfrak{r}}{r}}.$$

Introducing the ratio of the lever arms $\frac{r}{R} = \nu$, in this formula, we find that the expression for the efficiency can be written

$$\eta = \frac{1 - \nu \phi \frac{\mathfrak{r}}{r}}{1 + \phi \frac{\mathfrak{r}}{r}},$$

for which in most cases we may put approximately

$$\eta = 1 - (1 + \nu)\phi \frac{\mathfrak{r}}{r}.$$

The ratio $\frac{\mathfrak{r}}{r}$, that is to say, the ratio of the radius $\mathfrak{r}$ of the journal to the radius r of the smaller wheel, varies between $0·2$ and $0·4$ in windlasses; it is only in waggon-jacks which employ the smallest size of pinion that this ratio will exceed 0·4, and it is only in shafting that it will fall to 0·1 and below. In order to rapidly estimate the influence of the journal friction in windlasses and hoisting gear, we can refer to the following table, which gives for the ratio

$$\nu = \frac{r}{R} = \tfrac{1}{2}, \tfrac{1}{3}, \tfrac{1}{4}, \tfrac{1}{6}, \text{ and } \tfrac{1}{8},$$

and for

$$\frac{\mathrm{r}}{r} = 0\cdot5,\ 0\cdot4,\ 0\cdot3,\ 0\cdot2,\ \text{and } 0\cdot1,$$

the values of the efficiency,

$$\eta = 1 - (1 + \nu)\phi\frac{\mathrm{r}}{r},$$

the coefficient of journal friction being assumed to be $\phi = 0\cdot08$. According to this table the efficiency of the shaft ranges from 0·940 to 0·991, and for the common gears used in windlasses, corresponding to $\frac{r}{R} = \tfrac{1}{4}$ and $\frac{\mathrm{r}}{r} = 0\cdot3$, we may assume the efficiency to be about 0·97.

If we wish to determine the efficiency of the combination, considering both the friction of the teeth and that of the journal, we must put

$$\eta = \eta_1\eta_2 = \frac{1 - \phi\dfrac{\mathrm{r}}{R}}{1 + \zeta\left(1 + \phi\dfrac{\mathrm{r}}{r}\right)}.$$

TABLE FOR THE EFFICIENCY OF GEAR SHAFTS.

$$\eta = 1 - (1 + \nu)\phi\frac{\mathrm{r}}{r}.$$

$\nu = \dfrac{r}{R} =$	$\tfrac{1}{2}$	$\tfrac{1}{3}$	$\tfrac{1}{4}$	$\tfrac{1}{6}$	$\tfrac{1}{8}$
Ratio of $\dfrac{\mathrm{r}}{r}$ 0·5	0·940	0·947	0·950	0·953	0·955
0·4	0·952	0·957	0·960	0·963	0·964
0·3	0·964	0·968	0·970	0·972	0·973
0·2	0·976	0·979	0·980	0·981	0·982
0·1	0·988	0·989	0·990	0·991	0·991

Therefore for the most common proportions we may take a mean value of

$$\eta = 0{\cdot}95 \times 0{\cdot}97 = 0{\cdot}922,$$

or in round numbers $0{\cdot}92$. For proportions differing materially from the above we can find the exact value from the general formula. We may add that the efficiency (η) for the reverse motion is found from the same formula by prefixing the contrary sign to the terms containing ϕ, for, during the reverse motion, all the resistances act opposite to the directions they possess during the forward motion. Hence for the reverse motion

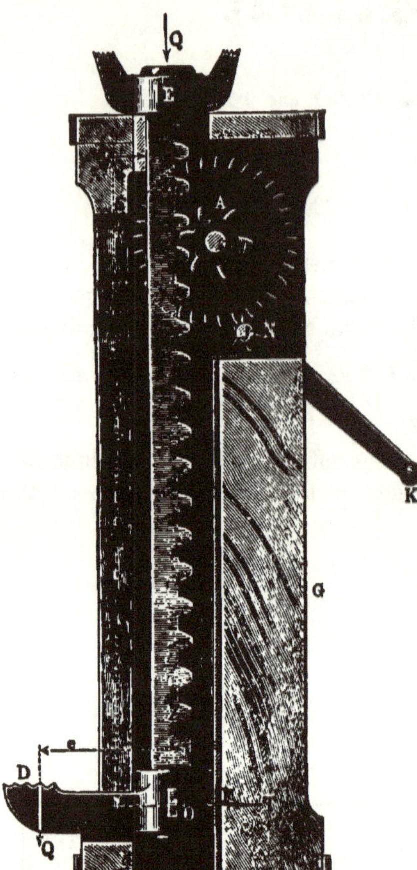

Fig. 8.

$$(\mathrm{P}) = (1 - \zeta)\mathrm{Q}\frac{r}{\mathrm{R}}\frac{1 - \phi\frac{\mathfrak{r}}{r}}{1 + \phi\frac{\mathfrak{r}}{\mathrm{R}}},$$

and the efficiency,

$$(\eta) = \frac{(\mathrm{P})}{\mathrm{P}_0} = \frac{(1 - \zeta)\left(1 - \phi\frac{\mathfrak{r}}{r}\right)}{1 + \phi\frac{\mathfrak{r}}{\mathrm{R}}}.$$

The values of (η) calculated from the formula differ so little from η that the efficiencies for the forward and reverse motion of such a train as we are now considering may be taken as equal. It will be shown in the following pages that such equality by no means exists in all mechanisms.

§ 4. **Rack and Pinion-Jack.**—For raising loads to small

heights the jack, worked by a pinion and rack, is largely used in practice. Its arrangement is evident from Figs. 8 and 9.

Fig. 8 represents a jack in the form in which it is employed in expediting building operations and in the erecting‑shops, while, for placing heavy pieces of work in the lathe the jack illustrated in Fig. 9 is used. In Fig. 8 the load Q acts either in the axis of the rack at E or on a claw D. In the latter case the action Q gives rise to friction at the bearing surfaces F and H. The pinion A, which has but a small number of teeth (5 to 8), meshes with the rack, and receives its motion directly from a crank on the shaft C when the load is small.

With greater loads the wheels B and N and a shaft J are inserted, and a crank JK is fixed to the latter. A ratchet wheel on this shaft, held by a pawl fixed to the frame, prevents the load from running down when the power is removed from the crank.

In order to ascertain the force P required to lift the load Q, we must first determine the friction at the bearing surfaces F and H, Fig. 8. Let T represent the equal reaction of the guides F and H, the point of applications being assumed at the middle of the surfaces in contact; let l represent the vertical distance between these horizontal forces; e the horizontal distance between the load Q and the lifting force Q_1, which acts

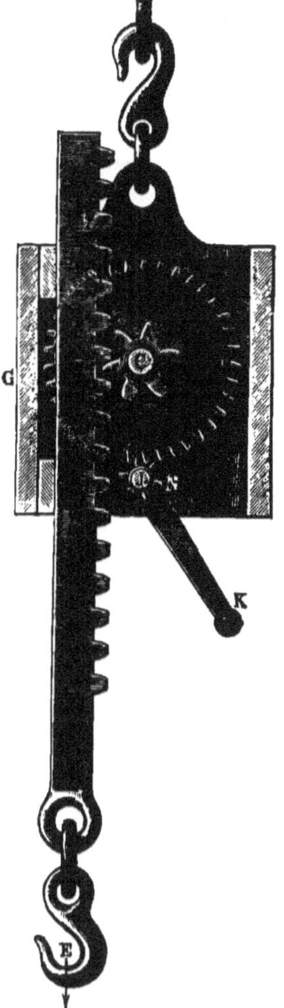

Fig. 9.

in the pitch-line of the rack; and, finally, let c denote the *algebraic* sum of the distances of the guiding surfaces F and H from the point O, which represents the intersection of the lines

of action of the force Q_1 and the lower reaction T, then the
equation of moments about O as a centre is

$$Qe = Tl \mp \phi Tc,$$

in which the upper sign corresponds to the lifting, and the
lower to the lowering of the load. Using the upper sign we
obtain

$$T = Q\frac{e}{l - \phi c},$$

so that the total friction of H and F equals

$$2\phi T = 2\phi Q\frac{e}{l - \phi c}.$$

This gives for the lifting force exerted by the rack

$$Q_1 = Q + 2\phi T$$

$$= Q\left(1 + 2\phi\,\frac{e}{l - \phi c}\right);$$

on the other hand, when the load is lowered the force exerted
by the rack on the pinion A is given by

$$(Q_1) = Q - 2\phi T$$

$$= Q\left(1 - 2\phi\,\frac{e}{l + \phi c}\right).$$

To balance this force Q_1, friction of teeth and journals being
neglected, would require the application of a force

$$Q_1\frac{r_1}{R_1}\frac{r_2}{R_2}$$

to the crank $JK = R_2$, where r_1 and r_2 denote the radii of the
pitch circles of the wheels CA and JN, and R_1 the radius of
the wheels CB. If η_1 denotes the efficiency of the rack and
pinion CA and the shaft C, and η_2 the efficiency of the pinion
JN and its shaft J, the required driving forces will be,

$$P = \frac{1}{\eta_1\eta_2}Q_1\frac{r_1}{R_1}\frac{r_2}{R_2} = \frac{1}{\eta_1\eta_2}\left(1 + 2\phi\,\frac{e}{l - \phi c}\right)Q\frac{r_1}{R_1}\frac{r_2}{R_2}.$$

Since, without wasteful resistances,

$$P_0 = Q\frac{r_1}{R_1}\frac{r_2}{R_2},$$

the efficiency of the jack for the forward motion becomes

$$\eta = \frac{P_0}{P} = \eta_1 \eta_2 \; \frac{1}{1 + 2\,\phi\,\dfrac{e}{l - \phi c}} = \eta_1 \eta_2 \eta_3,$$

if the value

$$\frac{1}{1 + 2\phi\,\dfrac{e}{l - \phi c}} = \eta_3$$

is regarded as the efficiency of the prismatic guides of the rack.

The force (P) which must be applied to the crank to prevent the running down of the load is determined in the same manner, and found to be

$$(P) = (\eta_1)(\eta_2)\left(1 - 2\phi\,\frac{e}{l + \phi c}\right) Q\,\frac{r_1}{R_1}\,\frac{r_2}{R_2} = (\eta_1)\,(\eta_2)\,(\eta_3)\,P_0,$$

where

$$1 - 2\phi\,\frac{e}{l + \phi c} = (\eta_3).$$

EXAMPLE.—Let $Q = 400$ kilograms [882 lbs.], $l = 0\cdot40$ metre [15·75 in.], $e = 0\cdot080$ metre [3·15 in.], and $c = 0\cdot060$ metre [2·36 in.], then assuming the coefficient of friction (sliding) to be $\phi = 0\cdot15$, the lifting force exerted by the rack is

$$Q = 400\left(1 + 0\cdot30\,\frac{80}{400 - 0\cdot15 \times 60}\right) = 424\cdot5\ \text{kg. [936 lbs.]}$$

and the force required for lowering is

$$(Q_1) = 400\left(1 - 0\cdot30\,\frac{80}{400 + 0\cdot15 \times 60}\right) = 376\cdot7\ \text{kg. [830 lbs.]};$$

hence the efficiency of the guides for lifting is

$$\eta_3 = \frac{400}{424\cdot5} = 0\cdot943,$$

and for lowering

$$(\eta_3) = \frac{376\cdot7}{400} = 0\cdot942.$$

Let each pinion have 6 teeth, and the wheel CB 36, then from the table on page 14 the efficiency of the rack is 0·948, and for the pair of wheels 0·940. If $r_1 = 30$ mm. [1·18 in.], $r_2 = 25$ mm. [0·98 in.], $R_1 = 150$ mm. [5·91 in.], and $R_2 = 200$ mm. [7·87 in.], then, for a ratio of the journals

$$\frac{r_1}{r_2} = \frac{r_2}{r_2} = 0\cdot4,$$

the efficiency of the shaft C, according to table, page 17, is

$$0\text{·}962 \qquad \left(\nu = \frac{r_1}{R_1} = \frac{30}{150} = 0\text{·}2\right),$$

and the efficiency of the shaft J is

$$0\text{·}964 \qquad \left(\nu = \frac{25}{200} = \tfrac{1}{8}\right).$$

Consequently the resulting efficiency of the shaft C, together with the rack and pinion, is

$$\eta_1 = 0\text{·}948 \times 0\text{·}962 = 0\text{·}912,$$

and that of the shaft J is

$$\eta_2 = 0\text{·}940 \times 0\text{·}964 = 0\text{·}906.$$

This gives for the effort required

$$P = \frac{1}{0\text{·}912} \times \frac{1}{0\text{·}906} \times 424\text{·}5 \times \frac{30}{150} \times \frac{25}{200} = 12\text{·}85 \text{ kg. } [28\text{·}3 \text{ lbs.}]$$

As the theoretical driving force is

$$P_0 = 400 \times \frac{30}{150} \times \frac{25}{200} = 10 \text{ kg. } [22 \text{ lbs.}],$$

the efficiency of the machine is

$$\eta = \frac{10}{12\text{·}85} = 0\text{·}778, \text{ or nearly } 78\%.$$

We also find

$$\eta = \eta_1 \eta_2 \eta_3 = 0\text{·}912 \times 0\text{·}906 \times 0\text{·}943 = 0\text{·}778.$$

This comparatively small value of η is mainly due to the small number of teeth (6), and the comparatively large size of the journals. To lower the load requires the application of a force

$$(P) = 0\text{·}912 \times 0\text{·}906 \times 376\text{·}7 \times \frac{30}{150} \times \frac{25}{200} = 7\text{·}78 \text{ kg. } [17 \text{ lbs.}].$$

to the crank.

The application of graphical methods for determining the force P and the efficiency η of hoists, and indeed of all mechanisms, is to be recommended as simpler than the numerical calculation, particularly when drawings of the machine are at hand.

The simple principles underlying such a graphical investigation follow directly from the character of the *angle of friction,* and are briefly stated in the appendix to vol. iii., part 1, Weisbach's *Mechanics* (see also *Zur graphischen Statik der Maschinengetriebe,* von Gustav Herrmann).

The diagram for the case under consideration is obtained in the following manner:—Let C and J, Fig. 10, be the axes of the shafts, a and n the points of contact of the pitch lines, b and e the surfaces for guiding the rack; let the load $Q = 1o_1$ act at D, and the required driving force at K. We draw, in the first place, the directions of the forces and the reactions of

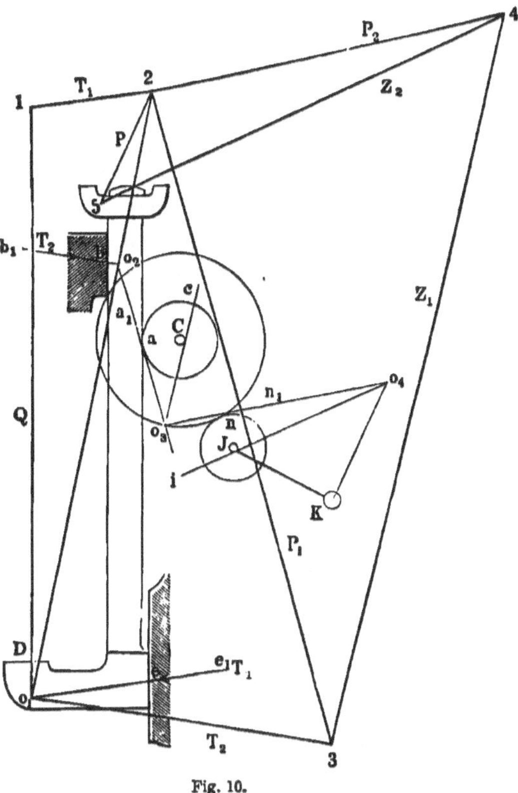

Fig. 10.

the guides and bearings. Assume the reactions T_1 and T_2 at e and b to make an angle with the normal to the guides equal to the angle of friction ρ—that is, draw them in the directions $e_1 e$ and $b_1 b$. It is here unnecessary to approximate by taking the direction of the pressures between the teeth at a and n normal to the line of centres—an approximation which is made in every calculation, but the direction of the pressure is to be drawn making an angle (about 75°) with the line of centres, equal to that made by the actual line of action of the teeth. But the effect of the friction of the teeth is to transfer the theoretical line of action (which passes through the points of contact a and n of the pitch circles) parallel to itself through a distance $\zeta = \phi \frac{t}{4}$, so that the leverage of the driving force relatively to the axis of the driver is increased by this amount. Then will $o_3 n_1$ represent the line of action of the pressure P_1 exerted between the teeth of the rack and pinion, and $o_4 n_1$ the line of action of the pressure P_2 between the teeth of the wheels. With the radius ϕr draw the friction-circles about the points C and J; then the directions of the reactions of the journals Z_1 at C and Z_2 at J are represented by the respective tangents $o_3 c$ and $o_4 i$ to these friction circles, and passing through the intersection o_3 of the forces P_1 and P_2 and the intersection o_4 of the forces P_2 and P. The shaft at c is acted upon by the three forces P_1, P_2, and the reaction Z_1, and the shaft J by P_2, P, and the reaction Z_2. After drawing these lines of action, the polygon of forces $o_1 12345$ is easily obtained from the load $1 o_1 = Q$ by drawing

$$12 \| e e_1 ; \quad o_1 2 \| o_1 o ,$$

(in the figure the latter lines coincide—that is to say, Q is measured from the point of intersection o_1); further draw

$$o_1 3 \| b_1 b ; \quad 23 \| o_2 o_3 ; \quad 34 \| o_3 c ; \quad 24 \| o_3 o_4 ; \quad 45 \| o_4 i, \text{ and } 25 \| o_4 K.$$

The length 25 represents the magnitude of the driving force P exerted on the crank K, and is drawn on the same scale as Q, and at the same time the sides of the force polygon represent the pressures P_1 and P_2 between the teeth of the gears, the reactions T_1 and T_2 of the guides, and the reactions Z_1 and Z_2 of the bearings. These forces can now be used for determining the dimensions of the teeth, journals, bearings, parts of the frame, etc.

The method here shown by a single example remains essentially the same for all kinds of mechanisms; hence we shall only exceptionally give such full details in the following pages. By constructing a diagram sufficiently large we obtain a degree of accuracy which suffices for every case. As regards the accuracy of this method, compared with that of computation, hitherto exclusively used in practice, we may say that with the graphical method it is unnecessary to make any approximate assumption, such, for example, as that the pressure of the teeth is perpendicular to the line of centres. It is only necessary to add that, for the determination of the vertical force P_0, the diagram is constructed in the same manner by taking the angle of friction, the radii of the friction circles, and the quantity ζ all equal to zero, consequently the reactions T normal to the bearing surfaces. For the *reverse* motion of the jack the diagram is altered only in this par-

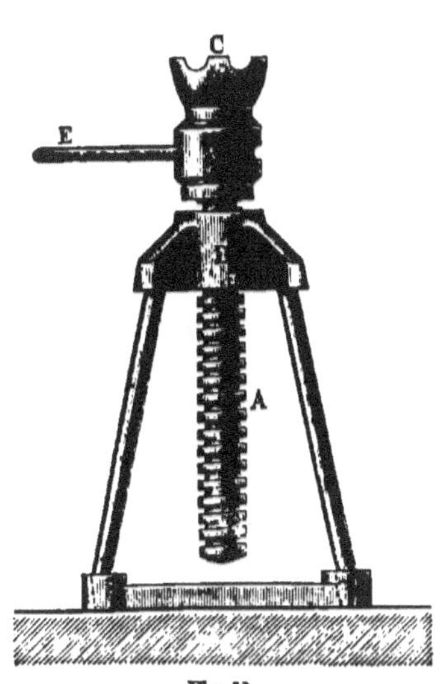

Fig. 11.

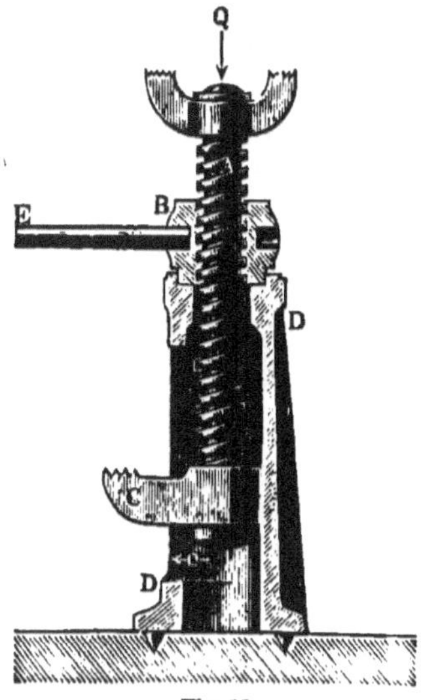

Fig. 12.

ticular; the directions of the reactions, etc., are inclined to the normals of the bearing surfaces at an angle equal to the angle of friction, but in a contrary direction to that drawn for the forward motion, and the reactions of the bearings are represented in direction by the other of the two possible tangents touching the friction circles.

§ 5. **Jack Screws.**—For lifting greater loads to small heights screws with square threads are frequently used, and by turning the screw or its nut, motion is imparted to the load. In the simple jack, Fig. 11, the screw A is made to advance in the fixed nut D by turning the lever E, thus lifting the load which rests on the head C, while in the lifting jacks represented in Figs. 12 and 13 the advance of the screw is produced by turning the nut. On the other hand, Fig. 14, which is used to lift a rail resting at B on the lever EC, is worked by turning the screw A, whose nut forms a part of lever EC. Also in the jacks employed for lifting locomotives for the purpose of removing or replacing axles, etc., the screw turns and the nut advances. Such jacks, which are always employed in pairs, consist, when constructed according to Fig. 15, of a vertical screw A connected with the wooden frame L by the box D and the step C. By employing two pairs of gears FG and HJ the rotation communicated to the crank K is imparted with reduced velocity to the screw, compelling the nut M to slide up between the guides on the frame L; the result is the lifting of the cross-beam T and its load (loco-

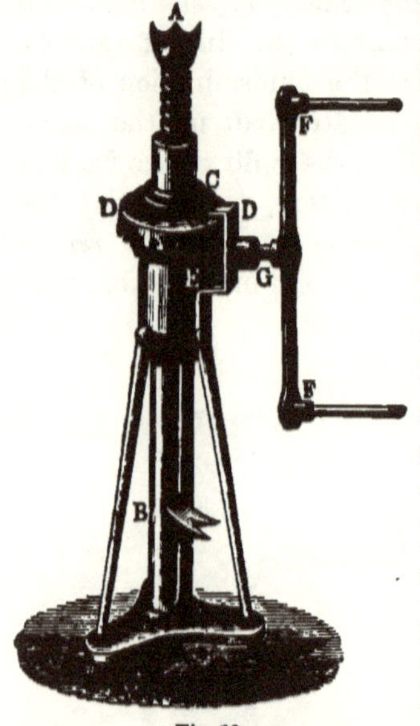

Fig. 13.

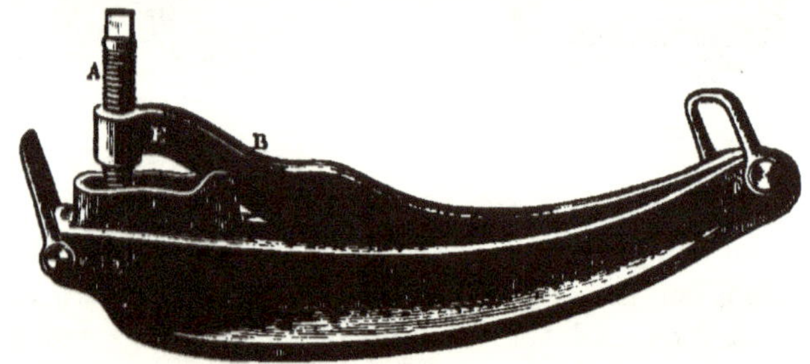

Fig. 14.

motive, boiler, etc.), the ends of the beam resting on the nuts M of this double jack.

The object in choosing a pair of spur and a pair of bevel wheels is not only to obtain the required increase of power by a reduction in speed, but also to make the working of the crank more convenient for the labourers. For very heavy loads the reduced velocity is sometimes obtained by using a

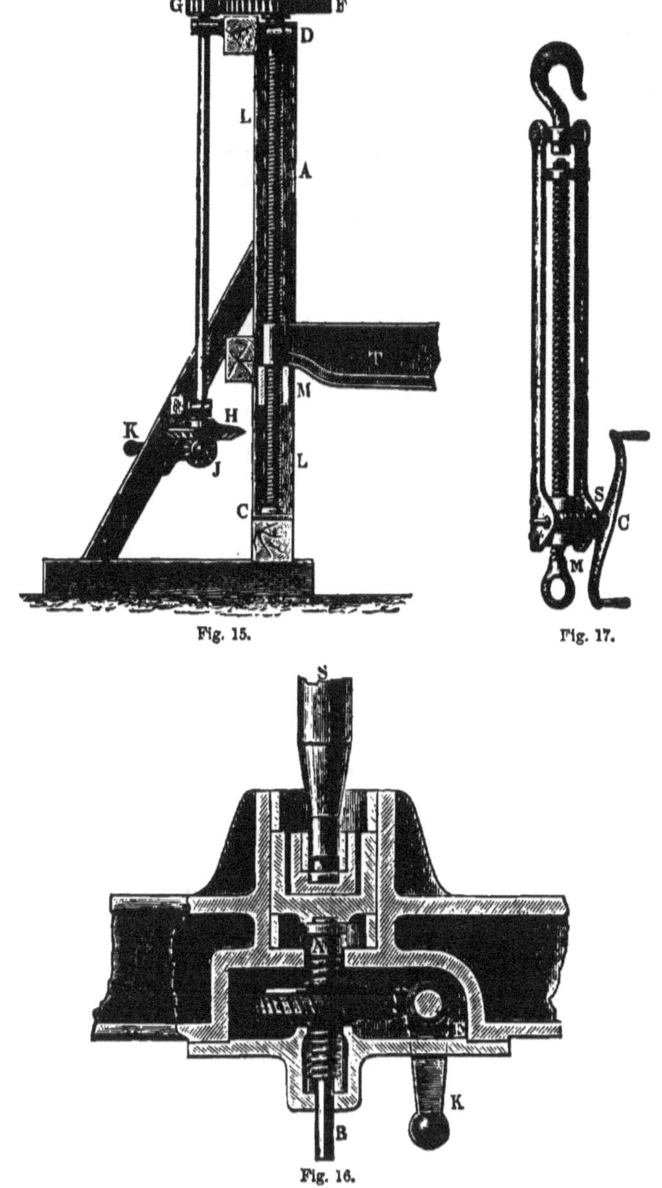

Fig. 15.

Fig. 17.

Fig. 16.

worm wheel and worm instead of spur wheels.   This is seen

in the apparatus, Fig. 16, used for adjusting the vertical millstone spindle with its runner. The screw A, which is prevented from turning by the square part B, receives its vertical motion by revolving a nut, which is made as a worm wheel gearing into a worm E, operated by the crank K. Again, in the jack, Fig. 17, the rotation of the nut M is effected by the screw S on the crank arbor C.

As regards the purchase of a screw or ratio borne by the resistance to the driving effort, it is shown in vol. iii. part 1, § 126, of Weisb., *Mech.*, that $P_1 = Q \dfrac{n + \mu}{1 - n\mu}$, where P is the driving force applied to a point at a distance r from the axis equal to the mean radius of the thread, and there, overcoming the load Q, acting along the axis of the screw. In this formula journal friction is neglected, μ represents the coefficient of sliding friction of the thread in its bearing, and the velocity ratio of the screw is $n = \dfrac{S}{2\pi r} = $ the tangent of the angle of the inclination of the helix. Furthermore let r_1 denote the lever arm of pivot friction due to the pressure acting in the direction of the axis of the screw, and r_2 the lever arm of friction between the neck-journal and its bearing, due to the force P acting at right angles to this axis ; then, the driving force acting with a leverage R was found to be

$$P = \frac{r}{R - \phi r_2}\, Q \left(\frac{n + \mu}{1 - n\mu} + \phi \frac{r_1}{r} \right),$$

where ϕ denotes the coefficient of journal friction.

The theoretical force required is $P_0 = \dfrac{r}{R} Q n$, therefore the efficiency is

$$\eta = \frac{P_0}{P} = \frac{R - \phi r_2}{R}\ \frac{n(1 - n\mu)}{n + \mu + (1 - n\mu)\phi \frac{r_1}{r}}.$$

For the reverse motion it was found that

$$(P) = \frac{r}{R + \phi r_2}\, Q \left(\frac{n - \mu}{1 + n\mu} - \phi \frac{r_1}{r} \right),$$

and hence

$$(\eta) = \frac{(P)}{P_0} = \frac{R}{R + \phi r_2}\ \frac{n - \mu - (1 + n\mu)\phi \frac{r_1}{r}}{n(1 + n\mu)}.$$

It was also noted in the article referred to, that for the same angle of inclination of the helix, the efficiency is materially diminished by increasing the radii r_1 and r_2. Therefore, in all apparatus of this class, in which the *nut is turned*, a smaller efficiency is to be expected than in those machines where the *spindle* of the screw is turned.

In the screw jacks as usually constructed, the velocity ratio n of the helix is seldom greater than $0\cdot1$, nor less than $0\cdot05$. Further, when the screw turns, we may assume as suitable the values $r_1 = 0\cdot5r$ and $r_2 = r$. When, however, the nut is rotated, it is provided with a collar-shaped bearing, having an inner radius r; we may then generally assume the leverage of the collar friction to be $r_1 = 1\cdot5r$, and that of the neck-journal $r_2 = 2r$. In the above formula the effect of the ratio $\dfrac{r_2}{R}$ upon the efficiency is but of secondary importance, for the friction of the neck-journal of radius r_2 depends upon the lever arm R of the driving force P, only this friction diminishing as R increases. For windlasses, R always considerably exceeds the radius r of the screw, so that $\dfrac{r_2}{R}$ seldom amounts to more than $\frac{1}{8}$.

The following table of the efficiencies η and (η) for different values of the pitch was computed on the supposition of a lever arm $R = 8r$, $\mu = 0\cdot1$, and $\phi = 0\cdot08$.

This table shows the important influence of the wasteful resistances upon the efficiency of the screw-jack, and that an estimate made without considering frictional resistances would not be even approximately true. Owing to their small efficiency, screws should not be employed in hoisting apparatus intended for *continuous and heavy service*. On the other hand, their application is often to be recommended for apparatus which run intermittently, owing to the great security which they insure against running down by virtue of their self-locking property. That this feature belongs to all the screws contained in the table follows from the fact that the values of (η) are negative throughout.

The values of η in the table, which refer to the case in which the *screw turns*, can also be employed for the efficiency of worm gears, as is shown in vol. iii. 1, § 132, Weisb. *Mech.*

TABLE SHOWING

THE EFFICIENCY OF SCREW GEARING.

$$\eta = \frac{R - \phi r_2}{R}\; \frac{n(1-n\mu)}{n+\mu+(1-n\mu)\phi\frac{r_1}{r}}; \qquad (\eta) = \frac{R}{R+\phi r_2}\; \frac{n-\mu-(1+n\mu)\phi\frac{r_1}{r}}{n(1+n\mu)}.$$

Values of	$n=$	0·04	0·05	0·06	0·07	0·08	0·10	0·125
The Screw turns	η	0·219	0·259	0·296	0·328	0·357	0·408	0·463
A. $r_1 = 0·5r$; $r_2 = r$; $R = 8r$	(η)	− 2·47	− 1·78	− 1·32	− 0·986	− 0·740	− 0·396	− 0·121
The Nut turns	η	0·151	0·183	0·210	0·236	0·260	0·304	0·352
B. $r_1 = 1·5r$; $r_2 = 2r$; $R = 8r$	(η)	− 4·406	− 3·328	− 2·611	− 2·097	− 1·713	− 1·777	− 0·747

When the rotation of the screw or its nut is not directly brought about by a lever, but is effected by means of a single pair of gears, as in Fig. 13, or several pairs, as in Fig. 15, the efficiency of the whole combination is to be determined as explained in the preceding pages, by taking the product of the efficiencies of the screw and gear, or gears. A few examples will make this clear.

EXAMPLE 1.—If in a windlass jack similar to the one described in Fig. 13, the bevel wheels have 10 and 30 teeth, and radii $r_2 = 50$ millimetres [1·97 in.], and $R_1 = 150$ millimetres [5·91 in.] respectively, the mean radius r_1 of the screw equal to 40 mm. [1·57 in.], the crank $R_2 = 300$ mm. [11·81 in.], and the tangent of the angle of inclination $n = \dfrac{S}{2\pi r} = 0·06$, what will be the driving force P required at the end of the crank to lift a load of 3000 kilograms [6615 lbs.] ? In this case the theoretical driving force will be

$$P_0 = Q n \frac{z_1}{z_2}\frac{r_1}{R_2} = 3000 \times 0·06 \times \frac{10}{30} \times \frac{40}{300} = 8 \text{ kg. [17·64 lbs.]}$$

The efficiency of the screw for $n = 0·06$, and for the suitable proportions $r_1 = 1·5r_1 = 60$ mm. [2·36 in.], $r_2 = 2r_1 = 80$ mm. [3·15 in.] is

$$\eta = \frac{150 - 0·08 \times 80}{150}\;\frac{0·06(1 - 0·06 \times 0·1)}{0·1 + 0·06 + (1 - 0·006) \times 0·08 \times \frac{60}{40}} = 0·957\frac{0·0596}{0·279} = 0·205.$$

(This value is $\frac{1}{2}$ per cent. less than that given in the table, 0·210, as $\dfrac{R_1}{r}$ was assumed to be $\dfrac{150}{40} = 3·75$ only).

Further, the efficiency of the teeth is

$$\frac{1}{1 + 0·33\;\sqrt{\left(\frac{1}{10}\right)^2 + \left(\frac{1}{30}\right)^2}} = 0·966,$$

and the efficiency of the crank shaft, supposing the radius of the journal to be $r = 15$ mm. [0·59 in.] is

$$\frac{1 - 0·08\dfrac{15}{300}}{1 + 0·08\dfrac{15}{50}} = 0·973 \text{ (see also table, page 17).}$$

Hence the efficiency of the machine is found to be

$$\eta = \eta_1\eta_2\eta_3 = 0·205 \times 0·966 \times 0·973 = 0·205 \times 0·940 = 0·193 \;;$$

thus the driving force required is

$$P = \frac{P_0}{\eta} = \frac{8}{0\cdot193} = 41\cdot5 \text{ kg. } [91.5 \text{ lbs.}]$$

In order to determine the force which must be applied to the crank to *lower* the load, we find for the screw

$$(\eta) = \frac{150}{150 + 0\cdot08 \times 80} \; \frac{0\cdot06 - 0\cdot1 - (1 + 0\cdot006)0\cdot08\frac{60}{40}}{0\cdot06(1 + 0\cdot006)} = 0\cdot959\frac{-0\cdot1607}{0\cdot0604} = -2\cdot55.$$

As the theoretical force, which acts in the periphery of the bevel wheel of radius 150 mm. [5·91 in.] is

$$3000 \times 0\cdot06 \times \frac{40}{150} = 48 \text{ kg. } [105\cdot8 \text{ lbs.}],$$

it follows that the actual force at this point is

$$48 \times 2\cdot55 = 122\cdot4 \text{ kg. } [269\cdot9 \text{ lbs.}]$$

The efficiency of the bevel gearing being $0\cdot966 \times 0\cdot973 = 0\cdot940$, there is evidently needed an exertion of

$$(P) = \frac{1}{0\cdot940} \times 122\cdot4 \times \frac{50}{300} = 21\cdot7 \text{ kg. } [47\cdot85 \text{ lbs.}]$$

on the crank, in order to cause the load to descend.

EXAMPLE 2.—If the nut, by the addition of a rim, is made into a worm wheel, having a radius of 100 mm. [3·94 in.], then the efficiency of the screw will be given by

$$\eta = \frac{100 - 0\cdot08 \times 80}{100} \; \frac{0\cdot06(1 - 0\cdot006)}{0\cdot1 + 0\cdot06 + (1 - 0\cdot006) \times 0\cdot08\frac{60}{40}} = 0\cdot200.$$

Now, let the mean radius of the worm be 40 mm. [1·57 in.], and the velocity ratio $n = 0\cdot08$, and let us choose a length of crank equal to 200 mm. [7·87 in.]; further, let us assume the lever arm of the pivot to be $r_1 = 10$ mm. [0·39 in.], and the radius of the journal of the worm shaft to be $r_2 = 15$ mm. [0·59 in.] Then (since the friction between the worm wheel and its bearings has already been included in determining the force required at the periphery of the worm wheel—that is to say, included in the expression $\eta = 0\cdot200$) the efficiency of the worm gear will be

$$\eta_2 = \frac{200 - 0\cdot08 \times 15}{200} \; \frac{0\cdot08(1 - 0\cdot008)}{0\cdot18 + (1 - 0\cdot008)0\cdot08\frac{10}{40}} = 0\cdot994\frac{0\cdot079}{0\cdot20} = 0\cdot393.$$

Hence the efficiency of the jack screw is

$$\eta = 0\cdot200 \times 0\cdot393 = 0\cdot079.$$

As the theoretical force required at the crank is

$$3000 \times 0\cdot06 \times \frac{40}{100} \times 0\cdot08 \times \frac{40}{200} = 1\cdot152 \text{ kg. } [2\cdot5 \text{ lbs.}],$$

the actual force will be

$$P = \frac{1 \cdot 152}{0 \cdot 079} = 14 \cdot 6 \text{ kg. } [32 \cdot 2 \text{ lbs.}]$$

EXAMPLE 3.—Finally, let us suppose that the screw, with the same dimensions, is driven by two pairs of gears as in Fig. 15, and let us assume the radius of the pivot to be 20 mm. [·79 in.], that of the end journal to be 40 mm. [1·57 in.], and that of the spur wheel on the screw shaft 450 mm. [17·72 in.], then the efficiency of the screw is found to be

$$\eta = \frac{450 - 0 \cdot 08 \times 40}{450} \quad \frac{0 \cdot 06(1 - 0 \cdot 006)}{0 \cdot 16 + (1 - 0 \cdot 006) \times 0 \cdot 08 \times \frac{20}{40}} = 0 \cdot 296.$$

Let the spur wheels have 15 and 75, and the bevel wheels 12 and 48 teeth, then the respective efficiencies of the teeth are

$$\frac{1}{1 + 0 \cdot 33\left(\frac{1}{15} + \frac{1}{75}\right)} = 0 \cdot 975,$$

and

$$\frac{1}{1 + 0 \cdot 33 \sqrt{\left(\frac{1}{12}\right)^2 + \left(\frac{1}{48}\right)^2}} = 0 \cdot 973.$$

If now we assume the journals of the vertical shaft on which the wheels are fixed to have a common radius of 30 mm. [1·18 in.], and the bevel and smaller spur wheel to have the respective radii 300 mm. [11·81 in.] and 90 mm. [3·54 in.], the efficiency of this shaft will be

$$\frac{1 - 0 \cdot 08 \times \frac{30}{300}}{1 + 0 \cdot 08 \times \frac{30}{90}} = 0 \cdot 966.$$

In like manner we determine the efficiency of the crank axle to be

$$\frac{1 - 0 \cdot 08 \times \frac{20}{400}}{1 + 0 \cdot 08 \times \frac{20}{75}} = 0 \cdot 976,$$

by taking its radius equal to 20 mm. [0·79 in.], the length of crank 400 mm. [15·75 in.], and the radius of the bevel wheel 75 mm. [2·95 in.] Consequently, the efficiency of the machine becomes

$$\eta = 0 \cdot 296 \times 0 \cdot 975 \times 0 \cdot 966 \times 0 \cdot 973 \times 0 \cdot 976 = 0 \cdot 268.$$

As the theoretical force required at the crank is

$$3000 \times 0 \cdot 06 \times \frac{40}{400} \times \frac{15}{75} \times \frac{12}{48} = 0 \cdot 9 \text{ kg. } [1 \cdot 98 \text{ lbs.}],$$

the actual driving force required is

$$P = \frac{0 \cdot 9}{0 \cdot 269} = 3 \cdot 34 \text{ kg. } [7 \cdot 36 \text{ lbs.}]$$

If we wish to apply the principles of graphical statics in order to determine the effort required to operate a jack screw, we may proceed as follows. Let A, Fig. 18, represent the axis of the screw, having a mean radius $AB = ab$; the nut of the screw carries a wheel of radius AJ; this wheel is driven by the spur wheel HJ fixed to the axle H of the crank HK. Let the bearing surface of the nut be represented by the ring having a radial width $BR = br$, and let

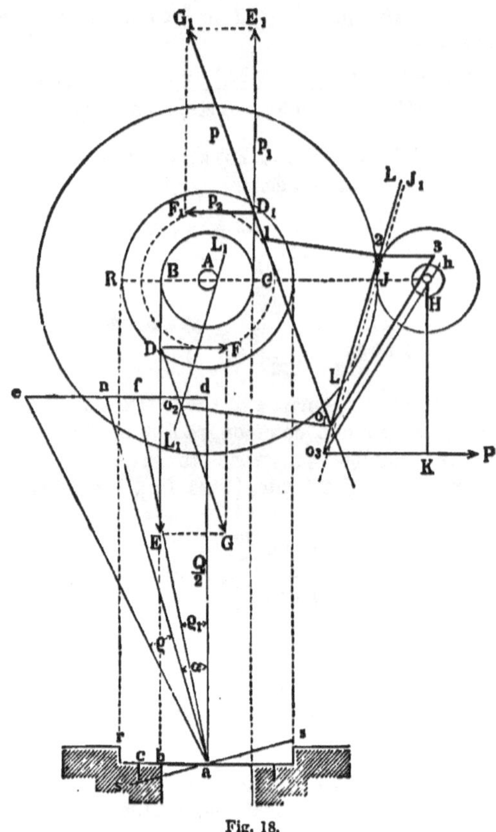

Fig. 18.

its mean radius ac be taken as the lever arm of friction. Assume ss to be the direction of the thread midway between the outer and inner edges at a, and a its angle of inclination to a plane normal to the axis. Further, if an is the normal to this line, and the angle nae is constructed equal to the angle of friction ρ, then, making $da = \frac{1}{2}Q$, and drawing the horizontal line de through d, we obtain in the length $de = \frac{1}{2}Q$ tang $(a + \rho)$ the forces p_1 of the couple acting at the circumference of the screw. The forces of this couple are represented by DE and D_1E_1. Now draw the line af, making an angle with the axis of the screw equal to the corresponding angle of friction ρ_1, belonging to the supporting surface c, then, similarly, the length df will represent the forces p_2 of the couple, due to the friction of the pivot, which couple acts in the circumference of the circle AC. Let the forces of this couple be represented by DF and D_1F_1. By combining the two couples p_1 and p_2 we find the resultant couple p, whose forces are represented by DG and D_1G_1. Let J be the point of contact of the wheels, and J_1J the direction of the pressure exerted by the teeth, whose inclination to the line of centres is about $75°$, then the actual pressure, taking friction into account, will be exerted along the line LL parallel to J_1J, and at a

distance $\zeta = \mu \dfrac{t}{2}$ (see Appendix III. 1, Weisb. *Mech.*) from the latter,

t representing the pitch and μ the coefficient of friction of the teeth. The pressure P_1 exerted by the teeth in the direction LL calls forth a reaction in the bearing of the nut, whose radius is AR, which reaction also acts in the direction L_1L_1 tangent to the friction circle of AR. But the two couples p and P_1 must be equal, a condition fulfilled only when the line o_1o_2 joining the points of intersection of each pair of forces coincides with their resultant. Therefore, making $o_11 = D_1G_1$, and drawing through 1 a parallel 12 to o_1o_2, the length $2o_1$ will give the force P_1 as the actual pressure exerted by the teeth of the wheel HJ on the wheel AJ. The force P required at the crank is now obtained simply by completing the triangle of forces of which $2o_1$ represents one side, the other two being respectively parallel to the direction o_3K, and to the tangent o_3h drawn from o_3 to the friction circle of the axle H, the tangent representing the direction of reaction of the bearing. The result gives as the driving force P the length of the line 23 in the polygon o_1123.

§ 6. **Differential Screw-Jacks.**—Another class of screw jacks has been constructed, which depends upon the principle of differential motion, so that a load can be lifted by imparting rotation to both screw and nut. As the motions are in the same direction, but differ in amount, the motion axially of the piece will be proportional to the difference of the rotations.

A modern example of this class of lifting jacks is shown in Fig. 19. Here the motion of the crank K is transmitted by means of the bevel wheels D and E to the nut M, and by means of C and B to the screw A; for this purpose the hole of the wheel B is provided with a feather, which fits a longitudinal key-way in the screw. Owing to the velocity ratios chosen for the bevel gears, the screw is rotated faster than the nut, although the reverse of this arrangement can be used. In order to lower the load at a greater velocity, the bushing J, in which the crank shaft turns, is made eccentric, so that a half turn will raise it sufficiently to throw the wheels C and B, as well as D and E, out of gear, leaving only the wheels C and B_1 in gear with each other to operate the spindle A, as in the ordinary jack screw. This mode of transmitting motion from C to B_1 is also adapted for lifting lighter loads, which require a smaller velocity ratio.

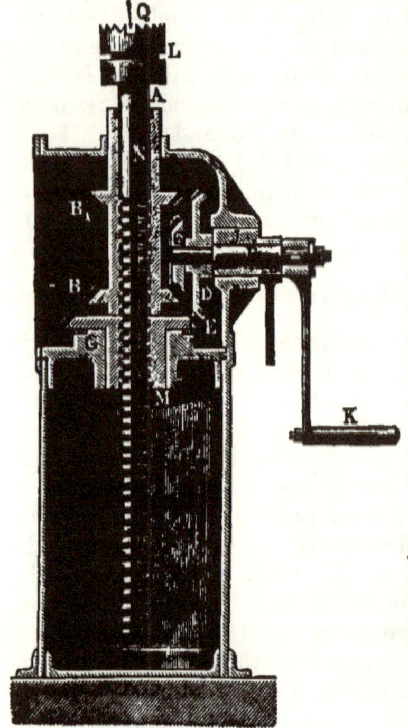

Fig. 19.

The action of this class of jacks has been fully investigated in vol. iii. 1, §130, Weisb. *Mech.*, and it was there found that a large efficiency can only be obtained by making the pitch of the screw as coarse as possible, and reducing the area of the supporting bearings as much as is practicable; two such bearings are employed in the present case, one for the spindle and one for the nut. Under this assumption the efficiency of the apparatus given in the example of the paragraph cited, for an angle of inclination $a = 45°$ or $n = 1$, was found to be 0·715. This represents an exceptionally large efficiency for screws, and the explanation lies in the unusual proportions assumed. It is easy to show that the arrangement of Fig. 19, using the ordinary proportions ($n = 0·05$ to $0·08$), can have but a very small efficiency, and that it is reduced by diminishing the difference between the velocities of the screw and nut.

In order to prove this, let the respective angles through which the screw and nut turn in a given time be denoted by ω_1 and ω_2, and again let r express the radius of the helix midway between the outer and the inner edges of the thread, the velocity ratio of this helix being $n = \dfrac{S}{2\pi r}$. Thus the load Q is lifted through a distance $(\omega_1 - \omega_2)\, rn$, and the useful work performed is expressed by $Q\,(\omega_1 - \omega_2)\, rn$. In addition, work has been performed in overcoming the friction between the threads, as well as in the journals and supporting bearings. For the purpose of simplifying the calculation, let the comparatively unimportant journal friction be entirely neglected, and let us only take into account the friction between the threads and that produced by the load Q at the support at L, and between the nut and its support at G. The friction generated at the two latter surfaces depends on Q, and is given by ϕQ. Letting the mean lever-arm of friction for the bearing at L be r_1, and for the nut r_2, we can express the useful and lost work by the following equation:

$$Q(\omega_1 - \omega_2)rn + \mu Q(\omega_1 - \omega_2)r + \phi Q(\omega_1 r_1 + \omega_2 r_2) = Q[(\omega_1 - \omega_2)r(n + \mu)$$
$$+ \phi(\omega_1 r_1 + \omega_2 r_2)].$$

Therefore, neglecting the friction due to transverse action of the driving force, the efficiency becomes:

$$\frac{\text{Useful work}}{\text{Energy expended}} = \frac{(\omega_1 - \omega_2)rn}{(\omega_1 - \omega_2)r(n + \mu) + \phi(\omega_1 r_1 + \omega_2 r_2)}.$$

If in this expression we put $n = 0{\cdot}06$, as in the examples of the preceding paragraph, and place $r_1 = 0{\cdot}5r$ and $r_2 = 1{\cdot}5r$, as being the smallest possible values; and if we further assume the velocity ratio to be $\omega_1 : \omega_2 = 4 : 3$, we shall obtain an efficiency:

$$\frac{(4-3)0{\cdot}06r}{(4-3)(0{\cdot}1 + 0{\cdot}06)r + 0{\cdot}08(4 \times 0{\cdot}5 + 3 \times 1{\cdot}5)r} = \frac{0{\cdot}06}{0{\cdot}16 + 6{\cdot}5 \times 0{\cdot}08}$$
$$= \frac{0{\cdot}06}{0{\cdot}68} = 0{\cdot}088,$$

that is not quite $0{\cdot}09$. The friction of the neck-journals further reduces the efficiency. When the difference between the angular velocities ω_1 and ω_2 is taken smaller than the above assumed values, the efficiency will be still smaller. Constructions of this kind, therefore, give a great efficiency only when the pitch of the screw is coarse and the radii of the bearings are small.

CHAPTER II

§ 7. **Pulleys.**—The hoisting arrangements thus far mentioned are suitable for small lifts only. For greater heights *ropes* or *chains* passing over pulleys or drums are made use of. The simplest arrangement of this kind is shown in Fig. 20, which illustrates the single *fixed* or *guide pulley* commonly employed in practice.

A rope sustaining the load Q upon the part BC passes over the pulley A, which turns in a fixed bracket or support G, and the load is lifted by applying a pull P to the other end of the rope in the direction DE.

As the distances through which the driving force and resistance act are equal, the theoretical force $P_0 = Q$.

The wasteful resistances in the present case are: stiffness of the rope, or friction connected with the use of the chain, and journal friction; taking these into account the actual pull P is determined as follows. Let

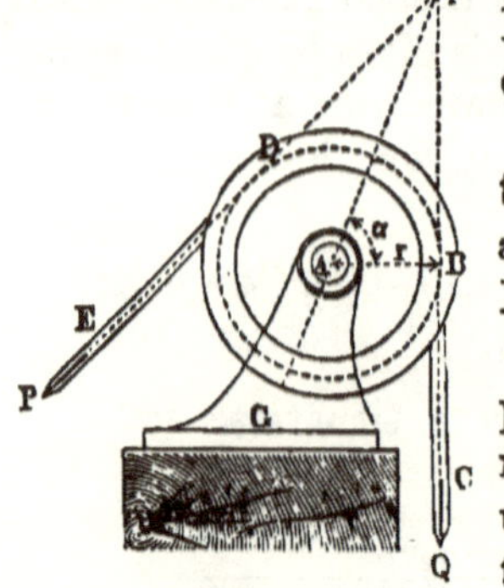

Fig. 20.

σ denote the coefficient of the resistance opposing the motion of the rope or chain as it moves on or leaves the pulley; according to vol. i. § 200, Weisb. *Mech.*, this resistance is proportional to the tension Q in the advancing part of the rope, and is to be taken equal to σQ; its effect is the same as if a force σQ resisted the motion of the rope at the points B and D at which the rope is bent. Denote by Z the pressure on the pin whose radius is r, and by r the effective radius of the pulley measured to the centre of the rope, then follows the equation:

$$P r = Q r + 2\sigma Q r + \phi Z r.$$

The pressure on the pin may be accurately determined from

$$Z^2 = P^2 - 2PQ \cos 2a + Q^2,$$

where $2a$ represents the angle BAD of the arc of contact of the rope. As this method of determining Z would involve unnecessary elaboration, it may be deemed sufficiently correct for practical purposes to put $P = Q$ in calculating the pin friction; we thus obtain $Z = 2Q \sin a$, which value of Z gives

$$P r = Q r + 2\sigma Q r + \phi 2Q \sin a r;$$

and thus the force

$$P = Q\left(1 + 2\sigma + 2\phi\frac{r}{r}\sin a\right).$$

For the sake of brevity putting

$$1 + 2\sigma + 2\phi\frac{r}{r}\sin a = k,$$

we have

$$P = kQ,$$

and accordingly the efficiency of the pulley

$$\eta = \frac{P_0}{P} = \frac{1}{k}.$$

When the two directions of the rope are parallel, *i.e.*, when $a = 90°$,

$$k = 1 + 2\sigma + 2\phi\frac{r}{r}.$$

For the reverse motion, that is, when the load is allowed to descend, the tension in the part ED is $\dfrac{S}{k}$, assuming the tension in BC $= S$; hence, while the work done by the load in descending a certain distance s is Ss, the useful portion in the rope DE will be only $\dfrac{S}{k}s$.

This gives for the efficiency of the reverse motion

$$(\eta) = \frac{\dfrac{S}{k}s}{Ss} = \frac{1}{k} = \eta,$$

as in the case of the forward motion.   When the two portions of the rope are parallel a more exact value for k is given by the equation

$$Pr = Qr + \sigma Qr + \sigma Pr + \phi(P + Q)r,$$

which gives

$$P = Q \; \frac{1 + \sigma + \phi\dfrac{r}{r}}{1 - \sigma - \phi\dfrac{r}{r}},$$

hence

$$k = \frac{1 + \sigma + \phi\dfrac{r}{r}}{1 - \sigma - \phi\dfrac{r}{r}},$$

a value which in most cases differs but little from the above given $1 + 2\sigma + 2\phi\dfrac{r}{r}.$

In fixing a value for σ we must make a distinction between chains and ropes.   It has been shown in vol. i. § 200, Weisb. *Mech.*, that for chains the friction at the joints where bending occurs, due to the tension Q, is given by

$$\phi_1 Q\frac{\delta}{2r},$$

where ϕ_1 represents the coefficient of friction for the links, and δ the diameter of the round iron in chain.   We therefore have for chains

$$\sigma = \phi_1 \frac{\delta}{2r}.$$

In the case of ropes, fuller information concerning the resistance due to stiffness is also to be found in vol. i. § 202 and following of the work referred to.   Basing our calculations on the formula of *Eytelwein*, the resistance due to bending the rope to the curvature of the pulley and then straightening it again, is

$$2\sigma Q = 0{\cdot}018\frac{\delta^2}{r}Q^*,$$

where δ and r are expressed in millimetres.

$$\left[\; 2\sigma Q = 0{\cdot}457\frac{\delta^2}{r}Q \text{ when } \delta \text{ and } r \text{ are in inches.}\;\right]$$

Therefore for ropes we have

$$2\sigma = 0{\cdot}018\frac{\delta^2}{r} \qquad\qquad \left[2\sigma = 0{\cdot}457\frac{\delta^2}{r}\right].$$

For the ordinary chain pulleys, as used in hoisting-tackle and windlasses, it is customary to take the radius r of the pulley not less than 10δ. Assuming this proportion and a co-efficient of friction $\phi_1 = 0{\cdot}2$, we find for *chains*

$$2\sigma = 2 \times 0{\cdot}2 \times \frac{\delta}{20\delta} = 0{\cdot}02,$$

that is, a value independent of the diameter of chain iron.

A suitable radius for rope pulleys is $r = 4\delta$, which gives

$$2\sigma = 0{\cdot}018\frac{\delta^2}{4\delta} = 0{\cdot}0045\delta \qquad \left[2\sigma = 0{\cdot}457\frac{\delta^2}{4\delta} = 0{\cdot}114\delta\right],$$

that is, a value directly proportional to the diameter of the rope. Assuming a diameter of pin

$$2r = d = 3\delta \text{ for chain pulleys,}$$

and

$$2r = d = \delta \text{ for rope pulleys,}$$

and a coefficient of journal friction $\phi = 0{\cdot}08$, the values of the efficiency of the fixed pulley

$$\eta = \frac{1}{k} = \frac{1}{1 + 2\sigma + 2\phi\dfrac{r}{r}\sin a},$$

---

* Redtenbacher gives, according to the experiments of *Prony*,

$$2\sigma = 26\frac{\delta^2}{a} \text{ (for metres) ;}$$

this would give

$$\sigma = 0{\cdot}013\frac{\delta^2}{a} \text{ (for millimetres)}$$

$$\left[\sigma = 0{\cdot}330\frac{\delta^2}{a}\text{(for inches)}\right].$$

have been calculated for arcs of contact $2a = 180°$, $120°$, and $90°$, and tabulated as follows:

TABLE OF THE EFFICIENCY OF THE FIXED PULLEY.

$$\eta = \frac{1}{k} = \frac{1}{1 + 2\sigma + 2\phi\frac{r}{r}\,\sin a}.$$

Diameter of Rope. $\delta =$	10 mm. [·39 in.]	20 mm. [·79 in.]	30 mm. [1·18 in.]	40 mm. [1·57 in.]	50 mm. [1·97 in.]	Chains.
$2a = 180°$	0·939	0·901	0·866	0·833	0·803	0·958
$2a = 120°$	0·942	0·903	0·868	0·835	0·805	0·960
$2a = 90°$	0·944	0·906	0·870	0·837	0·807	0·964

In the preceding calculation the weight of the pulley has been neglected, as its effect on the journal friction is slight in comparison with that of the forces acting on the rope or chain; in individual cases more exactness may be desirable, and then the pressure on the journal should be increased by this amount.

The influence of the weight of the ascending and descending ropes and chains will be examined later.

The ratio between the effort and resistance for the *movable pulley* can be easily derived from the relation $P = kQ$ already found for the fixed pulley. For this purpose let the pulley A, Fig. 21, having the two parts of the rope parallel, be suspended from a fixed hanger AF, and let the tension produced in the part BC by the load carried be denoted by S, then in order to bring about motion in the direction of the arrow, a force kS must be applied to the part DE, so that the hanger has to resist a total pressure $Z = S + kS = S(1 + k)$. In this case, when the pulling end of the rope passes over a certain distance s, the load Q at the other end must rise through the same distance. The mutual relations between the forces Q, P, and Z will not be altered, no matter what motion is given to the whole system, consisting of the pulley with its hanger and both ends of the rope, as the relative motions of the individual parts will remain the same as before. Therefore, if we introduce at each instant an additional motion equal to that of Q, only in opposite direction, that is, if the whole system be shifted vertically downward a distance s, then the load Q will come to rest, and the end C of the rope may be considered as attached to a fixed point. The pulley, together with the force Z, is shifted a distance s in a direction contrary to that in which Z acts, while the pulling part DE, besides its previous motion measured by s, receives an equal additional motion, so that the point of application of the force P moves through a distance $2s$ in the direction in which the latter acts. Through this reasoning we can pass from the fixed to the *movable pulley*, which in Fig. 22 is represented in

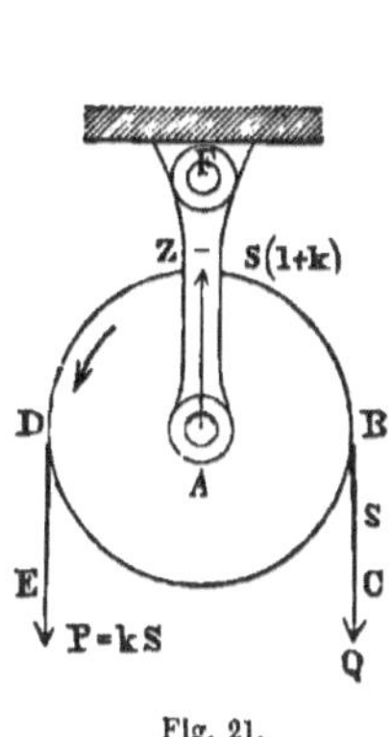
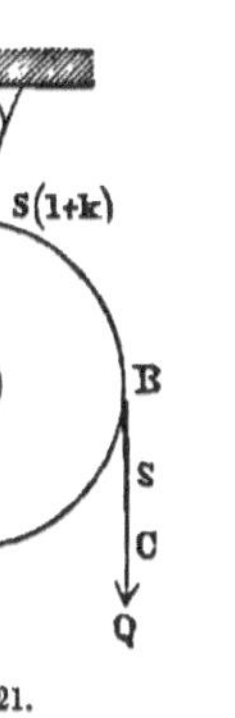

Fig. 21.

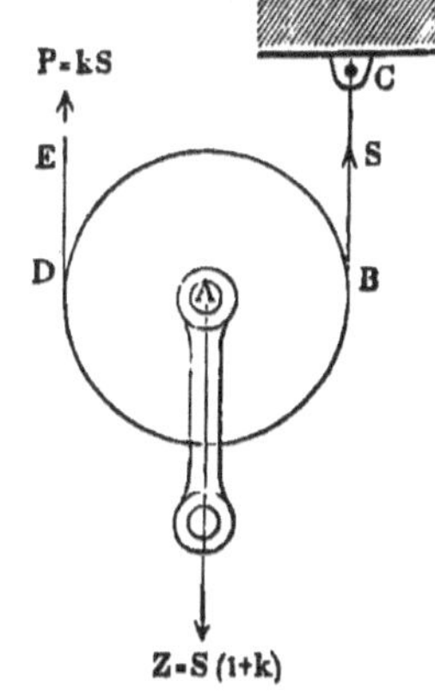

Fig. 22.

the usual inverted position. Here the end of part BC is attached to the fixed point C, which reacts with a force S′, while, as before, the pulling force $P = kS'$, tends to impart rotation to the pulley, and thus lift the load, which is now represented by the pressure $Z = S(1 + k)$ on the pin. The energy exerted is expressed by $kS \times 2s$, while the useful work performed is

$$S(1 + k) \times s.$$

Thus the efficiency of the movable pulley is found to be

$$\eta = \frac{1 + k}{2k},$$

which is greater than the efficiency $\dfrac{1}{k}$ of the fixed pulley, since k is always greater than unity.

By communicating motion to the movable pulley in the direction of the load Z, we can also determine the formulæ applicable to the reverse motion, during which the point of application of P is moved downward. In this case P represents the useful resistance, and Z the driving force. During the descent of the pulley the part DE is wound on, and the part BC unwound, so that the tension in BC becomes equal to kS, while that in DE will be denoted by S. Accordingly, the bearing pressure Z is again given by $S(1 + k)$. As the energy expended is expressed by $S(1 + k) \times s$, and the useful work performed by $S \times 2s$, the efficiency for the reverse motion of the movable pulley becomes

$$(\eta) = \frac{S2s}{S(1 + k)s} = \frac{2}{1 + k}.$$

This expression may be obtained directly from the value for the forward motion

$$\eta = \frac{1 + k}{2k},$$

by substituting $\dfrac{1}{k}$ for k, and taking the reciprocal of the result, inasmuch as

$$\eta = \frac{P_0}{P} \text{ and } (\eta) = \frac{(P)}{P_0}.$$

It also follows that the value $(\eta) = \dfrac{2}{1 + k}$ for the reverse motion must be less than the value $\eta = \dfrac{1 + k}{2k}$ for the forward motion, since $2 \times 2k < (1 + k)^2$.

For example, if for a particular pulley we had

$$k = 1 + 2\sigma + 2\phi\frac{r}{r} = 1 \cdot 1,$$

the efficiencies would be

$a.$ $\eta = (\eta) = \dfrac{1}{1 \cdot 1} = 0 \cdot 909$ for the fixed pulley.

$b.$ $\eta = \dfrac{1 + 1 \cdot 1}{2 \cdot 2} = 0 \cdot 955$ for the forward motion of movable pulley.

$c.$ $(\eta) = \dfrac{2}{2 \cdot 1} = 0 \cdot 952$ for the reverse motion of movable pulley.

The relation between effort and resistance in the pulley can be easily determined by graphical methods. Let BC and DE, Fig. 23, again represent the centre lines of the parts of the rope; then draw parallel to them and at a distance $\sigma = 0\cdot009\delta^2[\sigma = 0\cdot228\delta^2]$ for ropes, or $\phi\dfrac{\delta}{12}$ for chains, the lines of action bc and de of the forces, and from the point of intersection o draw the corresponding tangent oa to the friction circle of the journal, which circle is described with a radius ϕr. In this construction the lever arm Ab of the resistance must be taken larger, and that of the effort Ad smaller by the amount σ than the radius r of the pulley. Making $o1 = Q$, and drawing 12 parallel to de, the force P will be given by 12, and the reaction of the bearing by $2o$.

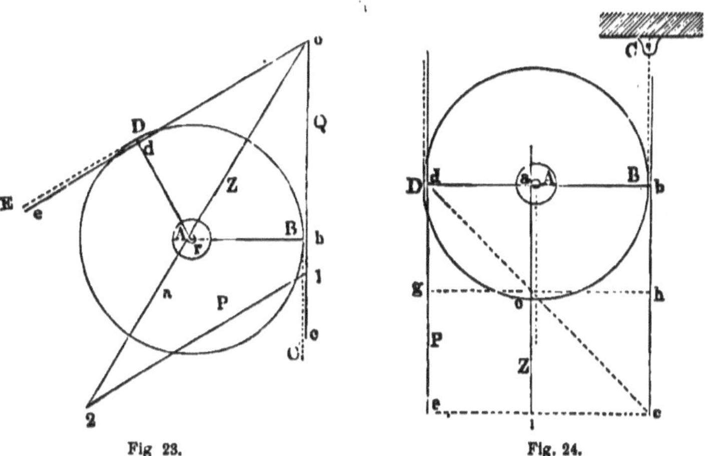

Fig 23.   Fig. 24.

For the movable pulley, Fig. 24, draw the lines of action cb and ed of the forces at a distance σ from and parallel to the centre lines of the rope, and assume the reaction oa of the bearing tangent to the friction circle of the journal and parallel to the parts of the rope. Make $a1 = Z$, draw through 1 the horizontal line $e1c$, and connect d and c; then the driving force P will be represented by $1o = eg$, and the reaction of the fixed point C by $oa = hb$.

In the following

TABLE OF THE EFFICIENCY OF THE MOVABLE PULLEY

$$\eta = \frac{1+k}{2k} = \frac{1+\sigma+\phi\frac{r}{r}}{1+2\sigma+2\phi\frac{r}{r}} \; ; \; (\eta) = \frac{2}{1+k} = \frac{1}{1+\sigma+\phi\frac{r}{r}}.$$

Diam. of Rope. $\delta =$	10 mm. ['39 in.]	20 mm. ['79 in.]	30 mm. [1'18 in.]	40 mm. [1'57 in.]	50 mm. [1'97 in.]	Chains.
$\eta =$	0·970	0·950	0·933	0·917	0·902	0·979
$(\eta) =$	0·968	0·946	0·928	0·909	0·891	0·978

are contained the results for the forward and reverse motion obtained under the same suppositions as previously for the fixed pulley, that is

$$r = 4\delta \text{ and } 2r = d = \delta \text{ for ropes.}$$
$$r = 10\delta \text{ and } 2r = d = 3\delta \text{ for chains.}$$

§ 8. **Hoisting-Tackle.**—With the aid of the preceding paragraphs it now becomes an easy matter to determine the relations between the forces acting in the many different systems of pulleys used in practice under the name of *tackle*. There is but one rule to be observed in such contrivances— namely, that *in every case where a rope encircles a pulley, the tension in the unwinding part is equal to k times the tension in the advancing part ;* k as before representing the coefficient of resistance for the pulley

$$k = 1 + 2\sigma + 2\phi\frac{r}{r}.$$

All combinations of pulleys are included under the head of tackle, but we may make a distinction, though not very marked, between the arrangement in which each *block* in the combination has but a single pulley or *sheave,* and that in which the several sheaves are mounted side by side in a frame or block.

A simple combination of pulleys of the former kind is shown in Fig. 25. Here the weight Q is suspended from the movable block A, which hangs in the bight of a rope, one end of which, called the *suspending part,* is attached to the fixed point F_1, while the other, representing the *hauling part*, is carried by the spindle of a second pulley B. This latter pulley hangs in the loop of a second rope, one end of which is made fast at F_2, while the other is secured to a third movable pulley C, whose rope is carried over a fixed pulley D, in order to allow of the application of a downward force P to the free end.

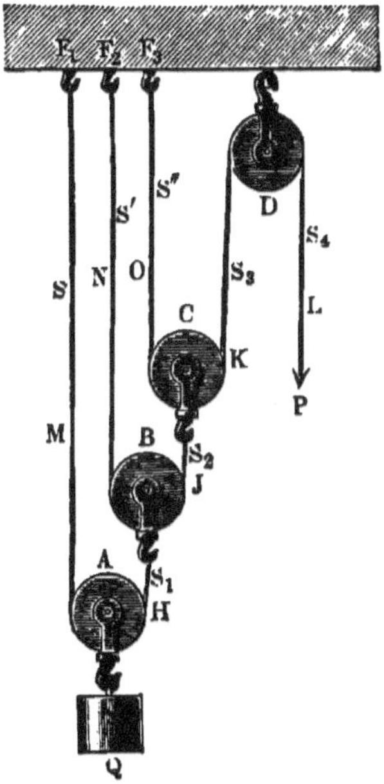

It is readily seen that, if the point of application of a force acting on the part L be moved through a distance s, the pulley C will be lifted through a height $\frac{1}{2}s$; the lift of B will be $\frac{1}{2}$ that amount, that is, $\frac{1}{4}s$, and finally the pulley A with its load Q will be lifted through a $\frac{1}{8}s$. Therefore, in the absence of wasteful resistances, the theoretical effort will be

$$P_0 = \tfrac{1}{8}Q,$$

Fig. 25.

or in general for n movable pulleys

$$P_0 = \frac{Q}{2^n}.$$

Assuming that all the pulleys used are of equal dimensions and all the ropes of equal size, the actual effort P required is obtained as follows. Let S represent the tension in the first part fixed at F_1, and S_1, S_2, S_3, and S_4, the tensions in the parts H, J, K, and L, then $S_1 = kS$; and since

$$Q = S + S_1 = S(1 + k),$$

it follows that

$$S = \frac{Q}{1 + k},$$

hence

$$S_1 = \frac{k}{1 + k}Q.$$

As the pull on the spindle of the pulley B is S_1, we find in a similar manner the tension in the part J to be

$$S_2 = \frac{k}{1+k}S_1 = \left(\frac{k}{1+k}\right)^2 Q,$$

and for the part K

$$S_3 = \frac{k}{1+k}S_2 = \left(\frac{k}{1+k}\right)^3 Q.$$

Finally, for the tension S_4 in the hauling part L of the fixed pulley we have

$$S_4 = kS_3 = k\left(\frac{k}{1+k}\right)^3 Q = P.$$

In general for one fixed and n movable pulleys the actual force required is

$$P = k\left(\frac{k}{1+k}\right)^n Q,$$

and the efficiency

$$\eta = \frac{P_0}{P} = \frac{1}{k}\left(\frac{1+k}{2k}\right)^n.$$

If the tackle shown in Fig. 25 were inverted and suspended by the pulley A, the fixed pulley D being omitted, and the driving force made to act directly on the part K, then the load may be carried on the three parts M, N, and O at the points F_1, F_2, F_3. For this case let S denote the tension in F_1M, and S' and S'' the tensions in F_2N and F_3O, then for the part H we have

$$S_1 = kS,$$

from which we deduce the tension in F_2N,

$$S' = \frac{1}{1+k}S_1 = \frac{k}{1+k}S.$$

Similarly for the part J we have,

$$S_2 = kS' = k\frac{k}{1+k}S.$$

hence the tension in F_3O,

$$S'' = \frac{1}{1+k}S_2 = \left(\frac{k}{1+k}\right)^2 S,$$

and in the ply K the tension

$$S_3 = kS'' = k\left(\frac{k}{1+k}\right)^2 S = P;$$

accordingly for the load Q we obtain

$$Q = S + S' + S'' = S\left[1 + \frac{k}{1+k} + \left(\frac{k}{1+k}\right)^2\right] = S\frac{1 + 3k + 3k^2}{(1+k)^2},$$

and consequently the driving force

$$P = k\left(\frac{k}{1+k}\right)^2 S = \frac{k^3}{1 + 3k + 3k^2}Q.$$

Without wasteful resistances we should have

$$P_0 = S_3 = S'' = \tfrac{1}{2}S' = \tfrac{1}{4}S;$$

therefore

$$Q = (4 + 2 + 1)S'' = 7S'' = 7P_0,$$

and the efficiency

$$\eta = \frac{P_0}{P} = \frac{1 + 3k + 3k^2}{7k^3}.$$

Another arrangement of tackle is shown in Fig. 26. Here the ropes, which are made fast to the blocks A, B, and C, are passed around the three pulleys, D, E, and F, from which the load Q is suspended. Supposing a force P to be applied to the part G, then, without hurtful resistances, the tensions in J and H will be P, and thus in each of the parts K, L, and M there will be a tension of 3P. Consequently the force acting in part N will be 9P, which also represents the tensions in O and T; by adding together the forces we obtain

$$Q = (1 + 1 + 3 + 3 + 9 + 9)P_0 = 26P_0.$$

Evidently the distances travelled by the points of application of the effort and load are also in the same ratio of 26 : 1. Owing to the varying dimensions of pulleys and sizes of

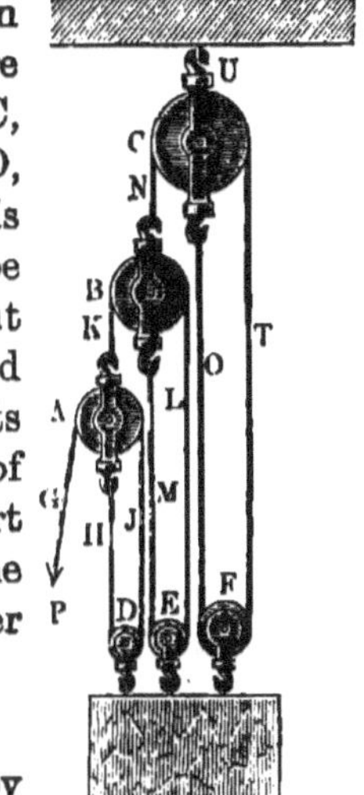

Fig. 26.

ropes, the values of k differ for the different pulleys. If a mean value of

$$k = \frac{1}{\eta} = \left(1 + 2\sigma + 2\phi\frac{r}{r}\right)$$

be assumed for all the pulleys, then the application of a force P to the part G will produce in J the tension

$$P\frac{1}{k} = P\eta,$$

further, in H, the tension

$$P\frac{1}{k^2} = P\eta^2,$$

and consequently the force acting in K will be expressed by $P(1 + \eta + \eta^2)$.

In the same manner we find the tension in the parts

L to be $P(1 + \eta + \eta^2)\eta$; and in M to be $P(1 + \eta + \eta^2)\eta^2$,

thus giving the pressure on the pin in the block B equal to $P(1 + \eta + \eta^2)^2$.

For the part T we find the value

$$P(1 + \eta + \eta^2)^2\eta,$$

and for O

$$P(1 + \eta + \eta^2)^2\eta^2a\ ;$$

therefore the force exerted on the hook U is

$$P(1 + \eta + \eta^2)^3\ .$$

For the load Q, which is equal to the sum of the tensions in the parts H, J, M, L, O, and T, we find

$$Q = P(\eta + \eta^2)[1 + (1 + \eta + \eta^2) + (1 + \eta + \eta^2)^2].$$

Such tackle find very little application in practice for raising heavy loads by means of a small effort, as even for moderate lifts the height required by the arrangement is considerable, which fact is apparent from Figs. 25 and 26. In Fig. 25, for example, a lift h of the load requires a space $2h$ between the pulleys A and B, and a space $4h$ between B and C, so that the height of the fixed pulley D above the lowest position of the load must be at least six times the lift h.

In Fig. 26 it appears that the arrangement is still more unfavourable in this respect. On the other hand, such systems of pulleys are frequently to be met with in the modern hydraulic hoists, where the stroke of the piston working under high pressure is greatly multiplied, and thus increases the range of motion of a comparatively small load. This is effected by inverting the system so as to carry the load at the free end of the rope, while the movable pulley, say A, Fig. 25, is acted upon by the effort P. More detailed information on this class of hoists will be found in the following.

On the other hand, the second class of tackle mentioned on page 46 is far more generally used, being employed for lifting heavy weights to considerable heights by the application of a small force, hand power for instance.

A clear idea of the mode of action is obtained from Fig. 27, in which the pulleys of each block are placed on separate pins, although in reality the arrangement generally preferred is to place the sheaves either side by side on a common stud in the block, or below each other on separate pins. The manner of suspending the load Q from the lower block which contains the sheaves C, E, G, and the mode of passing the rope over the sheaves, is evident from the figure.

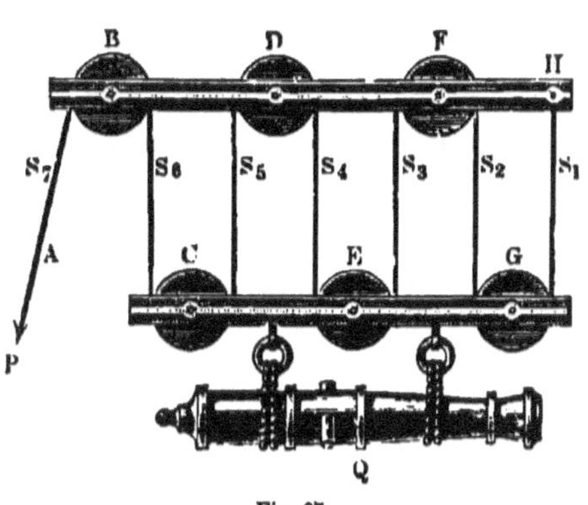

Fig. 27.

The most common arrangement is to place the same number of sheaves in each block, this number frequently being three, seldom or never more than four. One block may be given an extra sheave, however, for instance, by omitting the sheave G, and securing the rope to the lower block, or by leaving out the fixed sheave B, and allowing the driving force to act on the part rising from C.

In the arrangement represented in the figure the load Q is sustained by six plies of the rope, the tension in each, neglecting all wasteful resistances, being equal to the force P_0 in the hauling part BA. Therefore in this case $Q = 6P_0$; or in general for n carrying parts $Q = nP$. Taking friction and stiffness of rope into account, the tensions in the different parts will not be equal. Then if $S_1, S_2, S_3 \ldots S_7$ represent the tensions in the plies, and k the coefficient of resistance

$k = 1 + 2\sigma + 2\phi\dfrac{r}{r}$ (which is assumed the same for all the

sheaves), we have $S_2 = S_1 k$; $S_3 = S_2 k = S_1 k^2$, etc., and in general for the νth ply $S_\nu = S_1 k^{\nu-1}$. The tension S_7 in the pulling end A is given by $P = S_7 = S_1 k^6$, or in general for n sheaves, by $S_{n+1} = S_1 k^n$.

The load Q is found by means of

$$Q = S_1 + S_1 k + S_1 k^2 + \ldots S_1 k^5 = S_1 \frac{k^6 - 1}{k - 1};$$

or in general for n sheaves

$$Q = S_1(1 + k + k^2 + \ldots k^{n-1}) = S_1 \frac{k^n - 1}{k - 1}.$$

Since $P_0 = \dfrac{Q}{6}$, the efficiency of a tackle having six sheaves is

$$\eta = \frac{P_0}{P} = \frac{\frac{1}{6}Q}{S_1 k^6} = \frac{k^6 - 1}{6 k^6 (k - 1)},$$

and generally

$$\eta = \frac{k^n - 1}{n k^n (k - 1)}.$$

The pull on the upper block is given by

$$Z = S_1 + S_2 + S_3 + \ldots S_{n+1} = S_1(1 + k + k^2 + \ldots k^n) = S_1 \frac{k^{n+1} - 1}{k - 1}.$$

Therefore, when the lower block is fixed, and this force Z is employed to raise a load $Q = Z$, the efficiency is

$$\eta = \frac{P_0}{P} = \frac{\frac{1}{n+1}Q}{S_{n+1}} = \frac{\frac{1}{n+1}Q}{S_1 k^n} = \frac{k^{n+1} - 1}{(n+1) k^n (k-1)},$$

which expression, for the movable pulley, that is, for $n = 1$, becomes equal to the value already deduced, namely

$$\eta = \frac{k^2 - 1}{2k(k-1)} = \frac{k+1}{2k}.$$

In order to determine the efficiency for the reverse motion, with a view to calculating the force (P) which must be applied to the free end A of the rope so as to balance the load Q, we first substitute $\frac{1}{k}$ for k and find $(P) = \frac{S_1}{k^n}$, and as in this case

$$Q = S_1\left(1 + \frac{1}{k} + \frac{1}{k^2} + \cdots \frac{1}{k^{n-1}}\right) = S_1\frac{\frac{1}{k^n} - 1}{\frac{1}{k} - 1} = S_1\frac{k^n - 1}{k^{n-1}(k - 1)},$$

the efficiency becomes

$$(\eta) = \frac{(P)}{P_0} = \frac{S_1}{k^n}\ \frac{nk^{n-1}(k - 1)}{S_1(k^n - 1)} = \frac{n(k - 1)}{k(k^n - 1)}.$$

The graphical determination of the forces acting in a tackle is easily obtained by applying the methods previously employed.

Let A, Fig. 28, be the middle of, a horizontal diameter $BC = 2r$ of a sheave, measured from centre to centre of rope; lay off Bb and Cc equal to σ on the same side of the extremities of this diameter; make Aa_1 equal to the radius ϕr of the friction circle of the journal, so that $a_1b = r - \sigma - \phi r$, and $a_1c = r + \sigma + \phi r$. Now lay off upon the vertical through b the length $b1$ equal to the tension S_1 in the fixed part of the rope, and draw the straight line $1a_12$; then the portion $c2$ of the vertical through c will represent the tension S_2, as according to the construction

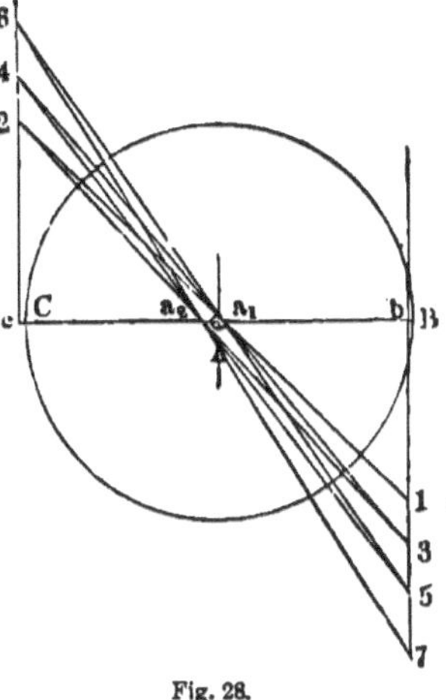

Fig. 28.

$$S_1(r + \sigma + \phi r) = S_2(r - \sigma - \phi r).$$

In order to find the tension in the remaining parts, we have only to make $ca_2 = ba_1$, and draw a line from 2 through a_2; then $b3$ will represent the tension S_3, and drawing $3a_1$ we shall have $c4 = S_4$ etc. Thus $b7$ will give the pulling force P for a tackle having six sheaves; and the load will be represented by the sum of the lengths :

$$b1 + c2 + b3 + c4 + b5 + c6 = Q.$$

This diagram also holds for the reverse motion; though in this case the pull exerted on the free end of the rope is represented by the length $b1$, when the running block descends under the action of a load

$$Q = b7 + c6 + b5 + c4 + b3 + c2.$$

The following table gives the efficiency of tackles both for the forward and reverse motions; the number of sheaves ranges from two to eight, and the chains and sizes of rope are those which occur most frequently in practice.

The relations of the radii of the pins and sheaves to the diameter of the rope or round iron composing the link (in the case of chains) are the same as those previously assumed for the fixed and movable pulley.

TABLE OF THE EFFICIENCY OF TACKLE.

$$\eta = \frac{k^n - 1}{n(k - 1)k^n}; \quad (\eta) = \frac{n(k - 1)}{k(k^n - 1)}.$$

Diam. of Rope.	$\delta =$	10 mm. ['30 in.]	20 mm. ['79 in.]	80 mm. (1'17 in.]	40 mm. (1'57 in.]	50 mm. [1'97 in.]	Chains.
2 sheaves .	η	0·913	0·856	0·808	0·764	0·723	0·930
	(η)	0·912	0·850	0·803	0·757	0·715	0·946
3 ,, .	η	0·834	0·817	0·754	0·702	0·656	0·917
	(η)	0·831	0·805	0·745	0·686	0·634	0·916
4 ,, .	η	0·858	0·776	0·706	0·647	0·597	0·900
	(η)	0·851	0·763	0·688	0·620	0·560	0·896
5 ,, .	η	0·833	0·739	0·663	0·598	0·544	0·880
	(η)	0·823	0·722	0·636	0·560	0·493	0·877
6 ,, .	η	0·807	0·706	0·624	0·555	0·496	0·863
	(η)	0·795	0·681	0·586	0·503	0·433	0·857
8 ,, .	η	0·762	0·645	0·552	0·479	0·422	0·827
	(η)	0·743	0·605	0·492	0·404	0·329	0·819

The table shows how in tackle operated by large ropes the resistances are considerably greater than in chain tackles, the efficiency of the latter being independent of the diameter of link iron, under the above assumption that the radii of pulleys and pins are directly proportional to the size of link iron. The preceding results also apply to the ordinary tackle, Fig. 29, page 55, in which sheaves of the same diameter are placed side by side in a block, and turn loosely on a common shaft or pin. Were the sheaves secured to the shaft, and the latter made to turn in bearings arranged in the blocks, it would be necessary to make the diameters of the sheaves proportional to the lengths of rope passing over them in order to avoid

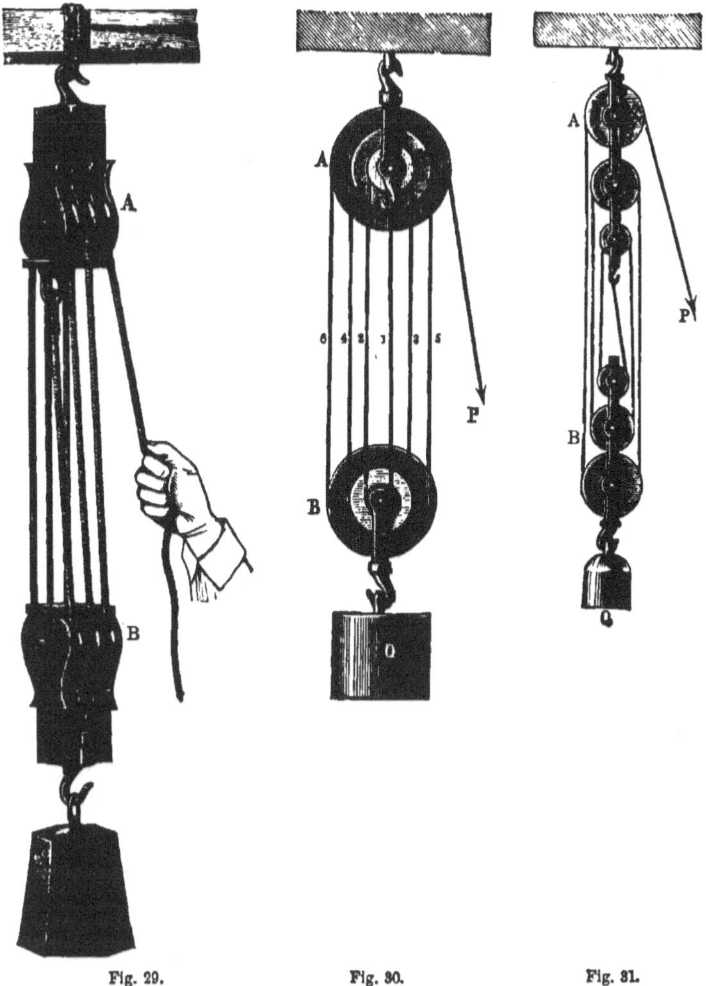

Fig. 29.   Fig. 30.   Fig. 31.

slipping; so accordingly, if the radius of the first sheave of the block B, Fig. 30, be placed equal to 1, the other sheaves of this block must have the radii 3 and 5, while those of the upper block must have the radii 2, 4, and 6, as the velocities of the plies of the rope vary as the numbers 1, 2, 3 4, 5, and 6. With this form of tackle, designed by *White*, a small gudgeon may be employed, thus reducing the friction, but it is not much used in practice, as the resistances due to stiffness of rope are largely increased when small sheaves are employed.

Besides, on account of the rope thickness it is impossible wholly to prevent slipping.

Sometimes the blocks are arranged one above another, Fig. 31, in which case the sheaves of each block have different diameters so as to prevent the plies of rope from rubbing against each other. The deductions already made are also applicable to this form of tackle, only we must here for each sheave substitute the value of

$$k = 1 + 2\sigma + 2\phi\frac{r}{r}$$

pertaining to it. This arrangement is not to be recommended, for here again the resistances due to stiffness of the rope are

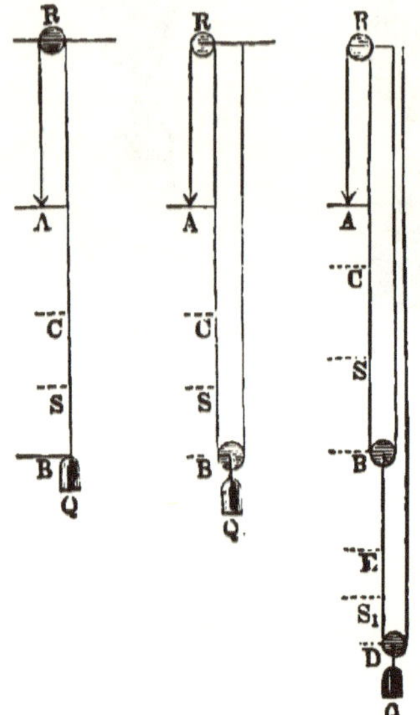

Fig 32.   Fig. 33.   Fig. 34.

unnecessarily large for the smaller sheaves, and moreover, owing to the increased length of the blocks, the efficient height to which the load can be raised is to some extent diminished.

So far the weight of the rope itself has been neglected. As regards its effect upon the pin friction and on the resistance due to stiffness of the rope, it is admissible in almost every case to neglect this factor on account of its insignificance. On the other hand, the work which must be expended for raising the weight of the rope, as also the work performed by the rope in its descent, demand special consideration in many cases.

Let the weight of a unit of length (metre) of rope be put equal to q, and let it be assumed that in the case of the fixed pulley R, Fig. 32, the point of application A of the effort, the position of the workman, for instance, is at a height $AB = a$ above the position of the load Q. In raising this load Q through a height $BC = h$, a portion of rope of the length h, and weight hq, and represented by its centre of gravity S, reaches the level of A. The energy expended in raising this weight of rope through a height $SA = a - \dfrac{h}{2}$, is

$$L = hq\left(a - \frac{h}{2}\right).$$

For $h = 2a$, $L = 0$, and for a greater lift L would even become negative; that is, the energy exerted by the portion of rope between R and A in its descent would be greater than the work performed in lifting the portion between Q and R; this case deserves attention when blocks are employed on high scaffoldings where the driving force is applied below.

In the movable pulley B, Fig. 33, when the load Q is raised through the same height $BC = h$, a weight of rope $= 2hq$ is lifted a distance $SA = a - \dfrac{h}{2}$; accordingly the corresponding work performed is expressed by $L = 2hq\left(a - \dfrac{h}{2}\right)$.

If a second pulley D, Fig. 34, is suspended from B at a distance a_1 therefrom, it will be seen that when the load Q has been lifted through a height $DE = h$, the pulley B will have passed over a distance $BC = 2h$; consequently the path of the lower piece of rope is expressed by $S_1S = a_1 + \dfrac{h}{2}$, and that of the upper piece by $SA = a - h$. Thus the work performed is found to be

$$L = 4hq(a - h) + 2hq\left(a_1 + \frac{h}{2}\right) = hq(4a + 2a_1 - 3h).$$

If a third movable pulley be suspended from D a distance a_2 below it, we shall in like manner find for the total work required to raise the three parts of rope:

$$L = 8hq(a - 2h) + 4hq(a_1 + h) + 2hq\left(a_2 + \frac{h}{2}\right)$$
$$= hq(8a + 4a_1 + 2a_2 - 11h).$$

In order to obtain the maximum effective lift in such arrangements, the pulleys must be brought as near together as possible in their lowest position; therefore, putting $a_1 = a_2 = 0$, we have

$$L = hq(8a - 11h),$$

and in general for n movable pulleys:

$$L = hq\left(2^n a - \frac{2^{2n-1} + 1}{3}h\right).$$

For an ordinary tackle with n sheaves which correspond to the fixed and movable pulleys in Fig. 33, we find the work consumed for lifting the rope to be

$$L = nqh\left(a - \frac{h}{2}\right).$$

EXAMPLE.—In an 8-sheaved tackle with a lift of 6 metres (19·68 ft.), and where the rope weighs 0·5 kilograms per metre (0·33lbs. per foot), the work required for lifting the rope is

$$L = 8 \times 0\cdot5 \times 6(5 - 3) = 48 \text{ metre kilograms (347 foot pounds)},$$

when the workmen are located 5 metres (16·41 ft.) above the load.

Assuming the workmen to stand at a height of 3 metres above the load, the work required for the same purpose would be reduced to zero, and were they placed on a level with the load, we should find

$$L = 8 \times 0\cdot5 \times 6(0 - 3) = -72 \text{ metre kilograms } [-522 \text{ ft. lbs.}]$$

Supposing the weight of the load $Q = 400$ kilograms [882 lbs.], the assistance rendered by the weight of the rope would be $\frac{72}{400 \times 6} = 3$ per cent of the total useful work. In determining the driving force, the weights of the running blocks are, of course, to be added to the load lifted. The effect of the weight of the sheaves upon the pin friction may be neglected, inasmuch as the action of this weight is to diminish the pressure on the pin of the lower block by the same amount as this pressure is increased in the upper block.

The use of tackle is confined chiefly to building operations and the rigging of ships. Such appliances are not adapted for heavy service on account of the inconvenience attending both the return of the running block when empty, and the handling of long ropes and chains. On the other hand, this form of hoisting gear is of

great value for temporary service in lifting moderate loads, which
accounts for its use in building operations, and also for manœuvring
sails, etc., on board ships. From the table on page 54 it appears
that, as a rule, the efficiency of a chain tackle is greater than that
of a large rope tackle.

§ 9. **The Differential Pulley-Block.** — The comparative
simplicity of this form of hoisting gear, as constructed by
Weston, has led to its extensive adop-
tion in machine shops and in building operations. Its name is due to
the fact that a movable pulley is
employed, which is caused to ascend
with a velocity proportional to the
difference of the motions of the two
parts of the chain passing over it.
The contrivance consists of two
pulleys running in separate blocks,
A and F, Fig. 35, the lower GH
being an ordinary movable pulley
carrying a hook J for attaching the
load Q. The upper pulley is provided
with two grooves for the chain, one
of these DE having a somewhat
larger diameter than the other BC.
An *endless chain* K connects the two
pulleys in a manner shown in the
figure, passing first over the smaller
pulley in the direction CB, then
downward to the movable pulley
which it supports in the loop GH,
and finally up over the large pulley
ED. The load is lifted by hauling on
the part DK, thus imparting motion
to the upper pulley in the direction
of the arrow. For if we imagine this
pulley to have moved through a
certain angle, say one complete

Fig. 35.

revolution, then a portion $2\pi R$ of the chain will be wound
on at E, while a portion equal to $2\pi r$ will be unwound on
the other side of the smaller pulley at B, when R and r

denote the radii of the two pulley grooves. In consequence, the portion of the chain which carries the movable pulley is shortened by the amount

$$2\pi(R-r),$$

so that the pulley F, together with its load Q, is lifted through a height $\pi(R-r)$ equal to one-half that distance. Since the effort P has travelled the path $2\pi R$, we obtain, when wasteful resistances are neglected, the theoretical force required

$$P_0 = Q\frac{R-r}{2R}.$$

As the tension in part CK of the chain is very slight, due as it is to the weight of this portion of the chain only, it is evident that the chain, unless checked, would slide down over the upper pulley on account of the far greater tension S, generated by the load Q in part BG; in order to prevent such sliding motion, the upper pulley is provided with pockets made to fit the links of the chain, Fig. 36. Here we find the reason why the use of ropes is not feasible in this type of hoists.

This hoisting gear, as commonly constructed, is capable of sustaining the load automatically, a feature of prime importance in many cases, although it is combined with the disadvantage of a small efficiency as previously referred to. In order to determine the latter as well as the actual effort P required, let R denote the radius of the larger pulley groove AD, r that of the smaller groove AB, and that of the movable pulley FG, the latter two most always being made of the same or nearly the same size; further, let r denote the radii of the pins

Fig. 36.

A and F, ϕ_1 the coefficient of friction for the chain, ϕ the coefficient of friction for the pin, and δ the size of the round iron of which the chain is made. Then, denoting the tension in the part BG by S, and that in HE by S_1, we shall have for the movable pulley F as before

$$S_1 r = Sr + \phi_1\frac{\delta}{2}S + \phi_1\frac{\delta}{2}S_1 + \phi r(S+S_1),$$

from which follows

$$S_1 = S\frac{1 + \phi_1\frac{\delta}{2r} + \phi\frac{r}{r}}{1 - \phi_1\frac{\delta}{2r} - \phi\frac{r}{r}} = Sk,$$

if, for the sake of brevity, we place

$$k = \frac{1 + \phi_1\frac{\delta}{2r} + \phi\frac{r}{r}}{1 - \phi_1\frac{\delta}{2r} - \phi\frac{r}{r}} \propto 1 + 2\sigma + 2\phi\frac{r}{r}.$$

Moreover, since

$$Q = S + S_1 = S(1 + k),$$

it follows that

$$S = \frac{Q}{1 + k} \text{ and } S_1 = \frac{k}{1 + k}Q,$$

as in the ordinary movable pulley.

For determining the equation of moments for the fixed pulley, we will regard the tensions S in the chain BG and P in the part DK as the driving forces, and the tensions S_1 in EH, together with all the wasteful resistances, as the opposing forces ; we then obtain the equation

$$PR + Sr = S_1R + \phi_1\frac{\delta}{2R}(P + S_1)R + \phi_1\frac{\delta}{2R}Sr + \phi(P + S + S_1)r,$$

from which, after dividing through by R, we get

$$P = S_1\frac{1 + \phi_1\frac{\delta}{2R} + \phi\frac{r}{R}}{1 - \phi_1\frac{\delta}{2R} - \phi\frac{r}{R}} - S\frac{r}{R}\frac{1 - \phi_1\frac{\delta}{2r} - \phi\frac{r}{r}}{1 - \phi_1\frac{\delta}{2R} - \phi\frac{r}{R}}.$$

Since R and r always differ by a small fraction only (ordinarily $r = \frac{9}{10}R$ to $\frac{14}{15}R$), we can put

$$1 - \phi_1\frac{\delta}{2R} - \phi\frac{r}{R} = 1 - \phi_1\frac{\delta}{2r} - \phi\frac{r}{r},$$

and thus place the coefficient of S_1 equal to k. As result we obtain

$$P = S_1k - S\frac{r}{R} = Q\frac{k}{1 + k}k - Q\frac{r}{R}\frac{1}{1 + k}.$$

Placing the ratio $\dfrac{r}{R} = n$, we have

$$P = Q\frac{k^2 - n}{1 + k},$$

and as we have found

$$P_0 = Q\frac{R - r}{2R} = Q\frac{1 - n}{2},$$

we obtain for the efficiency

$$\eta = \frac{P_0}{P} = \frac{1 - n}{2}\,\frac{1 + k}{k^2 - n}.$$

For the reverse motion of the block all wasteful resistances act in the opposite direction; hence, in order to obtain the formulæ applicable to this motion, we have only to give the terms containing ϕ and ϕ_1 opposite algebraic signs; that is, in place of the quantity

$$k = \frac{1 + \phi_1\frac{\delta}{2r} + \phi\frac{r}{r}}{1 - \phi_1\frac{\delta}{2r} - \phi\frac{r}{r}} \propto 1 + 2\sigma + 2\phi\frac{r}{r},$$

we substitute the value

$$\frac{1}{k} = \frac{1 - \phi_1\frac{\delta}{2r} - \phi\frac{r}{r}}{1 + \phi_1\frac{\delta}{2r} + \phi\frac{r}{r}} \propto 1 - 2\sigma - 2\phi\frac{r}{r}.$$

Thus we find for the reverse motion the force applied

$$(P) = Q\frac{\dfrac{1}{k^2} - n}{1 + \dfrac{1}{k}} = Q\frac{1 - nk^2}{k^2 + k},$$

and for the efficiency

$$(\eta) = \frac{(P)}{P_0} = \frac{2}{1 - n}\,\frac{1 - nk^2}{k^2 + k}.$$

In the differential block the radii R and r of the fixed chain pulley depend on the proportions of the links of the chain, as the inside length of the link must be contained an even number of times in the circumference of each pulley groove. The two circumferences are frequently made the length of 20 and 18 links, occasionally as many as 30 and 28, so that the ratios $n = \dfrac{r}{R}$ become equal to $\dfrac{9}{10}$ and $\dfrac{14}{15}$ respectively. For ordinary chains we can put the length of a link $l = 2\cdot6\delta$; thus, for 18 links the radius of the small pulley grove becomes

$$r = \frac{18 \times 2\cdot6}{2\pi}\delta = 7\cdot45\delta,$$

and for a radius of pin $\mathrm{r} = 1\cdot5\delta$,

$$\frac{\mathrm{r}}{r} = \frac{1\cdot5}{7\cdot45} = 0\cdot2 ; \text{ hence } 2\phi\frac{\mathrm{r}}{r} = 2 \times 0\cdot08 \times 0\cdot2 = 0\cdot032.$$

For 28 links, however,

$$r = \frac{28 \times 2\cdot6}{2\pi}\delta = 11\cdot6\delta ; \text{ hence } \frac{\mathrm{r}}{r} = \frac{1\cdot5}{11\cdot6} = 0\cdot13,$$

and

$$2\phi\frac{\mathrm{r}}{r} = 2 \times 0\cdot08 \times 0\cdot13 = 0\cdot021.$$

The resistances due to stiffness in winding the chain on and off the pulley, corresponding to a coefficient of friction $\phi_1 = 0\cdot2$, are found to be

$$2\phi_1\frac{\delta}{2r} = 0\cdot2\frac{\delta}{7\cdot45\delta} = 0\cdot027,$$

and

$$2\phi_1\frac{\delta}{2r} = 0\cdot2\frac{\delta}{11\cdot6\delta} = 0\cdot017.$$

Accordingly the value

$$k = 1 + 2\sigma + 2\phi\frac{\mathrm{r}}{r}$$

will be equal to

$k = 1 + 0\cdot027 + 0\cdot032 = 1\cdot059 \backsimeq 1\cdot06$ for pulleys holding 18 links.
$k = 1 + 0\cdot018 + 0\cdot021 = 1\cdot039 \backsimeq 1\cdot04$ for pulleys holding 28 links.

The slight difference between these values shows that the assumption made in the preceding pages, according to which the value of k for the large pulley was made equal to that for the small pulley, is approximately correct.

In the following table will be found the values of $\dot{\eta}$ and (η), corresponding to the ratios

$$n = \frac{r}{R} = 0{\cdot}75,\ 0{\cdot}80,\ 0{\cdot}85,\ 0{\cdot}9,\ 0{\cdot}933,$$

for a mean value of

$$k = 1 + 2\sigma + 2\phi\frac{r}{r} = 1{\cdot}06.$$

TABLE OF THE EFFICIENCY OF DIFFERENTIAL PULLEY-BLOCKS.

$$\eta = \frac{1-n}{2}\,\frac{1+k}{k^2-n}\,;\quad (\eta) = \frac{2}{1-n}\,\frac{1-nk^2}{k^2+k}\,;\quad k = 1{\cdot}06.$$

Ratio of Pulleys $n =$	0·75	0·80	0·85	0·90	0·933
$\eta =$	0·688	0·637	0·565	0·460	0·359
$(\eta) =$	0·575	0·462	0·272	− 0·106	− 0·668

This table plainly shows the slight efficiency of ordinary differential pulley-blocks with slightly differing radii, and it is therefore evident that they are also to be counted among the kinds of hoists which are not to be recommended for *continuous* service, on account of their want of economy of power. On the other hand, owing to their self-sustaining property, they must be regarded as useful apparatus for *occasional* use in machine shops, erecting shops, and in building operations. They possess the advantage over the screw-jack of being more easily adjusted for greater lifts, besides being simpler in construction and mode of application. Their disadvantage, in common with all chain pulleys, consists in the stretching of the chain under the stress to which it is subjected, the result being that the links no longer fit accurately in the pockets of the pulley groove.

The negative value of (η) signifies that the block is self-sustaining, and the limit of the ratio $n = \dfrac{r}{R}$, at which this self-locking property commences, is found from the equation

$$(\eta) = 0,\ \text{hence } 1 - nk^2 = 0,\ \text{or } n = \frac{1}{k^2}.$$

For example, if $k = 1\cdot06$, the limiting value of this ratio is given by $n = \dfrac{r}{R} = 0\cdot889.$

In order to determine graphically the pull P which is necessary to drive the differential pulley we may proceed as follows :

If GH, Fig. 37, is the movable pulley, then draw in the direction in which the load Q is assumed to act the vertical tangent ff_1 to the friction circle of the pin F, which circle has a radius ϕr. Further, draw the lines Mg and Oh, which represent the tensions S and S_1 in the chain parallel to the line of action of Q, and at distances $\sigma = \phi_1\dfrac{\delta}{2}$ from the tangents at G and H, the former at G away from the latter at H, toward the centre F. Now, divide the load Q = M1 at N, so that

$$MN : N1 = Of_1 : f_1M ;$$

for this purpose draw MO perpendicular to M1, connect O with 1, and through f_2, the point of intersection between the lines O1 and f_1f, draw the horizontal line f_2N ; thus MN

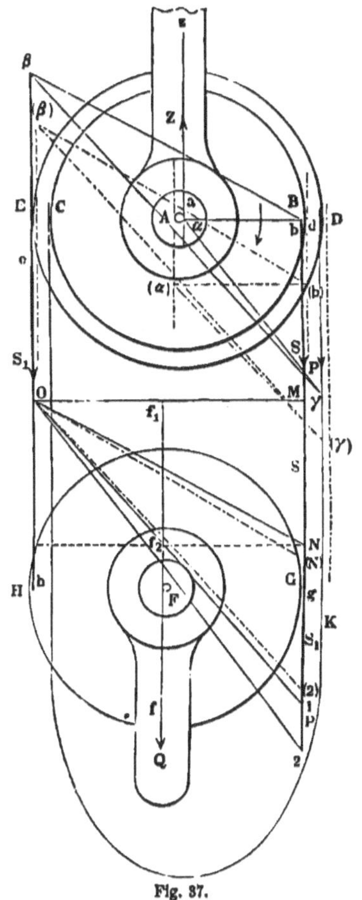

Fig. 37.

= S will give the tension in the chain BG, and N1 = S_1 that in EH. These forces are supposed to act on the fixed pulley along the verticles b and e, which lie at a distance σ from the vertical tangents at B and E, and are located inside the chain which winds off at B and outside that which winds on at E. Besides, the force P acts in the vertical d at a distance σ from the tangent at D. These three parallel forces are held in equilibrium by the reaction $Z = S + S_1 + P$ of the bearing A, which reaction is represented by az, located at a distance

Fig. 38.

$\phi_1 r$ from the centre A. The problem is therefore so to determine P that the resultant of S, S_1, and P shall coincide with za. Here we may use the polygon of forces MN1 with O as the pole. For this purpose draw through any point b of the line of action of S, the lines ba and $b\beta$ parallel to the pole rays OM and ON; further, draw $\beta\gamma$ parallel to O1, and join $a\gamma$. Then $a\gamma$ will give the direction in which O2 is to be drawn, which line together with O1 determines the length 12, which, measured according to the scale assumed, will represent the required force P.

If we wish to investigate the reverse motion by this method, S and S_1 must be interchanged; that is, make $S = M(N) = 1N$, lay off σ and ϕr in directions opposite to those of the preceding construction and, with the aid of the lines of action of the forces, complete the force and equilibrium polygons, as shown in the figure by the dotted lines. The length 1(2) represents the force (P), and the upward direction in which it is drawn shows that the tackle is self-locking, as during the reverse motion the upward force $(P) = 1(2)$ turns the pulley D in the same direction in which the load Q tends to turn it.

§ 10. **Other Forms of Tackle.**—A great variety of tackle have been constructed, a few of which we will now describe in detail. In the arrangement shown in Fig. 38, the load Q is suspended from the hook G of a running block F, one end of the chain being made fast to the upper block at C, while the other end is carried over the pulley AB and down to N, where it is secured to the chain CD by means of a ring. As in the differential tackle, the pulley B is provided with pockets for the links of the chain, in order to prevent slipping. Motion is imparted to this pulley B by a worm wheel M attached to it, and gearing with a worm S, which is operated by an endless chain K passing over the pulley J. For heaviei loads two or more pulleys can be connected side by side on the spindles A and F, as in the ordinary tackle. By the employment of worm gearing this hoisting gear is made self-locking, so that in order to lower the load it is necessary to apply power to the other end of the driving chain K. As has been previously explained, the efficiency of this arrangement is considerably reduced by the use of worm gearing, and for this reason the conclusions arrived at in the case of jack screws and differential pulley-blocks hold for this form of hoisting apparatus as well. In accordance with our previous deductions, the efficiency, as also the driving force, are found to be $\eta = \eta_1\eta_2$, where η_1 represents the efficiency of the tackle, and η_2 that of the worm gearing (see tables, pages 54 and 30). Take, for instance, the case of a chain tackle with two sheaves where we had found $\eta_1 = 0{\cdot}93$, combined with a worm and worm gear of efficiency $\eta_2 = 0{\cdot}35$, then we should have

$$\eta = 0{\cdot}93 \times 0{\cdot}35 = 0{\cdot}325.$$

Assuming the worm wheel M to have 15 teeth, and the radius of its pitch circle to be r, while that of the driving pulley J is equal to $2{\cdot}5r$, then the theoretical pulling force will be given by

$$P_0 = Q\frac{1}{2}\frac{1}{15}\frac{r}{2{\cdot}5r} = \frac{1}{75}Q = 0{\cdot}0133Q,$$

while the actual pulling force is

$$P = \frac{0 \cdot 0133}{0 \cdot 325} Q = 0 \cdot 0409 Q.$$

Thus a force of 4·09 lbs. would have to be exerted for each 100 lbs. lifted, whereas in the absence of wasteful resistances only 1·33 lbs. would be necessary.

In another form of hoisting apparatus, as constructed by *Collet* and *Engelhard* of Offenbach, and which properly comes under the head of windlasses, a worm driven by a chain wheel and chains is also made use of. This worm communicates opposite rotations to two worm wheels placed on the shafts of two chain-drums. Consequently the two chains, from which the load is directly suspended without the use of a movable pulley, are shortened by equal mounts, so that the load rises with double the velocity to what would be the case were a movable pulley em-ployed. With reference to ascer-taining the useful effect, we may regard this arrangement as a combination of two equal wind-lasses, each of which lifts half the load. The efficiency of the whole apparatus is then equal to the product of the efficiency of the worm gearing into that of the drum, the wasteful resistances of the latter consisting of the friction of the journals and the resist-ance due to friction in the chain as it winds on to the drum.

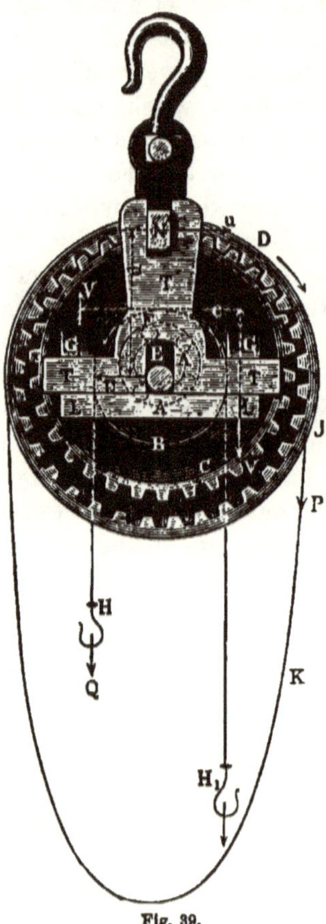

Fig. 39.

For information on this point we refer to the following paragraphs.

A peculiar tackle shown in Fig. 39 has been invented by

Eade.[1] Here the chain which sustains the load passes over a chain pulley B provided with pockets to prevent slipping, so that the load Q can be suspended from either end H or H_1 of the chain. The chain pulley B, which is loose on the stud A, has cast into it an internal gear D, which latter is driven by a spur wheel C, having one tooth less than D. The centre E of this wheel is therefore at a distance $AE = e = \dfrac{t}{2\pi}$ from the centre of A, t representing the pitch of the gears. The stud A is provided with an eccentric A$'$, on which the driver C turns loosely, friction rollers F being also introduced in order to reduce the friction of the eccentric. Rotation is communicated to the stud A by means of the chain pulley J, the latter being operated by the chain K as in the tackles just described. The centre E of the driver will then revolve about the stud A in a circle having the radius e, while the wheel C is prevented from turning about its own axis. The result is that, while the centre E moves in the aforesaid circle about A, a line connecting any two points of the wheel C will always remain parallel to itself. All points of C must move in circles described with radius e, and therefore the motion will not be one of rotation, but of translation in a circular path. In order to accomplish this movement a peculiar $\perp$-shaped piece is employed on the horizontal arms T of which the lugs G and L of the wheel C slide back and forth; while the upright arm is guided by the stirrup N, and the stud A is capable of a vertical motion, in which the driver C must take part on account of the lugs G and L. From this it is evident that by combining a horizontal and a vertical movement it is possible to impart a circular translation at each instant to the wheel C.

In consequence of this arrangement, it follows that for one complete revolution of the chain wheel J and shaft A the toothed wheel D is made to turn through the space of one tooth, as the following considerations will show. Let us first suppose *both* the toothed wheels and the shaft A to make one complete turn in the direction of the arrow, and then assume the wheel C to be turned in the opposite direction about its axis E through one complete revolution. The effect of the

---

[1] See *Engineer*, 1867, p. 135, and *Zeitsch. Deutsch. Ing.* 1868, p. 27.

latter motion is to cause the wheel D to turn back through a fractional part $\frac{z_2}{z_1}$ of a revolution, where z_2 denotes the number of teeth in the driver C, and z_1 the number in the internal wheel D.   Accordingly, for one complete revolution of the chain wheel J with its shaft A, the internal wheel D with the pulley B has been turned in the same direction as the shaft A through the fraction

$$+ 1 - \frac{z_2}{z_1} = \frac{z_1 - z_2}{z_1}$$

of a revolution, *i.e.* through the space of one tooth, if we assume z_2 to be one less than z_1.   From this we may also deduce the velocity-ratio or the relation between the distances traversed by the points of application of driving force and load.   Supposing the shaft A to have made one revolution, then the force will describe a path $2\pi R$, and the load a path $\frac{1}{z_2} 2\pi r$, when r denotes the radius of the pulley B, and R that of the driving wheel J.   In the tackle constructed by *Eade*, $z_1 = 31$, $z_2 = 30$, and therefore for a ratio $\frac{r}{R} = \frac{1}{2}$, the velocity ratio would be $\frac{1}{62}$, hence the theoretical effort

$$P_0 = \frac{z_1 - z_2}{z_1} \cdot \frac{r}{R} Q = \frac{1}{62} Q = 0 \cdot 0161 Q.$$

The actual driving force required is considerably larger than this, owing to the large wasteful resistances, which in addition to σQ and σP, due to bending the chain over the pulleys B and J respectively, consist of the following: the frictional resistances of the shaft A in the hanger N, those of the chain pulley B on A, and the wheel C on its eccentric E, the friction of the teeth of the wheels, and, finally, the resistances due to the sliding friction of the piece T against the lugs G and L, the stirrup N, and the shaft A.

These resistances are computed in the following manner : Let r be the radius of the pulley B upon which Q acts, and R the radius of the chain wheel J to which the force P is applied; further, let r_1 and r_2 denote the respective radii of the pitch circles of the internal wheel D having z_1 teeth, and of the driver C having z_2 teeth, and r the radius of the journal of the shaft A, thus making $r + e = r + (r_1 - r_2)$ equal to the radius of the eccentric AA$'$.   Then we find, in the first place, the pressure Q_1 between the teeth, from

$$Q(1 + \sigma)r + \phi(Q + Q_1)\mathfrak{r} = Q_1 r_1,$$

which gives

$$Q_1 = Q\frac{r(1 + \sigma) + \phi\mathfrak{r}}{r_1 - \phi\mathfrak{r}}.$$

Here we have again assumed the most unfavourable case for computing the journal friction, viz., that the point of contact of the wheels is diametrally opposite the point where the driven chain winds on to the pulley B, which gives the reaction of the bearing equal to $Q + Q_1$.   The resistance due to the friction of the teeth is given by

$$\zeta Q_1 = 0\cdot 33\left(\frac{1}{z_2} - \frac{1}{z_1}\right)Q_1,$$

hence there must be overcome at the point of contact of the gear a resistance

$$Q_2 = Q_1(1 + \zeta) = Q_1\left[1 + 0\cdot 33\left(\frac{1}{z_2} - \frac{1}{z_1}\right)\right].$$

Besides, at the centre E of the eccentric there is a force P_1 acting on the driver parallel to Q_2.   The tendency of these two opposite forces is to impart to the driver C a left-handed rotation, which is, however, resisted by the T-shaped piece, the latter applying two equal and opposite pressures VV upon the lugs L and G, forming a right-handed couple whose moment is Vc, where c is the lever arm of the couple, and the size of which is equal to the moment $Q_2 r_2$ of the circumferential resistances taken relatively to the centre E of the eccentric.   The two pressures $-$ V, exerted by the lugs on the piece T, give rise to two reactions U of the shaft A and the stirrup N, the moment of which $d \times$ U must be placed equal to cV.   By these four forces V and U four frictional resistances are produced at G, L, A, and N, which may be expressed by μV and μU respectively, their distances travelled for each revolution of the shaft A being equal to $4e$.   Now it is evident that the forces V and U will vary with the position of the eccentric, as in the slotted slider crank, consequently the force P_1 acting at E will not be strictly constant. But a mean value of same, during one revolution of the shaft A, is found from the equation pertaining to the work done by the various forces:

$$P_1 2\pi e = Q_2 \frac{z_1 - z_2}{z_2} 2\pi r_2 + \phi P_1 2\pi(r + e) + 2\mu V 4e + 2\mu U 4e.$$

If we wish to consider the influence of the friction rollers upon the magnitude of P_1, we must substitute $\nu\phi$ for ϕ in the term $\phi P_1 2\pi(r + e)$, which expresses the work performed in overcoming this friction, ν representing the ratio of the radius of the pin to the radius of the rollers.

From the above expression we obtain the force P_1 which must be exerted at the centre E of the eccentric, and which requires the application of a force P to the driving chain. Taking into account the friction of the chain and that of the shaft A in its bearings, we obtain the equation

$$P(1 - \sigma)R = P_1 e + \phi(P + P_1)r,$$

from which

$$P = P_1 \frac{e + \phi r}{(1 - \sigma)R - \phi r}.$$

Expressing P directly in terms of known quantities would involve the use of some inconvenient formulæ, and therefore an example has been chosen to elucidate the subject.

EXAMPLE.—In a differential tackle of this form let $z_2 = 30$, and $z_1 = 31$ be the number of teeth of the gears, $r_2 = 150$ mm. [5·91 in.] and $r_1 = \frac{31}{30} 150 = 155$ mm. [6·11 in.] the radii of the pitch circles, and thus $e = r_1 - r_2 = 5$ mm. [0·20 in.] Further, let $r = 80$ mm. [3·15 in.] be the radius of the pulley B, R = 160 mm. [6·30 in.] the radius of the chain wheel J, r = 15 mm. [0·59 in.] the radius of the shaft A, thus the radius of the eccentric $r + e = 20$ mm. [·79 in.] Then, for coefficients of journal friction $\phi = 0·08$, and sliding friction $\mu = 0·15$, and assuming a value $\sigma = 0·2 \frac{\delta}{20\delta} = 0·01$, we find the force Q_1 acting in the teeth of the internal wheel

$$Q_1 = Q \frac{80(1 + 0·01) + 0·08 \times 15}{155 - 0·08 \times 15} = 0·533 Q.$$

Taking friction of the teeth into account, we obtain the resistance in the circumference of the driver

$$Q_2 = \left[1 + 0\cdot 33 \left(\frac{1}{30} - \frac{1}{31} \right) \right] Q_1 = 0\cdot 534 Q.$$

If now the arms of the couples VV and UU be $c = d = 200$ mm. [7·87 in.], which supposition also makes $V = U$, and friction rollers be employed with a ratio of the journal $\nu = \frac{1}{3}$, then the force P_1 acting at the centre E of the eccentric will be obtained from

$$P_1 5 = 0\cdot 534 Q \frac{1}{30} \times 150 + \frac{1}{3} 0\cdot 08 P_1 20 + \frac{1}{2\pi} 4 \times 0\cdot 15 \times 4 \times 5 \times \frac{0\cdot 534 Q 150}{200},$$

which gives

$$P_1 = \frac{3\cdot 340}{4\cdot 47} Q = 0\cdot 747 Q.$$

Hence we obtain finally the driving force

$$P = 0\cdot 747 Q \frac{5 + 0\cdot 08 \times 15}{0\cdot 99 \times 160 - 0\cdot 08 \times 15} = 0\cdot 0294 Q,$$

which indicates that every 100 lbs. lifted requires an exertion of 2·94 lbs.

Since without friction

$$P_0 = \frac{1}{31} \frac{80}{160} Q = 0\cdot 0161 Q,$$

the efficiency is found to be

$$\eta = \frac{P_0}{P} = \frac{0\cdot 0161}{0\cdot 0294} = 0\cdot 548.$$

Assuming the above data, therefore,. *Eade's* tackle apparently gives a greater efficiency than the common differential tackle, about 45 per cent of the energy exerted being lost in overcoming the wasteful resistances, and 55 per cent being employed in lifting the load.   In order to ascertain whether the tackle is self-locking, we must give opposite signs to ϕ, μ, ζ, and σ.   If these substitutions give a positive value for (P), we must conclude that, for the above dimensions, the apparatus does not possess the self-locking feature. On the whole, this tackle deserves no special recommendation on account of its small efficiency, and, even for the case that it were made self-locking, the differential pulley block is to be preferred owing to its greater simplicity.

CHAPTER III

§ 11. **Windlasses.** — The various forms of tackle described above are employed chiefly for lifting moderate loads. For heavier service and greater lifts a *drum* is ordinarily made use of around which the rope or chain is coiled. The arrangement is then called a *windlass*, and is identical in principle with that shown in Fig. 40, though instead of rotating the drum by means of spoke wheels or hand spikes, a large toothed wheel is placed on the drum shaft and made to gear with a pinion on a separate crank shaft. The action of such gearing has already been explained in § 3.

Fig. 40.

A simple windlass worked by spokes is shown in Fig. 41. It consists of a tripod ABCD, two legs AD and BD of which are firmly joined in order to serve as a support for the drum E and guide pulley F, while the third leg C is jointed to the others by the pin D. When the requisite space or support is wanting, this third leg may be dispensed with, and the permanent legs stayed by one or two ropes, DH, termed guys, which are securely anchored to the ground. The resolution of the load Q into components acting in the directions of the legs AD and BD and the strut CD will give the pressures on the legs, or, when guys are used, will give the pull which the anchor stay must withstand.

In place of a tripod, a *derrick*, as illustrated in Fig. 42,

may be used. Here the post AB is mortised into a timber
platform S, and is held in its vertical position by several
ropes or chains radiating from the top B. The load Q is
suspended from a movable pulley E, one end of the rope being
secured to the cross-beam at K, while the other end is led
over the guide-pulleys F, G, and H to the drum of a geared
windlass W, which is mounted on the platform S. By employing the movable pulley the purchase is doubled, but its use

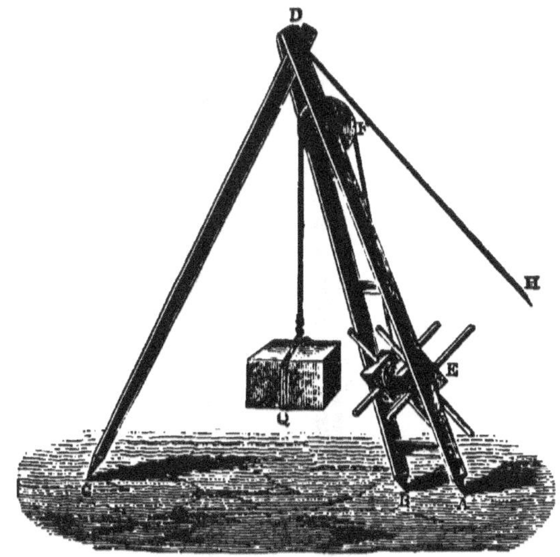

Fig. 41.

necessitates a correspondingly longer drum for holding the
increased length of rope.

For lifting very heavy loads, such as locomotives, marine
boilers, etc., tackle containing from four to eight sheaves are
frequently used in place of single blocks, not only for the sake
of gaining an increase of power, which might easily be effected
by inserting an additional pair of gears, but chiefly with a
view to avoiding excessively heavy chains, which in turn
would necessitate the use of very large winding drums.

Fig. 43 shows the arrangement of the windlass employed.
The drum D has its bearings in cast-iron standards ABC, and
carries the large spur wheel H, which gears with the pinion
G on the crank shaft E operated by the cranks F and F'.

The pawl *s*, which either engages with a special ratchet wheel on the crank shaft or directly with the teeth of the pinion,

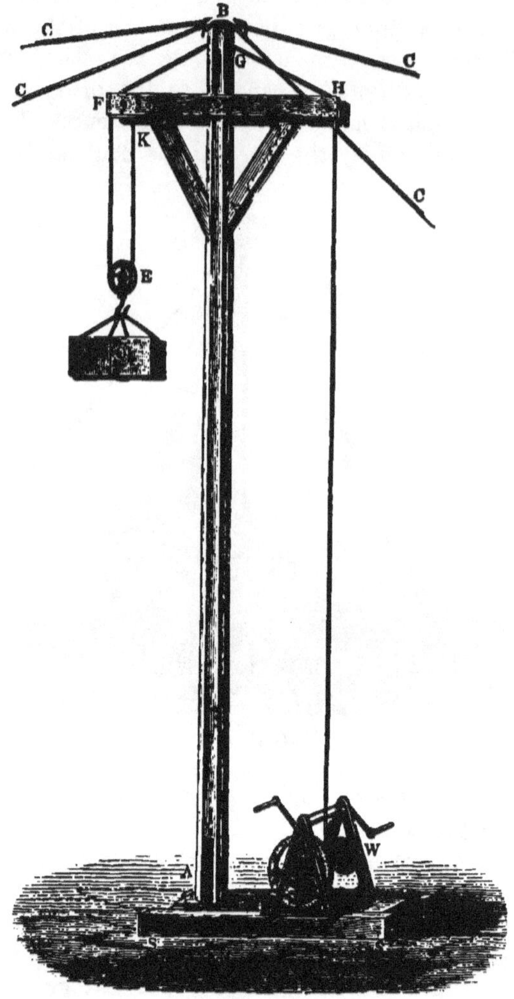

Fig. 42.

prevents the drum from reversing its motion under the action of the load when the application of power ceases — this device being necessary in all windlasses which are not self-locking. If it is desired to uncoil the rope from the drum, a too rapid rotation of the crank shaft may be avoided by shifting the latter in its bearings, thus throwing the pinion and wheel H out of gear. A *latch f* suspended from the cross-stay BB′ prevents any unintentional shifting of the crank shaft, and must be swung out of the position indicated when the shaft is to be shifted.

A double-geared windlass is shown in Fig. 44. Here the drum G receives its motion from the crank shaft A through the medium of two pairs of gears B, C and D, E, of which the

Fig. 43.

pinion D and the wheel C are keyed to an intermediate shaft H. With this arrangement, which is used when heavier weights are to be lifted, the velocity of rotation of the drum will be only a fractional part $\dfrac{BD}{CE}$ of that of the crank shaft.

For obtaining a more rapid rotation of the drum, when lighter loads are to be lifted, the windlass is also arranged to operate with a single pair of gears only, the wheels B and C being then disengaged by shifting the crank shaft A axially, thus allowing the pinion F on the shaft A to gear directly into the wheel E on the drum shaft. The ratchet wheel S again sustains the load until the pawl S_1 is raised clear of the teeth.

Unless provision be made to check the motion when the pawl
is released, the load would drive the mechanism backward with
an accelerated motion, which might easily lead to breakages,
and by the rapid rotation of the cranks endanger the lives of
the workmen.   To prevent such an occurrence, machines of
this kind are always provided with a brake, which generally
consists of a drum or pulley N encircled by an iron band or
strap operated by a lever L.   The arrangement and mode of

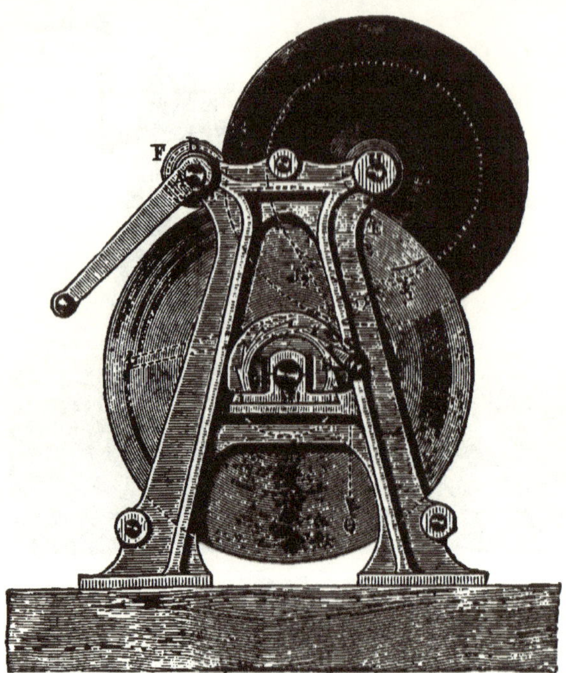

Fig. 44.

action of such friction brakes have been fully investigated in
vol. iii. 1, § 178, Weisb. *Mech.*, where attention is called to
the fact that the most advantageous results are obtained when
the brake is applied to a rapidly revolving shaft.   The friction
pulley is therefore generally placed either on the intermediate
or on the crank shaft, and only in rare cases is it attached to
the drum.   In the above-mentioned paragraph it is also
pointed out that the brake lever should be so arranged that it
will act directly on the driven end J_2 of the strap, and not

on the working end J_1, where the tension is considerably greater.

In smaller winches and windlasses operating with a rope the drum is frequently made of wood with wrought-iron gudgeons, while in hoisting apparatus of greater capacity it is usually made of cast iron, hollowed in the centre and secured to a wrought-iron shaft or provided with wrought-iron gudgeons only. Rope drums usually have a smooth cylindrical surface, while chain drums are more serviceable when made with a spiral groove for the upright links to drop into, the flat links obtaining their bearing on the cylindrical surface of the drum. This is evidently done for the same reason that chain wheels are given the cross-section shown in Fig. 45. When practicable the drum should be made of sufficient length to provide room for the full length of the rope or chain in a single layer. Allowing several layers to form on the drum is subject to various objections, as for instance, more rapid wear of ropes or chains, besides increasing the leverage of the load by an amount equal to the thickness of the rope or chain. It is only when the latter is very long, and thus would require inconvenient dimensions of the drum, that overlapping is allowed to take place, but in that case the drum must be provided with flanges in order to prevent the coils from running off; this end is sometimes attained by making the drum concave at the centre, thus forcing the coils toward the middle.

Fig. 45.

The determination of a suitable diameter for the drum is a matter of great importance. If it is made too small the resistances due to stiffness of rope or chain are disproportionally large, whereas an unnecessarily large diameter increases the lever arm of the load, thus leading to more inconvenience in obtaining the required velocity ratio, besides causing a waste of material. If δ denotes the thickness of rope or diameter of round iron when a chain is used, the diameter D of the drum should not be smaller than:

For hempen rope: $D = (7 \text{ to } 8) \, \delta$ (δ = thickness of rope),
For wire rope: $D = 1100 \, \delta$ (δ = diam. of wire),
For chains: $D = (20 \text{ to } 24) \, \delta$ (δ = diam. of iron).

In order not to subject the shaft of the drum to a torsional strain the large spur wheel is generally bolted directly to the drum instead of being keyed to the shaft, the drum being provided with a flange for this purpose. The gears are generally made of cast iron, and it is only for pinions or to secure additional safety that wrought iron or steel are used as material; the shaft should always be made of wrought iron or steel.

When a windlass is to be worked by several men, and to this end is provided with two cranks, it is advisable to place these at an angle of 180° in order to equalise the power exerted by the men operating the cranks. As the resistance to be overcome in a hoisting machine is uniform, fly-wheels need not be employed; in fact, their use would render difficult the exact stopping of the load, and under certain conditions might lead to breakages. In machines of this class, where owing to the varying resistance a fly-wheel cannot be dispensed with, for instance in the case of dredging machines, the wheel should be given a yielding connection with the hoisting drum, by the introduction of a friction-coupling, for instance.

For all windlasses the ratio of the theoretical driving force P_0 to the load Q is directly given by the velocity ratio, and therefore denoting the radius of the drum by r, and the length of the crank by R, we have

$$P_0 = Q\frac{r}{R}n_1 n_2 \ldots$$

where n_1, n_2 . . . represent the velocity ratios of the respective pairs of wheels; that is, the ratios of the number of teeth in the drivers to the number of teeth in the followers.

The actual driving force P is easily deduced from what has preceded. When the load Q acts with a lever arm equal to the radius r of the drum (measured to the centre of the rope or chain), the force P_1, which must be exerted at the radius R_1 of the large gear on the drum, taking into account friction of the journals and the resistance due to stiffness, is determined from the equation

$$P_1 R_1 = Q(1 + \sigma)r + \phi(P_1 + Q)r,$$

which gives

$$P_1 = Q\frac{(1 + \sigma)r + \phi r}{R_1 - \phi r}.$$

Here, as before, r denotes the radius of the journal of the drum shaft, and σ the coefficient due to stiffness of the rope or chain, the most unfavourable case being assumed, namely, that the forces Q and P_1 are parallel and acting each side of the centre of the journal. If the weight G of the drum were to be taken into account, which would be necessary only in the case of large winding drums for wire rope, we should have

$$P_1 = \frac{Q(1+\sigma)r + \phi(Q+G)r}{R_1 - \phi r}.$$

As the theoretical force in the circumference of the large gear of radius R_1 is given by $Q\dfrac{r}{R_1}$, the efficiency of the drum is

$$\eta_1 = \frac{Q\dfrac{r}{R_1}}{P_1} = \frac{\dfrac{r}{R_1}(R_1 - \phi r)}{(1+\sigma)r + \phi r} = \frac{1 - \phi\dfrac{r}{R_1}}{1 + \sigma + \phi\dfrac{r}{r}}.$$

Again, let η_2 and η_3 denote the respective efficiencies of the first and second pairs of wheels, each of these coefficients representing the product $\eta'\eta''$ of the efficiency of the teeth (see table, page 14), and that of the gear shaft (see table, page 17), then the total efficiency of the windlass is $\eta = \eta_1\eta_2\eta_3$, and consequently

$$P = \frac{1}{\eta}P_0 = \frac{1}{\eta}Q\frac{r}{R}n_1 n_2.$$

The coefficient σ for the stiffness is estimated as in the case of pulleys

$$\sigma = \phi_1\frac{\delta}{2r} = 0\cdot2\frac{\delta}{2r} \text{ for chains,}$$

and

$$\sigma = 0\cdot009\frac{\delta^2}{r} \text{ for hempen rope}$$

$$\left[\sigma = 0\cdot23\frac{\delta^2}{r}\right].$$

Assuming the most unfavourable case, namely, that the spur wheel is keyed to the drum shaft, thus subjecting this shaft to a twisting action, the radius r of the journal may be taken equal to

$$r = 0\cdot75\delta \text{ for ropes,}$$

or

$$r = 2\cdot5\delta \text{ for chains,}$$

as is shown by the following calculation.   According to vol. iii. 1, § 116, Weisb. *Mech.*, we find for hemp ropes

$$\delta = 1\cdot13 \sqrt{Q}, \text{ or } Q = 0\cdot785\delta^2$$
$$[\delta = \cdot03 \sqrt{Q} \text{ in inches, or } Q = 1116\delta^2 \text{ in lbs.}],$$

and according to vol. iii. 1, § 14, we find the diameter of a wrought-iron shaft, which is subjected to a twisting moment Qr, to be

$$2\mathfrak{r} = 1\cdot02 \sqrt[3]{Qr}   [2\mathfrak{r} = 0\cdot0907 \sqrt[3]{Qr}].$$

Substituting   in   this   expression $r = 4\delta$ and $Q = 0\cdot785\delta^2$ $[Q = 1116\delta^2]$, we obtain

$$\mathfrak{r} = 0\cdot51 \sqrt[3]{0\cdot785 \times 4\delta^3} = 0\cdot745\delta$$
$$[\mathfrak{r} = 0\cdot0454 \sqrt[3]{1116 \times 4\delta^3} = 0\cdot745\delta].$$

Similarly for the chain we find the size of link, according to vol. iii. 1, § 119, from

$$\delta = 0\cdot326 \sqrt{Q}, \text{ or } Q = 9\cdot42\delta^2$$
$$[\delta = \cdot00858 \sqrt{Q}, \text{ or } Q = 13395\delta^2];$$

hence for a radius of the drum $r = 12\delta$ we have

$$\mathfrak{r} = 0\cdot51 \sqrt[3]{9\cdot42 \times 12\delta^3} = 2\cdot47\delta$$
$$[\mathfrak{r} = 0\cdot0454 \sqrt[3]{13395 \times 12\delta^3} = 2\cdot47\delta].$$

Thus the ratio $\dfrac{\mathfrak{r}}{r}$ of the diameter of journal to that of the drum may be taken equal to $0\cdot2$ whether rope or chain is used, since in the case of ropes we have

$$\frac{\mathfrak{r}}{r} = \frac{0\cdot75\delta}{4\delta} = 0\cdot19,$$

and for chains

$$\frac{\mathfrak{r}}{r} = \frac{2\cdot5\delta}{12\delta} = 0\cdot208.$$

Substituting this value for $\dfrac{\mathfrak{r}}{r}$ in the above expression for η_1, and assuming a mean value for $\dfrac{r}{R_1} = \dfrac{1}{4}$, we can compute the efficiency of the windlass corresponding to different sizes of rope and chain ; these values are contained in the following table.   It may here be noted that the ratio $\dfrac{r}{R_1}$ varies from $\frac{1}{3}$

to $\frac{1}{8}$ for the ordinary windlasses; as this proportion has but a slight influence on the efficiency of the machine, however, it is sufficiently exact, when making estimates, to assume $\frac{r}{R} = \frac{1}{4}$, as in the table. These proportions are to be regarded as approximations only which are near enough to the truth in the ordinary windlass working with rope or chain; but for proportions departing from the above, as for instance the drums of winding engines working with wire rope, the efficiency in every case must be computed according to the general formulæ. This matter will be more fully treated in connection with the latter class of machines.

$$\text{TABLE OF THE EFFICIENCY OF HOISTING DRUMS.}$$

$$\eta = \frac{1 - \phi\frac{r}{R}}{1 + \sigma + \phi\frac{r}{r}} ; \quad \frac{r}{r} = 0{\cdot}2 ; \quad \frac{r}{R} = 0{\cdot}25.$$

Diameter of Rope. $\delta =$	10 mm. [·39 in.]	20 mm. [·79 in.]	30 mm. [1·18 in.]	40 mm. [1·57 in.]	50 mm. [1·97 in.]	Chains.
$\eta =$	0·959	0·939	0·920	0·901	0·883	0·972

EXAMPLE.—A load $Q = 3000$ kg. [6615 lbs.] is suspended from a chain made of iron 18 mm. [0·71 in.] in diameter by means of a double-geared windlass. What force must be exerted by the workmen at the cranks, if these have a length of 400 mm. [15·75 in.], and we assume the radius of the drum to be $r = 0{\cdot}20$ metres [7·87 in.], that of the spur wheel on the drum shaft $R_1 = 0{\cdot}75$ metres [29·53 in.], and the velocity ratios of the two pairs of gears to be $\frac{1}{6}$ and $\frac{1}{5}$?

The theoretical force is

$$P_0 = 3000 \frac{200}{400} \frac{1}{6} \frac{1}{5} = 50 \text{ kg. } [110{\cdot}25 \text{ lbs.}]$$

Assuming the radius of the journal to be $r = 40$ mm. [1·57 in.], the efficiency of the drum is

$$\eta_1 = \frac{1 - 0{\cdot}08\frac{40}{750}}{1 + 0{\cdot}2\frac{18}{400} + 0{\cdot}08\frac{40}{200}} = \frac{0{\cdot}996}{1{\cdot}025} = 0{\cdot}97$$

According to the tables on pages 14 and 17 let us assume the efficiency of the first pair of wheels to be $\eta_2 = 0.96 \times 0.97 = 0.93$, and that of the second pair to be $\eta_3 = 0.95 \times 0.96 = 0.91$, then we have for the efficiency of the windlass $\eta = 0.97 \times 0.93 \times 0.91 = 0.82$, hence the force

$$P = \frac{50}{0.82} = 60.98 \propto 61 \text{ kg. [134·5 lbs.]}$$

If it is desired to determine the resistance (P) which must be produced by a brake in order to prevent the acceleration of the load during its descent, let us assume the brake-wheel to have a diameter of 0·5 metre [19·69 in.] and to be located on the gear shaft; then, without wasteful resistances, (P) becomes

$$3000 \frac{200}{250} \frac{1}{6} = 400 \text{ kg. [882 lbs.]}$$

Owing to the wasteful resistances, however, which of themselves oppose the backward motion, the actual resistance required of the brake is only $(\eta_1)(\eta_2)$ 400 kg. $[(\eta_1)(\eta_2)$ 882 lbs.] where (η_1) and (η_2) denote the efficiencies of the drum and the first pair of wheels; these values in the present case differing but slightly from η_1 and η_2. Hence the resistance to be offered by the brake is to be taken at $0.97 \times 0.93 \times 400 = 361$ kg. [796 lbs.] How to find the tensions in the brake strap required for producing this resistance is fully explained under the head of Brakes in vol. iii. 1, § 178.

In order to determine by graphical methods the force P required to drive a double-geared windlass, let us first draw the line ee_1, Fig. 46, of the load Q parallel to the centre line EQ of the sustaining rope, and at a distance $Ee = \sigma$ therefrom. If through the points of contact D and B of the pitch circles we lay off the directions DD' and BB' of the pressures between the teeth at an angle of 75° with the lines of centres HG and AH, then the lines of action of the actual pressures Z_1 and Z_2 will be parallel to these directions, and at the distances Dd and Bb equal to ζ from them. The direction of the force P is to be taken perpendicular to the crank AK at the point K. If now through the points o_1, o_2, and o_3—representing the respective intersections between the lines of action of the forces Q and Z_1, Z_1 and Z_2, and Z_2 and P—we draw the corresponding tangents o_1g, o_2h, and o_3a to the friction-circles of the journals G, H, and A, then these lines will give the directions of the corresponding reactions R_1, R_2, and R_3 of the bearings. Using the lines of action of the forces thus established, we obtain the polygon of forces as follows : make $o_1 1 = Q$, draw 1 2 parallel to $o_1 o_2$, 2 3 parallel to $o_2 h$, and 1 3 parallel to $o_2 o_3$, and finally draw 1 4 parallel to $o_3 K$ and 3 4 parallel $o_3 a$.   The length 1 4 will give the driving force P, and in the remaining sides of the force polygon o_1 1 2 3 4 we obtain the respective forces Z and R exerted between the teeth and at the bearings, from which the dimensions of these parts together with the proportions of the framework may be

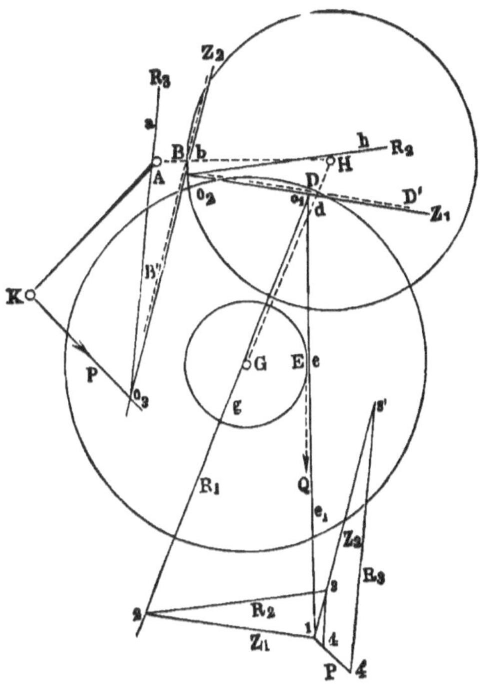

Fig. 46.

computed.   For the sake of clearness the part 1 3 4 of the polygon is magnified five times in 1 3′ 4′.   Were we to assume the quantities σ, ζ, and the radii of the friction-circles equal to zero, that is to say, were the lines of action of the forces drawn through the points E, D, and B, as also through the centres G, H, and A, we should obtain the force P_0, while for the reverse motion we must lay off the quantities σ and ζ in directions contrary to those for the forward motion, and then draw the lines of action of the actual forces parallel to the theoretical directions. In a similar manner the reactions of the bearings are determined by the tangents drawn from the points of intersections *o* to the other side of the friction-circles.

§ 12. **Windlasses operated by Steam Power.**—All the hoisting machines thus far mentioned have been assumed to be driven by hand power. When some other source of energy, such as steam or hydraulic power, is employed for the purpose of performing a greater amount of work in less time, the only alteration in the arrangement of the windlass is to replace the crank by a suitable mechanism for receiving the power. Thus, in the case of hoists used in shops and warehouses, the driving shaft is usually operated by means of a tight and a loose pulley driven by a belt from a continuously rotating shaft. By shifting the belt when in motion from the loose to the tight pulley, or the reverse, the hoisting apparatus can be engaged or disengaged. This arrangement is applied to *sack-hoists*, for instance, used in flour-mills for lifting sacks of grain, and is also employed in saw-mills for hauling logs. In vol. iii. 1, § 170, is shown a form of hoist also used in flour-mills, which is directly driven by a belt running over a pulley on the winding drum. The same article also describes the method of effecting a uniform descent of the load by means of a brake. Where a windlass is used for hoisting only, not for lowering, the brake is usually dispensed with, as the hook then is not lowered by power, being instead brought to its lowest position either by means of a light weight or by a direct pull on the chain. Concerning the various forms of hoisting apparatus arranged to run both forward and backward, a full description will be given under the head of *Lifts*.

Windlasses are frequently combined with a special small steam-engine, where it is desired to hoist heavy loads quickly, and no other source of power is available. This is the case in loading and unloading vessels, for which purpose we find steam hoisting gear extensively used. The engine is usually of the simplest possible construction, since in cases where occasional service only is required, simplicity of construction and safety in operation are matters of greater importance than economy of fuel. Condensing engines are therefore rarely employed for this purpose, and expansion of the steam is only used so far as it can be obtained by means of the ordinary slide-valve. A reversing gear should always be employed, however, as the windlass is used equally often for lowering and lifting. Two cylinders with cranks at right angles are generally applied to this type of hoisting gear, notwithstanding the small amount of power required on account of the ease and safety with which the engine can be reversed with this arrangement.

Oscillating engines are particularly adapted to such purposes on account of their simplicity and the small space they occupy.

Steam hoists are mostly single-geared, the crank shaft transmitting motion by means of a pinion which gears into a larger wheel on the drum shaft. The application of two pairs of gears is unnecessary for ordinary loads, as the pressure of the steam on the piston is far greater than the force which could be conveniently exerted by hand power. Consequently the chain is wound upon the drum with greater speed than in windlasses operated by hand, where the velocity is naturally very slight, on account of the small driving force. The velocity of the load is determined in all cases from the available amount of power. If, for an illustration, L denotes the energy expended per second, expressed in kilogram-metres (foot-pounds), then for an efficiency η of the windlass, the velocity v of the load Q can never exceed that deduced from the equation

$$\eta L = Qv.$$

Assuming, for example, that four men exert a pressure of 12 kg. [26 lbs.] each at the cranks, and turn them with a velocity of 0·8 metres [2·6 ft.] per second; then, for a value of $\eta = 0·75$, they are able to lift a load of 2000 kg. [4410 lbs.] at a velocity not exceeding

$$v = \frac{0·75 \times 4 \times 12 \times 0·8}{2000} = 0·014 \text{ m. } [0·046 \text{ ft.}]$$

per second, and consequently the corresponding velocity ratio must be secured in the train of gears employed. The application of a steam-engine of four horse power would, under the same conditions, give the load a velocity of

$$\frac{0·75 \times 4 \times 75}{2000} = 0·112 \text{ m. } [0·37 \text{ ft.}] \text{ per second.}$$

As a rule the limit of velocity of the load in steam hoists is 0·15 m. [0·5 ft.] per sec. A steam hoist for use on board ships is illustrated in Fig. 47.[1] The piston rods of the two oscillating cylinders A are connected with the driving shaft B, and the reduced motion of the drum is obtained by means of the pinion C and gear E. The brake pulley F is fixed to the spur wheel E, and the flexible strap is applied by means of a lever as heretofore explained. The distribution of the steam takes place through the trunnions Z on which the cylinders oscillate,

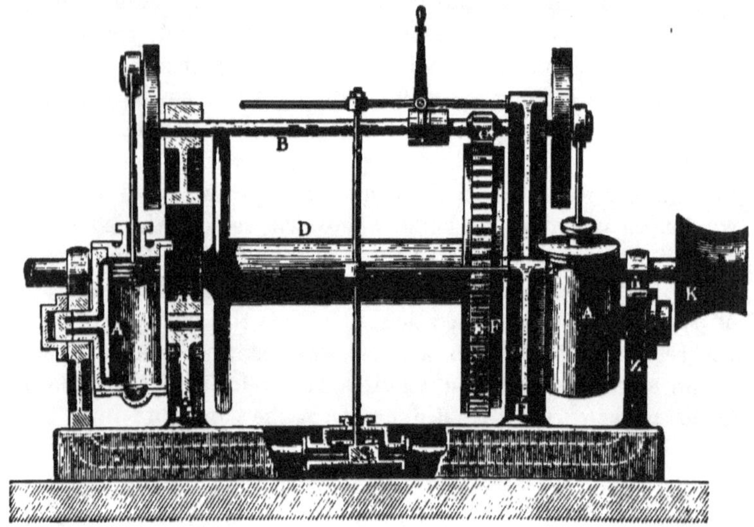

Fig. 47.

and the engine is reversed by the action of a suitable valve, S. For the purpose of operating the hoisting drum by hand when required, a special shaft, not shown in the figure, is provided, having square ends for receiving the crank handles. By shifting this shaft in its bearings, a pinion keyed to it is made to gear with the wheel E. Drums K are also attached to each end of the drum shaft for winding rópe when desired.

The steam-cylinders are 0·15 m. [5·9 in.] in diameter; the stroke is 0·25 m. [9·84 in.], and the crank shaft makes 100 revolutions per minute. The number of teeth of the gears are

[1] See Oppermann, *Portefeuille économique des Machines*, 1868, p. 18, and also Rühlmann, *Allgem. Maschinenlehre*, vol. iv.

11 and 68 respectively, and the diameter of the drum shaft
0·20 m. [7·87 in.], and thus the velocity of the chain is

$$\frac{100}{60}\ \frac{11}{68}\ 0\cdot200 \times 3\cdot14 = 0\cdot169 \text{ m. } [6\cdot65 \text{ in.}],$$

and the energy expended by the steam-engine in lifting a load
of 1800 kg. [3970 lbs.] with this velocity is

$$1800 \times 0\cdot169 = 304 \text{ kg.-m. } [2198 \text{ ft. lbs.}] = 4\cdot05 \text{ h.-p.}$$

As due allowance must be made for wasteful resistances, the
engine evidently must develop more than five horse power.

§ 13. **Other Forms of Hoists.**—In the apparatus shown
in Fig. 48, and known under the name of the *Differential*

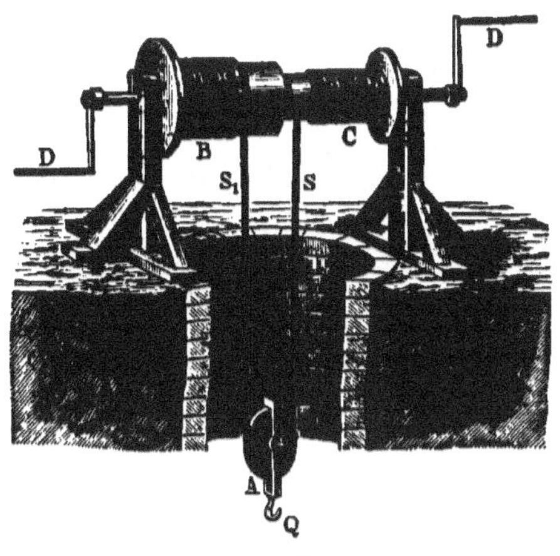

Fig. 48.

or *Chinese windlass*, the action is the same as in the above-
mentioned differential pulley blocks. The load is suspended
from the movable pulley A, both ends of the rope being
fastened to the drum BC, so that when this is turned one end
is unwound while the other is wound on. Owing to the
differing diameters of the drum at B and C, more rope is
wound on than is unwound, and consequently the load rises
with a velocity corresponding to the difference in the diameters.
Letting R denote the radius of the larger portion of the drum
B, and r that of the smaller part C, then, after one revolution
of the drum, the portion of the rope which hangs below the

windlass will have been shortened by an amount $2\pi(R-r)$, and the load accordingly has been hoisted a distance equal to one-half of this length, that is $\pi(R-r)$. Assuming that no wasteful resistances exist, the effort at the crank D of length l would be found from

$$P_0 2\pi l = Q\pi(R - r), \text{ which gives } P_0 = Q\frac{R - r}{2l}.$$

The stiffness of the rope, however, as well as the friction in the bearings, make a far greater effort necessary, and essentially reduce the efficiency of this apparatus, as was also the case in the differential pulley block. Denoting by σ the coefficient due to stiffness of the rope in winding on or off, and assuming a mean value for this coefficient in common for both the drum and the sheave in the block, we obtain the following equation, with reference to the tensions S in the unwinding part at C and S_1 in the part which winds on at B:

$$S_1 = S\left(1 + 2\sigma + 2\phi\frac{r_1}{r_1}\right) = Sk \quad \text{(see § 8),}$$

where r_1 is the radius of the sheave, and r_1 that of the pin on which it turns. We also have

$$Q = S + S_1 = S(1 + k).$$

If r denotes the radius of the journal of the drum BC, we can now deduce the equation

$$Pl + S(1 - \sigma)r = S_1(1 + \sigma)R + \phi(P + S + S_1)r,$$

or, after substituting

$$S = \frac{Q}{1 + k} \text{ and } S_1 = \frac{k}{1 + k}Q,$$

$$Pl + \frac{Q}{1 + k}(1 - \sigma)r = \frac{k}{1 + k}Q(1 + \sigma)R + \phi(P + Q)r.$$

The value of P may thus be derived from

$$P(l - \phi r) = Q\left(\frac{k}{1 + k}(1 + \sigma)R - \frac{1 - \sigma}{1 + k}r + \phi r\right).$$

The efficiency of a hoisting apparatus of this description is very small, and when ropes of large size are used it is even smaller than that of the differential pulley block, owing to the slight amount of friction due to the use of a chain in the latter.

EXAMPLE.—Taking a rope 20 mm. [0·79 in.] in diameter, and using a sheave of a radius $r_1 = 100$ mm. [3·94 in.] with a pin having a radius $r_1 = 10$ mm. [0·39 in.], then

$$k = 1 + 2\sigma + 2\phi\frac{r_1}{r_1} = 1 + 2 \times 0·018\frac{20^2}{2 \times 100} + 2 \times 0·08\frac{10}{100} = 1·088.$$

For a mean radius of the drum of 0·120 m. [4·74 in.] we have $\sigma = 0·018\dfrac{20^2}{2 \times 120} = 0·03$. Further assuming R = 150 mm. [5·91 in.], $r = 120$ mm. [4·74 in.], the length of crank $l = 0·40$ m. [15·75 in.], and the radius of the journal of the drum r = 20 mm. [0·79 in.], we shall obtain the effort P required at the crank from

$$P(400 - 0·08 \times 20) = Q\left(\frac{1·088}{2·088}1·03 \times 150 - \frac{0·97}{2·088}120 + 0·08 \times 20\right),$$

which equation gives

$$P = \frac{26·34}{398·4}Q = 0·0661Q.$$

Neglecting wasteful resistances the result would be

$$P_0 = \frac{150 - 120}{2 \times 400}Q = 0·0375Q,$$

and thus the efficiency $\eta = \dfrac{0·0375}{0·0661} = 0·567$.

Evidently the efficiency would be still smaller for larger ropes and a smaller difference in the diameters of the winding drum.

For this reason, and also on account of the great lengths of rope required necessitating the use of long drums, this apparatus must be considered a very imperfect hoisting mechanism. In order to lift a load through a height h, for instance, assuming a ratio of the radii of the drum $\dfrac{r}{R} = \dfrac{4}{5}$, it would be necessary to wind a length of rope equal to $5h$ on to the larger end of the drum.

The preceding remarks also apply to the geared differential windlass shown in Fig. 49. Here the two drums B and C are of different diameters, and driven simultaneously in opposite directions by means of the crank shaft with pinion J, the resistance to be overcome in this case being the difference between the pressures at the circumference of the pinion J, due to the tensions S and S_1 in the ropes. As may be concluded from the preceding calculation, this arrangement is not to be recommended, and for this reason a detailed analysis of the same will not be presented.

A peculiar hoisting mechanism has been designed by *Long*,[1] Fig. 50. A scroll-shaped cam ABC, projecting from a disc S on the crank shaft, operates an intermediate shaft by means of a star wheel J provided with friction rollers mounted on studs

Fig. 49.

E in its circumference. A pinion F on the intermediate shaft drives the gear G on the wind drum H. For every revolution of the disc S the star wheel J is evidently moved through one division in a manner similar to the action of a worm and worm gear. The ratio of the forces at work may also be determined similarly, the theoretical effort required at the crank of a length l being

$$P_0 = Q \frac{r}{l} \cdot \frac{r_1}{R_1} \cdot \frac{1}{n},$$

---

[1] See *Civil Engineer and Architect's Journal*, July 1852, and Dingler's *Journal*, vol. cxxv.

where r_1, R_1, and r are respectively the radii of the gears F and G and the rope drum, the number of studs E being denoted by n. As may be readily seen, however, the efficiency is greatly reduced by friction, as is also the case in worm gearing. For let a be the mean radius of the scroll, which may be shaped like an *Archimedean spiral* with a radial pitch equal to the pitch s of the star wheel, then the work required for overcoming the friction between the cam and the studs when in

Fig. 50.

operation will for every revolution of the disc be expressed by

$$\mu Q_1 2\pi a,$$

when Q_1 denotes the resistance acting through the space s in the circumference of the star wheel. Taking for mean radius a of the scroll as low a value as $2s$, and assuming a coefficient $\mu = 0\cdot1$, the work spent in overcoming friction will still be

$$0\cdot1 \times 2 \times 3\cdot14 \times 2sQ_1 = 1\cdot256Q_1s,$$

which exceeds the useful work Q_1s; thus we obtain the efficiency of the scroll disc alone to be $\dfrac{1}{1 + 1\cdot256} = 0\cdot443$ only.

It is possible to reduce the friction to one-half of the above, amount by the use of friction rollers, but in spite of this fact a very small efficiency is always obtained from the windlass as a whole. Consequently it would be advisable to employ a more efficient mode of driving—another pair of gears, for instance.

In single windlasses, as used in building enterprises, the winding drum is sometimes operated by a lever AC (Fig. 51), provided with a pawl B, engaging a ratchet C on the drum, the load thus being raised by depressing the lever AC, while a second pawl D attached to the framework prevents it from running down. Besides comparative simplicity and efficiency this type of windlass possesses the advantage of being less restricted as to the length of the lever arm CA than is the

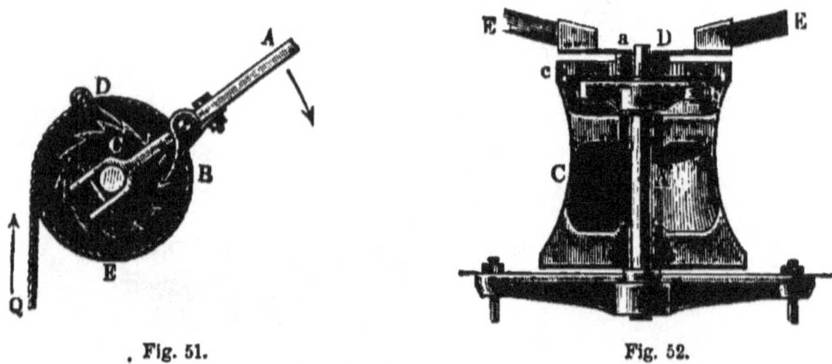

. Fig. 51.   Fig. 52.

case with the diameters of gears and radii of cranks in other constructions. Thus the length of the lever may always be chosen to suit different loads. The non-continuous movement must be counted as a disadvantage, however.

When the lift is considerable it is necessary to make the winding drum of large dimensions, in order to be able to wind on the great length of rope. In such cases the rope is not made fast to the drum, only placed around it in a few coils, one end being allowed to unwind while the other is winding on, thus leaving always the same number of coils on the drum. A sufficient number of coils must evidently be left on the latter to prevent the rope from slipping.

One of the simplest hoisting machines arranged according to this plan is the *capstan* (Fig. 52). Here several coils of rope are placed around the vertical drum C, which is rotated by means of *handspikes* E through the pinion *a* attached to the cover D and the intermediates b. As these intermediates are held rigidly by the upright spindle A, and engage the internal gear c, which forms part of the drum C, it is evident that for every revolution of the cover D the drum will be rotated in the ratio $\dfrac{a}{c}$. If the wheels a and b are made equal, which is the usual arrangement, then $c = 3a$, and the ratio of the gearing will be $\tfrac{1}{3}$.

The drum C is frequently provided with pockets for the handspikes, so as to be able to obtain a more direct application of the power when the load is light. A pawl at the lower circumference of the drum, engaging a ratchet wheel cast on the bed plate G, prevents the load from running down. The drum is made dishing, so as to cause the advancing coils to move towards the centre, while the unwinding rope is pulled away by the workman.

Let S_1 denote the tension in the latter end of the rope, then, according to vol. i. § 194, Weisb. *Mech.*, the tension S in the tight part may be taken as

$$S = S_1 e^{\phi\gamma},$$

before slipping can begin, when γ is the arc encircled, of unit radius, ϕ the coefficient of friction, and e the base of the hyperbolic logarithms. Thus, in case the rope surrounds the drum in n coils, we have $S = S_1 e^{\phi n 2\pi}$, that is, for $n = 3$, and assuming a coefficient of friction between the rope and the drum of $\phi = 0{\cdot}28$, we should get

$$S = S_1 \times 2{\cdot}7183^{\,0{\cdot}28 \times 3 \times 6{\cdot}28} = 198 S_1,$$

which indicates that the tension required in the slack part of the rope is only $\tfrac{1}{2}$ per cent of the load Q.

In order to avoid the lengthwise sliding of the coils, which must necessarily take place if the rope is to remain on the drum, the rope is sometimes carried around two drums G and H in succession (Fig. 53), the latter being provided with grooves for the reception of the former. The part which supports the load is carried to the first groove of the drum G at a, then leaving it at c reaches the first groove of the drum H, which it embraces in a semicircle db; passing further to the second groove of G at *a*, etc., it finally leaves the last groove of the drum H at *b*.

As before, the object to be gained by repeatedly carrying the rope around the drums is to prevent slipping, and therefore we have also for this case the general equation $S = S_1 e^{\phi \gamma}$, when

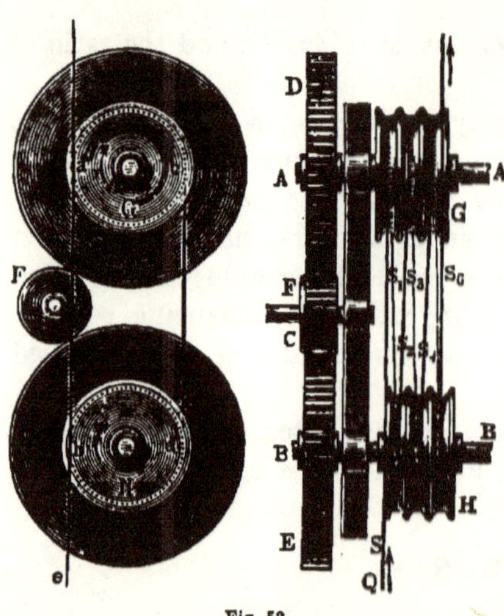

Fig. 53.

γ denotes the sum of all the encircled arcs of the two drums. If *n* is the number of grooves half encircled, which may be an even or odd number, then $\gamma = n\pi$. For hoisting the load the drums G and H must be rotated in the same direction, at the same velocity, which is accomplished by the pinion F on the driving shaft C gearing into the two equal drum gears D and E.

In order to determine the relation between driving force and load for the hoists under consideration, let $S = Q$ denote the tension in the tight part *ae*, S_1 that in the part *cd* which passes from drum G to H, likewise S_2 the tension in the second part *ba* winding on to G, etc., so that S_n will denote the tension in the final, leaving part, when in all *n* grooves are half encircled. Then, for the limit when the rope is on the point of slipping, we have

$$S = S_1 e^{\phi \pi} = S_2 e^{2\phi \pi} = S_3 e^{3\phi \pi} = \ldots \; S_n e^{n\phi \pi}.$$

After one revolution of the drums, of radii *r*, in the direction of the arrow, the resistance $S = Q$ has been overcome through a distance $2\pi r$, the driving force P during the operation evidently being aided by the tension S_n in the last, receding part *bf*. Neglecting the influence of friction and the stiffness of the rope, the tensions $S_1, S_2, S_3 \ldots S_{n-1}$ in the respective parts between the two drums will then have neither performed nor absorbed work, as each tension furthers the motion of one drum equally as much as it impedes that of the other. Thus, in the absence of wasteful resistances, the theoretical driving force P_0 at the circumference of the drums is $P_0 = S$.

To find the actual effort required at the circumference of the drums, let σ denote the coefficient of stiffness and r the radius of the journals. The pressure on the journals of the shaft A will then be

$$Z = S + S_1 + S_2 + S_3 + \ldots S_{n-1},$$

and of the shaft B

$$Z_1 = S_1 + S_2 + S_3 + \ldots S_n.$$

Under the assumption that the tension S_n is just sufficient to prevent slipping, thus $S = S_n e^{n\phi\pi}$, or $S_n = S e^{-n\phi\pi}$, we shall obtain

$$Z = S(1 + e^{-\phi\pi} + e^{-2\phi\pi} + e^{-3\phi\pi} + \ldots e^{-(n-1)\phi\pi})$$

$$= S\frac{e^{-n\phi\pi} - 1}{e^{-\phi\pi} - 1} = Q\frac{1 - e^{-n\phi\pi}}{1 - e^{-\phi\pi}},$$

and

$$Z_1 = S(e^{-\phi\pi} + e^{-2\phi\pi} + e^{-3\phi\pi} + \ldots e^{-n\phi\pi})$$

$$= S e^{-\phi\pi}\frac{e^{-n\phi\pi} - 1}{e^{-\phi\pi} - 1} = Q e^{-\phi\pi}\frac{1 - e^{-n\phi\pi}}{1 - e^{-\phi\pi}} = Z e^{-\phi\pi}.$$

The resistances due to the stiffness of rope at the drum G will be

$$U = \sigma(S + S_1 + S_2 + \ldots S_{n-1}) = \sigma Z = \sigma Q\frac{1 - e^{-n\phi\pi}}{1 - e^{-\phi\pi}},$$

and at the drum H

$$U_1 = \sigma(S_1 + S_2 + S_3 + \ldots S_n) = \sigma Z_1 = \sigma e^{-\phi\pi}\frac{1 - e^{-n\phi\pi}}{1 - e^{-\phi\pi}}.$$

Consequently, the force P required at the circumference of the drums is obtained from the equation :

$$Pr = (S - S_n)r + \sigma(Z + Z_1)r + \phi(Z + Z_1 + P)r,$$

or with the above values of Z and Z_1 :

$$P(r - \phi r) = Q(1 - e^{-n\phi\pi})r + (\sigma r + \phi r)Q(1 + e^{-\phi\pi})\frac{1 - \rho^{-n\phi\pi}}{1 - e^{-\phi\pi}}.$$

Hence we may derive P, and from $P_0 = Q$ the efficiency $\eta = \dfrac{P_0}{P}$ may be obtained. From the force P at the circumference of the drums, we may further calculate the effort P_1 acting with a lever arm R_1 on the crank shaft C, when due attention is paid to the journal friction of the shaft C, and the friction between the teeth of the gears CF, AD, and BE. Further explanation will be obtained from the following example :—

EXAMPLE.—In a hoisting apparatus of the kind shown in Fig. 53, the rope runs around the two drums three times, the load is 300 kg. (661·5 lbs.), the radii of the drums are 0·10 m. (3·94 in.), and the ratio of the gearing is 0·2. What is the driving effort required at the end of a crank 0·36 m. (14·17 in.) in length ?

Here $n = 6$, hence, assuming $\mu = 0·28$, we obtain

$$e^{-0·28 \times 3·14} = 0·4152$$

and

$$e^{-6 \times 0·28 \times 3·14} = 0·00512,$$

which gives a tension in the slack end of the rope of

$$S_6 = 300 \times 0·00512 = 1·54 \text{ kg. [3·4 lbs.]}$$

Assuming the size of rope $\delta = 1·13 \sqrt{300} = 20$ mm. (0·79 in.), we shall find from the formula given by *Eytelwein* :

$$\sigma = 0·018 \frac{\delta^2}{2r} = 0·018 \frac{400}{2 \times 100} = 0·036$$

$$\left[\sigma = 0·457 \frac{\delta^2}{2r} = 0·457 \frac{(0·79)^2}{2 \times 3·94} = 0·036\right],$$

and for a radius of the journal $r = 15$ mm. [0·59 in.] we have

$$\phi \frac{r}{r} = 0·08 \frac{15}{100} = 0·012.$$

Hence we get

$$P(1 - 0·012) = Q(1 - 0·0051) + (0·036 + 0·012)Q(1 + 0·415) \frac{1 - 0·0051}{1 - 0·415}$$

$$= Q(0·9949 + 0·1155),$$

which gives

$$P = \frac{1·1104}{0·988} Q = 1·124 Q = 337·2 \text{ kg. [743·5 lbs.]}$$

As $P_0 = Q$, when wasteful resistances are neglected, the efficiency of the two drums will be

$$\eta_1 = \frac{P_0}{P} = \frac{1}{1·124} = 0·889.$$

Taking for the gears an efficiency $\eta_2 = 0\cdot94$, in accordance with § 3, we obtain for the effort required at the crank

$$P_1 = \frac{1}{0\cdot94} \times \frac{100}{360} \times 0\cdot2 \times 337\cdot2 = 19\cdot95 \text{ kg. [44 lbs.]}$$

The total efficiency of the apparatus will thus be

$$\eta = 0\cdot889 \times 0\cdot94 = 0\cdot836.$$

With a view to avoiding the use of a winding drum in connection with chain hoists, various constructions have been made. The leading principle in all these designs consists in

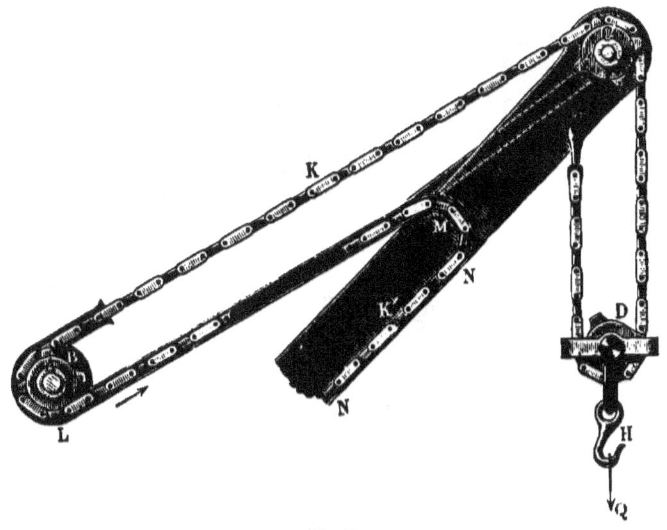

Fig. 54.

allowing the chain to pass over a roll or a small wheel, the circumference of which, as in the differential block, is provided with spurs or recesses which prevent the chain from slipping, or in giving to the chain wheel the shape of a prism on the sides of which the links of the chain are carried.

In the crane constructed by *C. Neustadt*,[1] a chain wheel together with a *Gall's chain* is made use of. The chain wheel B on the shaft A, Fig. 54, is substituted for the chain drum, the pins of the *Gall's* chain K resting in the spaces between the spurs of the former. The chain further passes over the fixed pulley C on the jib of the crane, and carries the load by

[1] Armangaud, *Public Industr.* vols. xii. and xiv.

means of the hook H attached to the movable pulley D. The slack end K' of the chain is guided by a closed casing LMN,

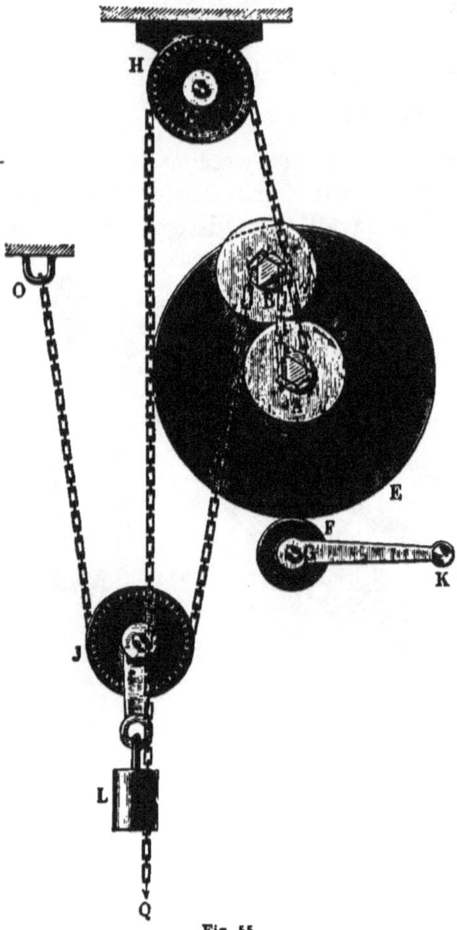

Fig. 55.

in which it slides back and forth under the action of the driving wheel B, thus preventing entanglement of the links, the chain being always kept tight by the weight of its lower end. This construction is not used in practice to any great extent, however, chiefly on account of the large amount of friction generated at the pins of the link chain, and besides, in common with all machines employing chains for driving, it is liable to the disadvantages due to the gradual elongation of the links. The resistances generated by the chain in running on and off the pulleys may be calculated in the same manner as that employed for ascertaining the friction between a rack and its pinion.

In *Bernier's*[1] windlass the chain which carries the load Q passes over the fixed guide-pulley H, Fig. 55, and is then brought around two shafts A and B, which are of triangular cross-section at the points where the chain is carried. Slipping of the chain is made impossible in this manner, and the load

---

[1] See Rühlmann, *Allgem. Maschinenlehre*, vol. iv. p. 402.

is therefore raised if the two shafts A and B are rotated in opposite directions. The opposite rotation is accomplished by means of two gears of equal size placed on the shafts and gearing together, the motion being transmitted from the crank shaft GK to the notched shaft A through the gears E and F. The slack end of the chain might be allowed to drop freely from the shaft B, but in order to secure a uniform tension and prevent entanglement, it is usually suspended, as at O, and carries in its beight a movable pulley J together with a tension weight L.

This hoist, besides requiring no chain drum, is evidently a very powerful contrivance, inasmuch as the lever arm of the load is equal to the small radius of the chain shaft only. The friction as well as the wear of the chain is considerable, however, owing to the small radius. This evil is of great consequence in this machine, as it is capable of smooth running only as long as the links retain their correct lengths.

Fig. 56.

In *Stauffer's* windlass this requirement of links of a constant length, which can scarcely be depended on in the long-run, is a matter of little moment. Here a *chain lock* is also made use of, Fig. 56, with depressions for the links, but since the chain is not carried around the shaft, only between the shaft and a drum B, very few links are grasped at the same time, and for this reason a slight elongation will do no harm. This hoisting machine, besides, offers several advantages, as for instance with reference to safety in lowering a load and keeping it suspended. The above-mentioned drum B is cast in one piece with the pinion C, and is operated from the crank K through the clutch coupling D. Thus, while the chain shaft is slowly revolved by means of the gears C and E, the chain is at the same time pulled through the lock between A and B.

The ratchet wheel H on the crank shaft and the pawl J prevent a reversing of the motion as long as the drum is connected with the crank by means of the clutch D.   By pressing the crank slightly *backwards* the clutch may be disengaged from the drum, however, and the load Q thus released.   The drum being loose on the shaft, will then under influence of the load run backwards with an increasing velocity depending on the ratio of the gearing.   By simply quitting hold of the crank the clutch is automatically again thrown into gear, and the motion arrested by the pawl J.   In order to obtain a uniform velocity in lowering, a braking resistance is, by the motion of the drum shaft, produced in the following ingenious manner.   In the circumference of the drum a number of leaden segments are applied in such a manner as to be forced outward by the action of the centrifugal force during the rotation of the drum, thereby producing the necessary frictional resistance.   This resistance, increasing with the velocity, soon reaches a point where no further acceleration of the load is possible.   To avoid shocks when the load, after the crank has been dropped, is brought to a sudden stop, the clutch is made yielding to a certain extent. This hoist undoubtedly is a perfectly safe contrivance, since the load is brought to a stop merely by releasing the hold of the crank, an operation which requires no effort whatever on the part of the workman.

§ 14. **Lifts.**—This class of hoisting apparatus is used for lifting building materials, grain, coal, ore, etc., and receives names to correspond, such as brick, grain or coal elevator, and furnace-lift.   We may divide these machines into two classes— those in which the hoisting action is continuous, being produced by means of an endless chain, and those in which the load is carried at the end of a rope or chain in the manner made use of in the windlasses just described.

In lifts which employ an endless chain the latter is either provided with special receptacles for the load, or with hooks for its reception, the material being occasionally placed in a so-called cage.   In either case the continuous motion of the chain and the lifting of the load are accomplished by revolving a shaft on which the wheel or wheels are fastened, around which the endless chain is carried.

The second class of lifts either operates with a drum upon

which the rope is wound, or with a plunger which is put in motion by steam or water pressure, and pulls the rope with it. Motion may be imparted to the drum either by hand or by the application of steam or water power, etc. In most cases only one hauling rope is employed, which makes it necessary to return the rope to the first position before another load can be hoisted. For controlling the motion of the descending rope, more especially when it carries an extra load in the shape of an empty cage or platform, a counterpoise or a brake is used. Occasionally, as in the shafts of mines, two ropes, carrying two cages, are employed in such a manner that the empty cage descends while the load ascends. For reversing the motion an engaging and disengaging gear must be applied. When loose material is to be hoisted, the lift may be arranged with a bucket chain. To this class of machinery belong the elevators used in flour-mills for raising grain, and also, in a measure, the dredging machines for excavating or removing obstructions in the beds of rivers or in harbours.

Fig. 57 illustrates a *furnace-lift* operated by an endless chain. A and C are two pairs of sprocket wheels measuring at least 2 metres ($6\frac{1}{2}$ ft.) in diameter; the spurs entering between the links of the chains ABCD carry the latter along in their motion. The two chains are connected by

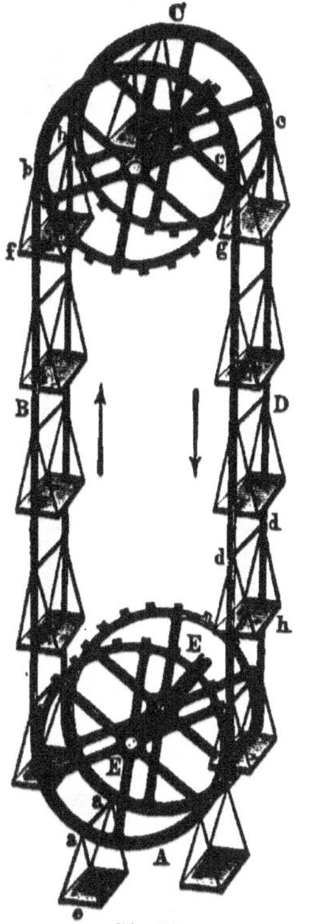

Fig. 57.

wrought-iron rods, *aa*, *bb*, *cc*, . . . from which the swinging platforms *f*, *g*, *h*, etc., are suspended. The lower shaft EE is driven by spur gearing, receiving its motion from a water wheel or a steam-engine, the velocity of the chains being about 0·15 metres ($\frac{1}{2}$ foot).

When the load, a car containing ore for instance, has been placed on an ascending platform *e*, it is slowly lifted until it reaches a certain level, say at *f*, where it is removed. After the car has been emptied it is returned to a platform which descends on the other side of the lift, and upon reaching the bottom it is again removed and refilled. Should the removal of a full or empty car be neglected at any time, no loss worth mentioning would result, since the car would then only complete another revolution, which would require practically no extra expenditure of energy, owing to the fact that the work performed by the empty car in its descent would nearly equal that absorbed during the ascent.

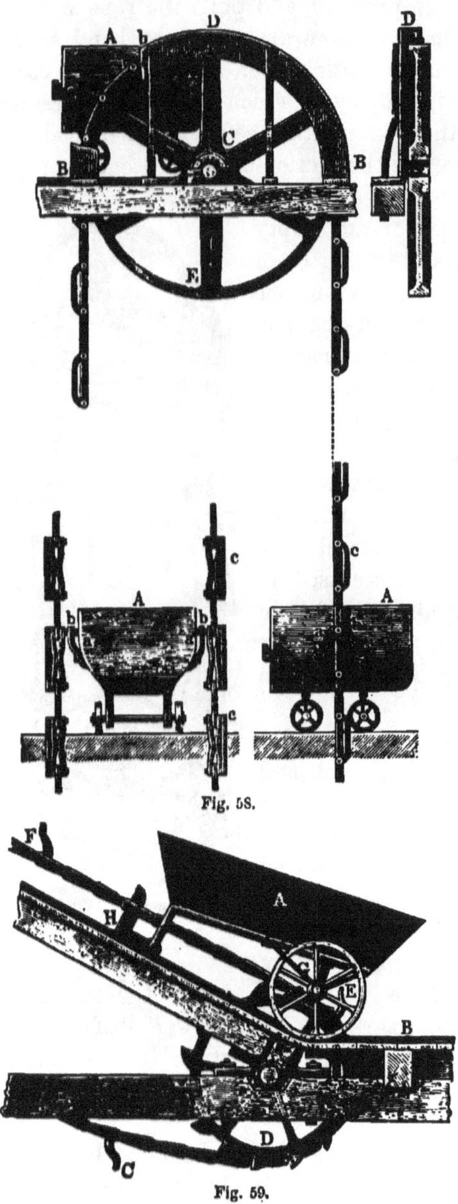

Fig. 58.

Fig. 59.

By providing the endless chain with studs or hooks for receiving the load, the platforms may be dispensed with. In the furnace-lift, partly illustrated in Fig. 58, the truck A which is to be raised has its sides fitted with hooks *aa* which are sustained by the studs *bb* of the chain until the truck has reached the top. The trucks are brought to the hoist by a railway at the bottom, and removed by the same means at the top. The upper railway is sufficiently inclined to convey the truck from the lift to the mouth of the furnace by the action of gravity, the empty cars being returned to the shaft by a second railway. To prevent any deviation which, owing to the one-sided action of the load, might cause them to run off the pulleys, the chains are provided with stiffening links *cc*, and the wheels CE are surrounded by fixed guides BDB for these links.

An inclined lift operated by an endless chain is shown in Fig. 59, which represents its lower portion, and makes clear the method employed for taking hold of the trucks by means of hooks attached to the chain. The truck A is brought to the hoist on a railway B, the endless chain CDEF passing over a sprocket wheel engaging the links of the former. At intervals of 3 metres (10 ft.) the links, which are about 0·3 m. (12 in.) in length, are made in the shape of hooks (*dogs*) CEF . . . which grasp the rear axle G of the car, and thus raise it to the top, where it unhooks itself, and is conveyed to the mouth of the furnace on an inclined railway. The upper sprocket wheel is located above the charging platform, and through suitable gearing it is driven by water or steam power.

In case the chain should break, the cars are prevented from running down the incline by means of small elbow-shaped levers H attached at several points along the railway. These levers, while they permit the axles of the car to pass on the upward journey, are so arranged as to arrest any backward motion. The empty cars are lowered on a second railway by means of a windlass controlled by a brake.

Neglecting wasteful resistances, the power required for the operation of a lift with an endless chain can easily be computed as follows. Let Q denote the weight of the material to be raised on the truck, h the height to which it is to be lifted, and n the number of trucks to be hoisted per minute. Then the useful work expended in hoisting each truck will be Qh, and the work performed per minute will be nQh, or per second

$$L = \frac{n}{60} Qh.$$

This expression, however, is true only when the descending

platform or truck G, as in Fig. 57, is entirely counterbalanced by the ascending one.   In every other case we must place

$$L = \frac{n}{60}(Q + G)h.$$

The velocity of the load in the vertical direction is $v \sin a$, where v denotes the speed of the chain, and a the angle of inclination of the railway to the horizon.

The arrangement of an *elevator*, as used in flour-mills for

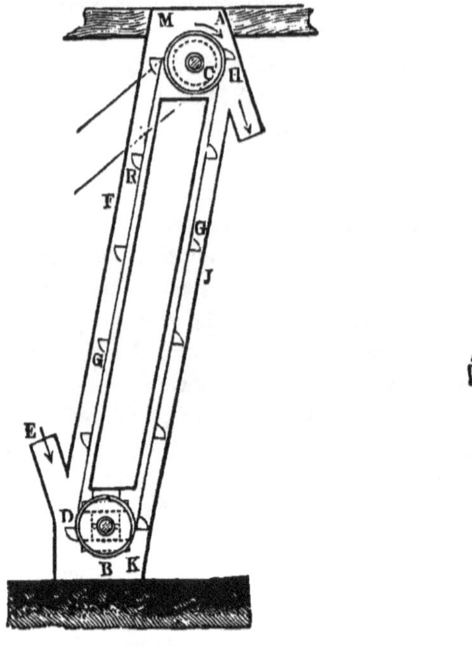

Fig. 60.

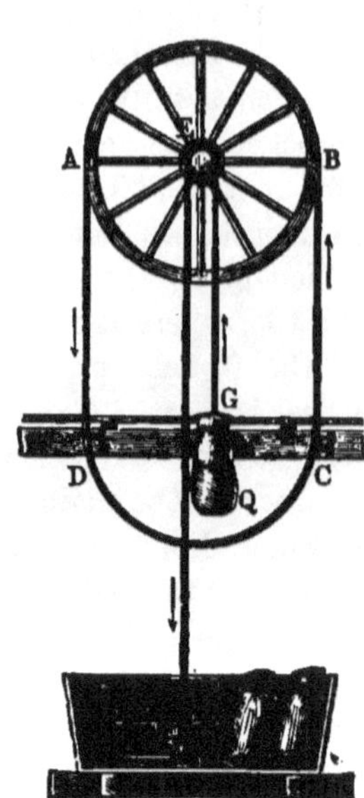

Fig. 61.

carrying grain to the upper story, is shown in Fig. 60.   In place of an endless chain, an endless belt carried by the two smoothly finished pulleys A and B is here employed, continuous motion being communicated to the upper pulley by the driving-belt pulley C.   At equal distances along the hoisting-belt small sheet-iron buckets G are placed which, after the grain has been coarsely ground at D, carry it to the top story, where it is finally discharged at H.   In order to give to the belt a sufficient degree of tension to prevent slipping, the bearings for the lower pulley shaft are usually fitted with an adjustment in the shape of wedges. The belt has an inclination of 12° or 15° to the vertical, which keeps the grain from dropping back into the trough at the point where it is discharged. The wooden troughs F and J, and the box-shaped casings K and M, which surround the belt and pulleys, prevent the scattering of dust from the grain.

The arrangement of dredging machines will be described later.

NOTE.—Elevators operated by an endless chain suffer from the disadvantages already referred to, and which attend all use of chains for transmitting motion, namely, that the links gradually stretch, and then fail to work with the accuracy necessary for smooth running, and, besides, that the frictional resistances between the links and their pins are considerable. *Cavé* has recommended the use of a hoist like that shown in Fig. 58 for hoisting in the shafts of mines (see Armangaud's *Genie industriel*, Dingler's *Polytech. Journal*, vol. cxxvi., or *Polytechn. Centralblatt*, 1852). In order to make possible the ascent of the loaded and the descent of the empty truck on the same endless chain, he suggests that a platform car be used on a railway for moving the trucks to and from the endless chain. For hoisting in very deep shafts a machine of this kind would hardly be suitable.

A simple *hand-lift* is shown in Fig. 61. A grooved pulley AB, about 2 m. (6½ ft.) in diameter, is rotated to the right or left by the endless rope ABCD, the rotation causing another rope which is wound several times around the shaft E to unwind on one side and wind on to the shaft on the other side. A load Q attached to one end of this rope will rise during the rotation, while the other unloaded end will be lowered. When the object hoisted has reached its destination a new load may be attached below at the other end of the rope, and then lifted by turning the pulley in the opposite direction.

When water power can be had on a level with the charging platform, a very simple form of *furnace-lift* may be constructed, as illustrated in Fig. 62, no actual driving mechanism being in this case required. A is a large pulley around which a rope is carried two or three times, the hoisting platforms B and C being suspended from the ends of this rope. The platforms are made with double bottoms, forming reservoirs which can be filled with water at the top from tanks D and E, and emptied below by means of the valve *b*.

All that is required when a loaded platform is to be hoisted is to fill the reservoir in the empty platform with water, the weight of which causes it to descend, and at the same time lifts the load. When the descending platform reaches the lowest point, a release valve is opened by striking against a projection K, and thereby causes the water to run out. This platform may now be loaded, and then hoisted by filling the

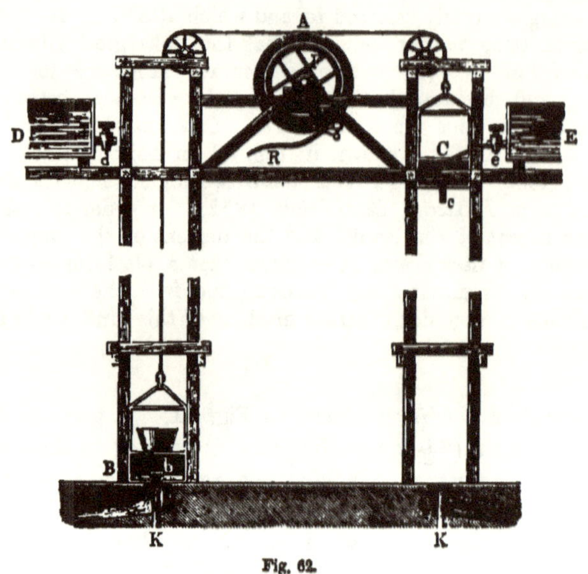

Fig. 62.

upper reservoir in turn with water. A brake - wheel F, operated by a treadle R, controls the speed during the ascent and descent.

Fig. 63 shows a *large furnace-hoist* driven by water or steam power. It consists of two parallel railways A and B, inclined at an angle of 30° to 40°, and having a length corresponding to the height of the furnace. On each railway runs a car, C or D, mounted on wheels of unequal diameters, and provided with a horizontal platform for receiving baskets, buckets, or trucks containing ore, coal, etc. Both cars are attached to the same rope EFG, which passes over the drum

F, the rotation of the latter thus causing one of the cars to ascend while the other descends. In order that the ascent of the loaded car may take place alternately with the descent of the empty one, the drum must be arranged to rotate in either direction, and for this purpose a reversing motion must be employed. This may be operated either by gearing or by belts running on tight and loose pulleys. In the above hoist the latter arrangement is used. One of the two belts, H and K, which transmit motion to the drum shaft, is open, and the other is crossed; by means of a forked lever

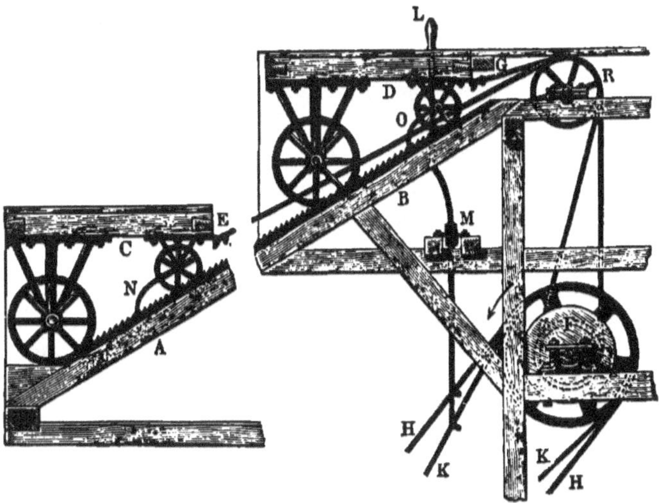

Fig. 63.

LMK either belt may be shifted to its corresponding tight pulley, thus causing the drum shaft to rotate in either direction. Should the rope break at any time the car will be brought to a stop by the pawl N or O engaging with the toothed rail along the track.

Let G be the weight of a car, including the empty receptacle which is placed upon it, and Q the load placed in the other receptacle, then for the state of rest the tensions in the portions E and G of the rope are $S_1 = (G + Q) \sin a$, and $S_2 = G \sin a$ respectively, where a is the angle of inclination of the plane to the horizon. Let η' denote the efficiency for the ascent of the mechanism consisting of the car C, the rope E, the pulley

R, and the drum F, and let (η') be the efficiency of the corresponding parts belonging to the car D for the descent; then the resistance to be overcome at the periphery of the drum is expressed by

$$W = P = \frac{S_1}{\eta} - (\eta')S_2 = \frac{1}{\eta}(G + Q)\sin a - (\eta')G\sin a.$$

In the absence of wasteful resistances, that is to say, for $\eta' = (\eta') = 1$ we have $W = P_0 = Q\sin a$, and therefore the efficiency of the whole hoist becomes

$$\eta = \frac{P_0}{P} = \frac{Q}{\dfrac{Q}{\eta} + G\left(\dfrac{1}{\eta} - (\eta')\right)} .$$

From this we see that the dead weight of the car, including the receptacle, may have considerable influence on the efficiency of the hoisting machine, which proves the fallacy of the supposition upon which the calculation is often based, namely, that the dead weight can be left out of account, because the two cars balance each other.

EXAMPLE.—In an inclined furnace-lift, similar to the one in Fig. 63, $a = 30°$ is the angle of inclination, the useful load of a car is 1000 kg. (2200 lbs.), and the weight of the car, including the empty receptacle, is taken at 800 kg. [1764 lbs.]

Assuming the efficiency of the car, the guide-pulley, and the drum, to be $\eta = 0\cdot90$, and to be the same for the ascent and descent, under which supposition the frictional resistances of the journals of the car and pulley and the resistance due to stiffness of the rope consume 10 per cent of the energy exerted, then we shall have for the resistance at the periphery of the drum

$$P = \frac{1800}{0\cdot90}\sin 30° - 0\cdot90 \times 800\sin 30° = 1000 - 360 = 640 \text{ kg. [1410 lbs.]}$$

whereas without wasteful resistances $P_0 = 1000\sin 30° = 500$ kg. [1100 lbs.]; hence the efficiency of the hoist is $\eta = \dfrac{500}{640} = 0\cdot781$.

If we assume the circumferential velocity of the drum to be $0\cdot5$ metres [$1\cdot64$ ft.] per second, the power to be transmitted by the belt will be $640 \times 0\cdot5 = 320$ kg.-m. [2315 ft.-lbs.] per second, corresponding to $\dfrac{320}{75} = 4\cdot27$ horse powers [$4\cdot21$ h. p.] The time required to lift a load to a height $h = 16$ metres [$52\cdot5$ ft.] is

$$t = \frac{16}{0\cdot5 \times \sin 30°} = 64 \text{ seconds.}$$

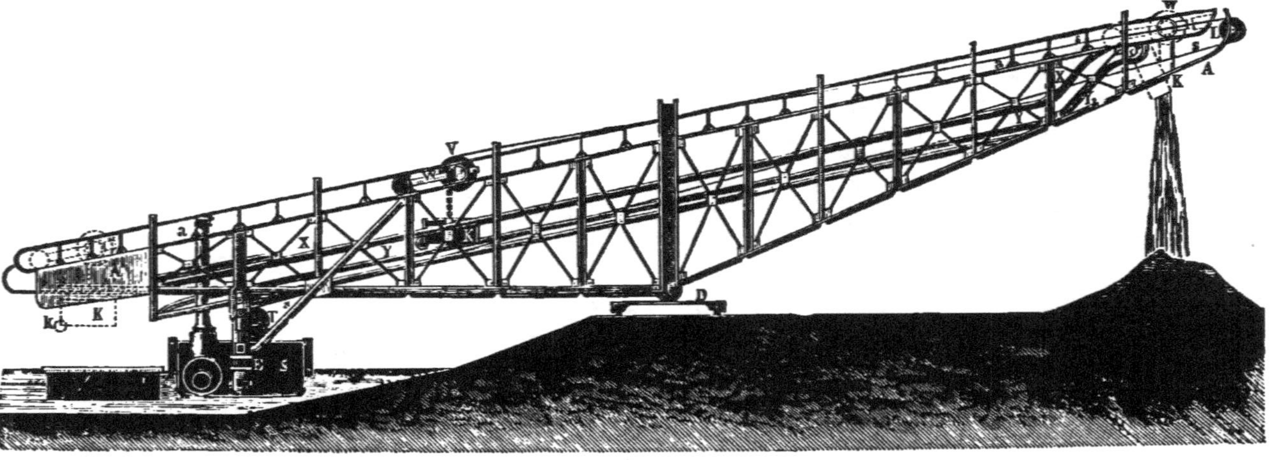

Fig. 64.

A very interesting form of hoist [1] was employed in building the Suez Canal, the material raised by the dredges being carried to the banks on inclines. The essential features of this elevator are shown in Fig. 64. Two large iron girders A, securely braced to form a strong frame, carry the rails *a* of an inclined track upon which a four-wheeled car W travels. This car supports the drum on which two chains are wound, from the ends of which a box K, filled with the dredged material, is suspended. This framework rests at C upon a movable platform running on a track along the canal bank, and also at E upon the dredge-boat S, which contains the steam-engine for operating the windlass. The framework A and the boat are connected by a kind of universal joint, enabling the frame to adjust itself to changes in the water level, and the platform D is made movable about a vertical axis, as in the case of a turntable. The steam-engine drives the drum T of a windlass, thus winding on two parallel wire ropes *s*, which are carried over fixed pulleys L at the apex of the frame, and ultimately secured to two large drums V carried by the front axle of the car W. Two smaller drums are cast on to the drums V, and the box K is suspended from chains attached to the former. As the wire rope is wound upon the drum T the drum V is made to rotate with its smaller chain drums, thus causing the box K to rise vertically. This vertical motion continues until the box, by means of two rollers *k* fixed to its hind end, strikes against the guard rails X arranged at the sides of the frame. These prevent any further vertical motion of the box, and hence any further rotation of the drum V. Any additional pull exerted by the wire rope on V causes the car W with its suspended box K to travel along the rails *a* ; during this motion the box is kept in a horizontal position by guiding the rollers *k* between the parallel rails X and Y. At the apex of the frame the rails X and Y are curved in such a manner as to tip the cage K, which thus automatically discharges its contents. At this point the engine is stopped, and then reversed after the box has been emptied, thus allowing the empty car W to roll back under the action of its own weight, whereupon another loaded car may be lifted in the same manner. ·

[1] See Oppermann, *Portefeuille des Machines*, 1869, p. 28.

In building the Suez Canal eighteen of these elevators were employed; the average slope of the inclines was about 0·23, the extreme ends of the frame were 3 metres [10 ft.], and 14 metres [46 ft.] above the water level, and each cage had a capacity of about 3 cub. m. [4 cub. yds.]

§ 15. Hydraulic Hoists.—Recently a form of hoisting apparatus has come into use, founded on the principle of the *hydrostatic press;* that is, water under great pressure acting on the surface of a piston, which fits water-tight in a cylinder, is employed to raise a load resting upon the piston. The essential arrangement of a *hydrostatic* or *Bramah press* is

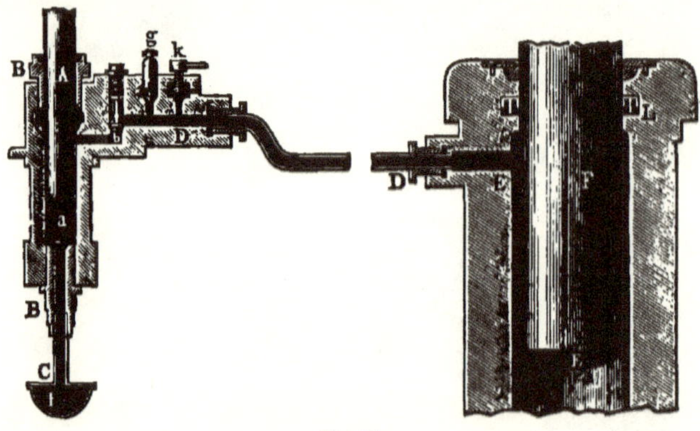

Fig. 65.

shown in Fig. 65. By means of a small force-pump, whose plunger A receives a vertical reciprocating motion by the application of muscular or steam power, the water, drawn from a reservoir through the suction pipe BC is forced through the tube DD into a strong cast-iron cylinder E. During this motion the suction valve *a* and the delivery valve *b* act like the valves of a common force-pump. The loaded safety valve g prevents the water from reaching a pressure that is unsafe for the parts ; the discharge valve k worked by means of a screw, allows (when open) the water to escape from the press cylinder E.   In the latter cylinder the accurately turned plunger F is guided through the water-tight packing L, and its end is suitably arranged either to exert a desired pressure, or, as in a hoist, to receive the load that is to be lifted. When water is forced from the pump AB to the cylinder E it pushes the plunger F out of the cylinder with a force Fp, where F denotes the cross-section of the plunger, and p the pressure of water per unit of area.   The pressure p which, in the absence of hydraulic resistances, is the same in the press as in the pump cylinder, is given by $p = \dfrac{P}{f}$, where f denotes the cross-section of the pump plunger A, and P the force necessary to drive this plunger when the friction of the stuffing box is neglected.   Hence the resistance which can be overcome by the plunger F is

$$Fp = \frac{F}{f}P,$$

that is, by adopting this arrangement the driving force P transmitted to the pump plunger is increased in the ratio of the cross-sections $\dfrac{F}{f} = \dfrac{D^2}{d^2}$.   It is obvious that the distances moved through by the plungers are diminished in the same ratio.   The force exerted by the plunger F is reduced by the amount of the wasteful resistances.   In the present case the friction between the plunger and packing is the principal wasteful resistance, for the motion of the water in the apparatus is generally so slow that the loss of work due to this cause can be neglected in comparison with the friction of the packing.   At least this is true for presses worked by hand.

In presses driven by steam power the resistance of the water passing through the pipes, etc., may be determined by the rules given in vol. i. § 7, Weisb. *Mech.*   Experiments show that the friction of the plunger may be taken proportionally to the pressure upon it, so that, assuming ϕ to be the corresponding coefficient of friction, the force exerted by the press plunger will not exceed $(1 - \phi)\dfrac{F}{f}P$.   Further information concerning the values of ϕ will be given later.

Fig. 66 represents an *hydraulic lifting jack*, in which F is the press cylinder and K the lifting plunger projecting through the upper end of the cylinder, the enlarged base G serving as a reservoir for the liquid.   The pump cylinder B forms a part

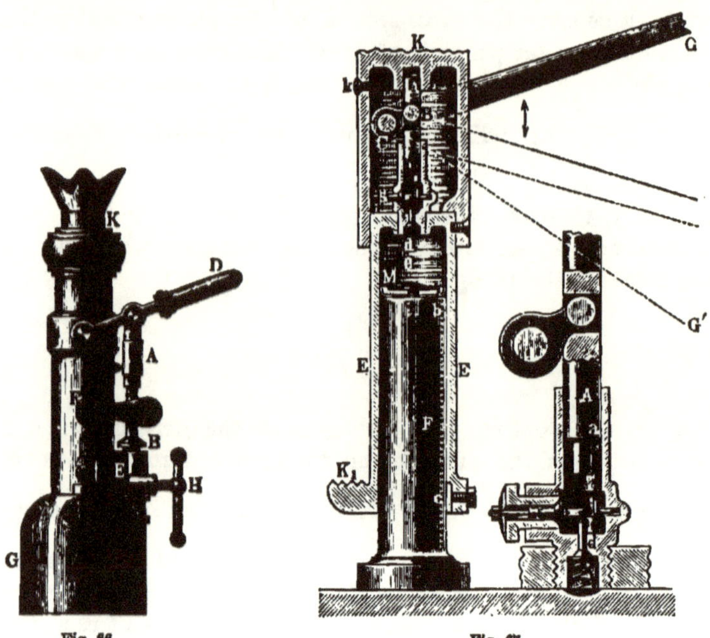

Fig. 66.   Fig. 67.

of the frame, and the figure shows how the plunger A receives its motion from the lever CD oscillating about C.   The reservoir is filled through an opening S, and, to prevent freezing, oil or glycerine is generally used.   When the load is to be lowered the communication between the cylinder F and the reservoir G is effected by a valve worked by a screw H.

Owing to their exposed position the pump cylinder and the valve chamber are liable to injury and fracture.   To avoid these dangers various arrangements[1] have been devised in

[1] See *Zeitschr. deutsch Ing.* 1866, p. 707.

which the sensitive parts are placed in the interior of the jack.   By this means greater security against injury has been obtained, but the parts are less accessible.   We will only mention one of these arrangements, the apparatus of *Tangye Bros.*, Fig. 67.   The jack is provided with a projecting claw K_1 and the tube EE sliding on the cylindrical standard F, being fitted with a water-tight joint in the shape of a leather cup M; a feather c, sunk in the cylinder E, and fitting in a groove cb, guides the cylinder and prevents it from rotating. The reservoir K is above the cylinder, the oil is admitted through a screw-hole k, and by means of the reciprocating motion of the plunger A it is drawn through the valve s, and forced through the valve d into the space e bounded by the plunger F and cylinder E.   The manner of operating the pump plunger by the lever CBG is evident from the figure. When the load is to be lowered by the action of its own weight, the lever G is depressed into its lowest position G', when the plunger A opens the delivery valve, at the same time pulling back the suction valve s by means of a projection a.   To enable the end of the plunger A to strike the valve d, the spindle of the valve s is made ring-shaped.   It is calculated that this jack will raise a maximum load of 30,000 kilograms [66,150 lbs.] when the cylinder E has an inside diameter of 89 mm. [3·5 in.], and the pump plunger A has a diameter of 19 mm. [0·75 in.]   For raising the tubular girders of the Britannia Bridge,[1] very powerful hydraulic presses were applied.   These were placed in recesses purposely constructed in the towers, 40 feet above the permanent bed of the tubes, and the force-pumps were operated by two steam-engines of 40 h.-p. each.   Each tube was 460 ft. long, and weighed, with the additional weights raised, 1726 tons.   Each was provided at the ends with cast-iron frames, to which two lifting chains made up of sets of eight and nine links alternately, were attached.   These chains were suspended from the cross-head of the plungers of the hydraulic presses.   The manner of arranging the apparatus and securing the chains which carry the ends of the tubes will be understood from Figs. 68 and 69.   In both illustrations A is the press plunger, B the cylinder, and C the pipe through which water

---

[1] See Clark, *The Britannia and Conway Tubular Bridges.*

is forced by the engines.   DD are the walls of the tower,
E F and G cast-iron girders, and H is a cast-iron support
for the press cylinder B.   Furthermore, K represents the
cross-head from which the lifting chains are suspended, and
N the cylindrical guide rods, which pass through the cross-
head, and are secured above to the cast-iron girder O, while
below they are firmly attached to the cylinder.   The cross-

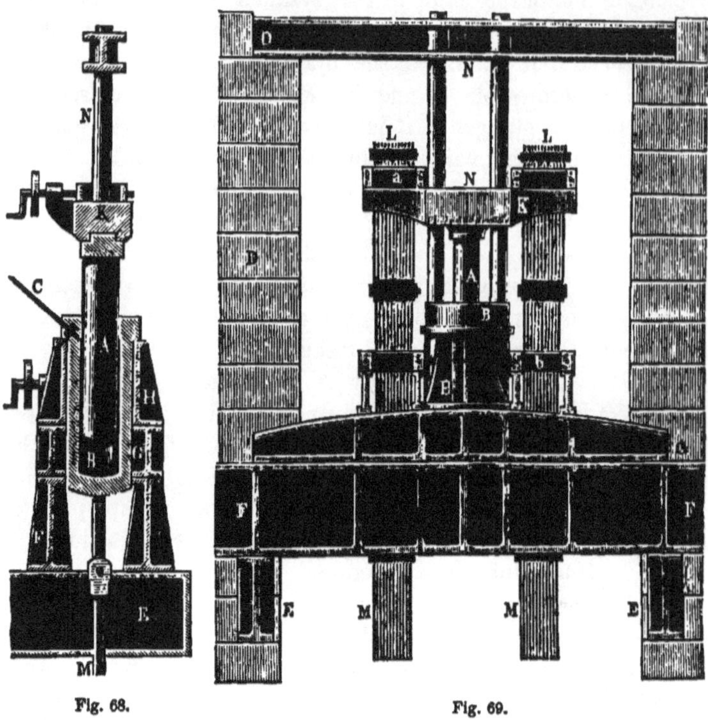

Fig. 68.   Fig. 69.

head is provided with clamps *a* which hold the ends of the
chains and are screwed up firmly against the links.   When
the lifting piston has completed its stroke two other sets of
clamps *b* are screwed up tight and hold the chains while a set
of links is removed, thus permitting the cross-head to be
lowered for the purpose of commencing a new stroke.   To
prevent the tube from falling in case the press should burst
or the chains break, the ends of the tube were followed up by
masonry during the ascent.

Three hydraulic presses were used at the erection of the Britannia Bridge. The largest was 10 ft. long, and had a cylinder 11 inches in thickness and 20 inches in internal diameter. The smaller presses had lifting pistons 18 inches in diameter. The two smaller presses were applied at one end of the tube and the larger press at the other end. The plunger of the force-pump was $1\frac{1}{16}$ inches in diameter, and was directly secured to the rod of the steam piston, whose diameter was 17 inches. The length of stroke of both the steam piston and pump plunger was only 16 inches, the lift of a press plunger, however, was 6 feet.

The time occupied for a single stroke of 6 feet varied between 30 and 40 minutes, and the tubes were raised over 100 feet. The water was delivered to the press cylinder through a $\frac{1}{2}$-inch pipe, the outside diameter of which was 1 inch.

EXAMPLE.—Supposing the weight of one of the above-mentioned larger tubes of the Britannia Bridge to be 1726 tons, then a force of $\dfrac{1726}{2} = 863$ tons must be exerted by the large press alone, and by the two smaller ones together, in order to raise the tube. A cross-section of $\left(\dfrac{20}{2}\right)^2 \pi = 314\cdot16$ sq. in. corresponds to a diameter of 20 inches of the press plunger. Hence the pressure of the water in the interior of the press must be $p = \dfrac{863}{314\cdot16} = 2\cdot747$ tons per sq. in., or assuming a ton of 2240 lbs., $p = 2\cdot747 \times 2240 = 6153$ lbs.; taking the atmosphere $= 14\cdot706$ lbs., the pressure $p = \dfrac{6153}{14\cdot706} = 418$ atmospheres. The pump plunger must therefore exert at least a force of $\left(\dfrac{17}{16}\right)^2 \dfrac{\pi}{4} 6153 = 5455$ lbs., and since the area of the piston of the steam cylinder is $8\cdot5^2 \times \pi = 227$ sq. in., the pressure of the steam must be $p_1 = \dfrac{5455}{227} = 24\cdot03$ lbs. per sq. in., *i.e.* not quite two atmospheres.

By suitably assuming the sectional areas of the pump and press plunger of the preceding hydraulic hoists a great increase of force can be attained, for example, the purchase of the hydraulic jack, Fig. 67, is expressed by $\left(\dfrac{89}{19}\right)^2 = 21\cdot94$, and of the hoisting apparatus for the Britannia Bridge by $\left(\dfrac{20}{1\frac{1}{16}}\right)^2 = 354\cdot3$. With the usual rigid combinations, wheels, screws, etc., such large purchases can only be produced by employing additional mechanisms of the same kind, and this greatly reduces the efficiency. In comparison with these combinations the wasteful resistances of the hydraulic hoists are trifling. These consist principally of the friction between the press plunger and its stuffing box, and of the hydraulic resistances of the pump. By omitting all pipes for conducting the fluid as in the jack, Fig. 67, a pump efficiency of at least $0\cdot80$ is realised, and assuming in addition a loss of about 5 per cent for the friction of the press plunger (see below), we conclude that $\eta = 0\cdot75$ is a suitable value for the efficiency of such hydraulic hoists. It is of course understood that these hoists sustain the load automatically when the supply of water into the press cylinder ceases, for the delivery valve is then closed and acts as a pawl. It is also unnecessary to provide this class of hoists with brakes, for by regulating the opening of the escape valve the descent of the load is perfectly controlled and acceleration made impossible. Nevertheless, when worked only at irregular intervals they are apt to get out of order, and in winter to be destroyed, unless oil or some other fluid which is not liable to freeze is employed in place of water.

EXAMPLE.—If in the lifting jack of Fig. 67, the diameters of the plungers are assumed to be 19 mm. [0·75 in.] and 89 mm. [3·5 in.], and the maximum load to be lifted is 30,000 kg. [66,150 lbs.], then, basing our calculations on an efficiency $\eta = 0\cdot75$, the force to be exerted by the pump plunger must be $\dfrac{19^2}{89^2} \times \dfrac{30,000}{0\cdot75}$ $= 1823$ kg. [4020 lbs.]

If the lever arm of the pump plunger is taken equal to 40 mm. [1·57 in.], and the effort of the workmen is applied at the end of a lever 1 metre [3·28 ft.] long, this effort becomes $0\cdot040 \times 1823 = 72\cdot9$ kg. [160·7 lbs.] The ratio borne by the resistance to the effort is given in this case by $\dfrac{89^2}{19^2}\dfrac{1}{0\cdot04} = 548\cdot5$.

§ 16. **Pressure Reservoirs.**—Instead of forcing the water into the cylinder of an hydraulic hoisting apparatus by means of a pump, an elevated reservoir may be employed to furnish the necessary head. To lift the load it is necessary that the natural head should exert on the lifting piston a pressure $FH\gamma$, greater than the weight Q; here F again denotes the area of the lifting piston, γ the weight of water per unit of volume, and H the height of the pressure column measured from the surface of the water in the reservoir to the bottom of the lifting piston. It is understood that Q represents the weight to be lifted, including the friction of the piston and other wasteful resistances. In cases where the necessary head of water is available, for example, in places where the water mains give sufficient pressure, hydraulic hoists may be conveniently employed, as they do not then require a special prime mover. This has led to their use in many of the larger *hotels* and *warehouses* for lifting persons and goods from one story to another. A well-known example of this kind is the hoisting apparatus (constructed by Edoux [1]) employed to carry visitors to the top of the exhibition building at the Universal Exposition at Paris in 1867, and used in Vienna in 1873 for a similar purpose.

This lift is illustrated in Fig. 70. The hollow lifting plunger

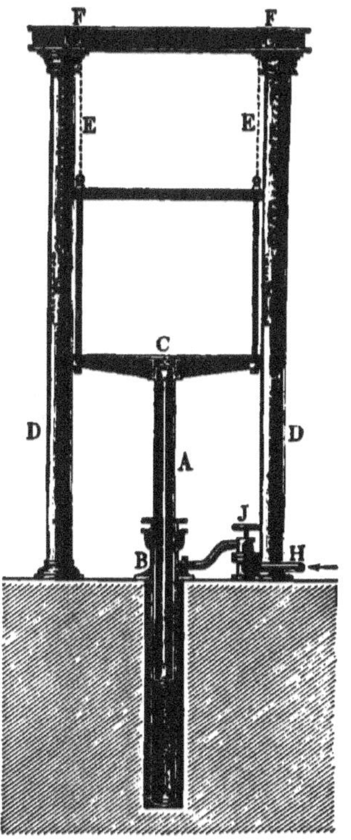

Fig. 70.

A, made of cast iron in several pieces (see vol. iii. 1, Fig. 330), passes water-tight through the cylinder cover B, and supports the cage C, which is arranged to carry passengers. The greater portion of the cylinder is sunk in the ground. The cage is guided by four vertical standards D, and attached to chains E, which are carried over the pulleys F, and loaded at the free ends with counter-weights G placed within the standards. By this means the weight of the cage and its load

---

[1] See Lacroix, *Études sur l'Exposition de* 1867, 6 série.

is partly counterbalanced, the excess of weight of the cage being sufficient to ensure its descent. The water is admitted through the pipe H, and to control the descent of the cage a regulating valve J is introduced, which, after shutting off the supply, allows the water to be discharged from the lifting cylinder. In the present case the diameter of the lifting plunger A was 0·25 m. [9·8 in.], and the maximum pressure exerted upon it was 1500 kg. [3300 lbs.], which corresponds to an effective head of water of

$$\frac{1500}{0·25^2\frac{\pi}{4}1000} = 30·57 \text{ m. } [100·3 \text{ ft.}]$$

Fig. 71 shows a similar lift, and represents an arrangement used in the locomotive works of Borsig in Berlin for lifting the locomotives in the erecting-shops to the height of about 2 metres [6½ ft.] to the track of the Stettin railroad. Here the platform C, constructed of iron beams, is provided with rails upon which the engines run. It is then lifted by three lifting pistons A working in cylinders placed side by side, the water being conducted through the pipe H from a supply reservoir lying about 18 metres [59 ft.] above the cylinders. The platform is guided vertically by two standards D fixed to it, each of which is held by three rollers attached to brackets E projecting from the walls. To support the platform when in its highest position, there are four pillars G and two sliding shoes which may be introduced under the ends of the lifted pillars. The platform is lowered as in all such contrivances by discharging the water from the lifting cylinder. To shut off the supply of water at the right time, a dog K fixed to the platform strikes against a lever F when the platform reaches its highest position. The apparatus is so arranged that for light loads the two outer cylinders can be disengaged, and therefore their lifting plungers are not fastened to the platform. Each lifting plunger has a diameter of 1·1 metre [3·61 ft.]

When the available head of water is insufficient, a special motor may of course be employed to drive a set of pumps, to force the water into an elevated reservoir from which the lifting cylinder may then be supplied. Although this arrangement may not at first appear desirable on account of its indirect action, nevertheless, under certain circumstances, it is so advantageous that its use is constantly increasing. These advantages may be stated as follows: in hoisting machines, with few exceptions, the work is done *periodically*, and usually during short intervals, these being separated by *periods of rest* which are required for removing the load, running back the machine, reloading, etc. The time required for these latter operations

Fig. 71.

often largely exceeds the time actually employed in lifting. Suppose, for example, that the time of one complete operation of an hydraulic crane is on an average about two minutes; perhaps not more than one quarter of this time is actually employed in lifting, while the remaining three-quarters are spent in attaching and removing the load, swinging back the jib, paying out the chain for a new lift, etc. If therefore the hoist were driven by the direct action of a motor, it would be necessary to have the driving force sufficiently powerful to perform the required work in the short period during which lifting occurs, while in the interval following the motor would have to stand idle. The disadvantage thus attending the use of powerful motors for intermittent work is avoided by employing the indirect mode of action mentioned above. In this case the motor can be employed during the whole time of an operation to pump water into an elevated reservoir, where it may be at once available, during the intervals of actual lifting. For this reason these elevated reservoirs, especially when made in the form in which they are represented in the following article, have received the name of *accumulators*. It is evident from what precedes that a motor, which is unable to lift the load directly, can easily perform the work when a reservoir is employed. For let T denote the time of a complete operation, t the time during which the crane is rising, and N the horsepower of a motor which is capable of directly performing the work of lifting, then the continually and indirectly working motor needs only to exert energy to the amount of $\frac{t}{T}N$ horsepowers, provided the additional wasteful resistances arising from the indirect action are neglected.

Moreover, it is evident that *any* prime mover already employed for other purposes, as a steam-engine driving the machinery of a shop, can be used to work a pump to supply the reservoir.

Another advantage of the indirect system, and under certain circumstances a very important one, is that we have a convenient means of easily distributing a large amount of mechanical energy to considerable distances; such a case arises when a large number of hoists are to be driven at once by a single prime mover. For, since the water pressure can be conducted to the hoisting apparatus through a series of pipes adapted to the local features of the ground, we have a method of transmitting motive power to long distances which is free from the disadvantages of having many bearings, bevel wheels, and rope-pulleys, all of which must be estimated in considering the cost of a transmission of power by shafting or wire rope, other things being equal. Nor would it be advantageous to employ steam-hoisting machinery on account of the great loss of heat due to the condensation of steam in the long pipes.

These circumstances explain satisfactorily the advantage of employing an hydraulic hoist driven by an elevated reservoir in all cases where the machine works intermittently, and where a large amount of work must be performed in a short time.

§ 17. **Accumulators.**—The pressure $FH\gamma$ exerted upon a lifting plunger, increases in direct proportion to the height H of the free surface of the water in the reservoir above the plunger, hence by increasing the natural head of water H, the cross-section F can be correspondingly reduced. But the erection of elevated reservoirs is generally attended with considerable expense and great difficulties as soon as the height becomes considerable, and where they are employed it is seldom possible to obtain a head of water of more than 30 m. [100 ft.] A small head will generally involve large dimensions for the lifting cylinder, and, as will be shown hereafter, will involve a relatively large loss of work due to wasteful resistances, so that such a construction cannot be recommended as an economical one.

For this reason Armstrong's invention, an *accumulator* which replaces the elevated reservoir, is of great importance.

An accumulator, Fig. 72, consists essentially of a strong

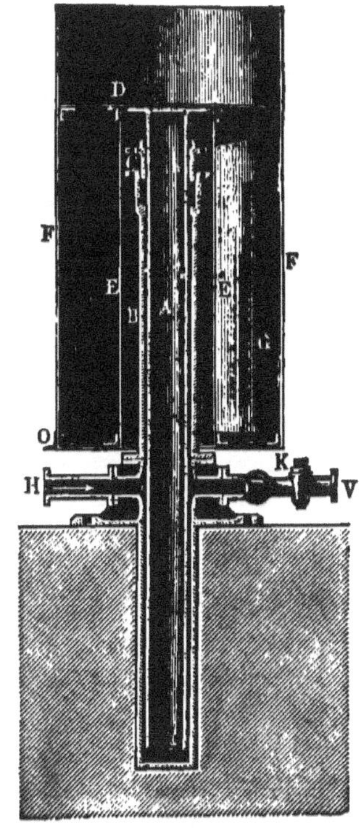

Fig. 72.

cylinder B, whose plunger passes, water-tight, through the stuffing box C. This lifting plunger is heavily loaded by a weight receiver which consists of two wrought-iron cylinders E and F suspended from a cross-head D. Let us now suppose water to be pumped into the accumulator cylinder through the pipe H, during which operation the pipe K is closed; then the lifting plunger A is forced upward by the pressure of the water upon its lower surface, and in this manner the weight G is raised as in a hydraulic hoist. Here again we have $G = Fp$, where G denotes the fixed weight, F the area of the lifting plunger A, and p the water pressure per unit of area; this expression does not take into account the friction of the stuffing box. The water contained in the cylinder B is therefore subjected to a pressure p per unit of area expressed by $p = \dfrac{G}{F}$.

This water pressure could be obtained by replacing the loaded piston A by a pressure column of the same cross-section F and weight G, and its height would therefore be found from $G = FH\gamma$, to be $H = \dfrac{G}{F\gamma}$. Thus the water in the accumulator cylinder B is under the same pressure as though the cylinder were connected by means of a pipe with an elevated tank, the height of the free surface of the water above the bottom of the lifting piston being $H = \dfrac{G}{F\gamma}$. If, therefore, by opening the cock K, water is conducted from this cylinder through the pipe V to any hydraulic hoist, the machine will perform its functions in the same manner as if the reservoir with the above head of water were employed. In other words, with reference to its action on the hoist the accumulator replaces a pressure column of $H = \dfrac{G}{F\gamma}$. The available quantity of water in the accumulator is likewise given by its dimensions, and is equal to $V = Fl$ where l denotes the distance through which the lifting plunger A can fall when the cock K is opened; that is, it denotes the height to which the plunger was previously raised by the pump. Furthermore, we see that while the water is being conducted through the pipe V to the hoist, the pump may replenish the accumulator continuously through the pipe H. If the water which is forced by the pump is just sufficient to drive the hoist, the lifting piston will remain stationary; on the other hand, it will fall or rise according as the hoist uses a greater or smaller quantity of water than the pump can supply during the same time. We may therefore regard the accumulator as a regulator which stores up or gives off the difference between the energy supplied and that consumed.

This will enable us to determine the necessary capacity of the accumulator in every case in which the data are known. Let V denote the quantity of water required by the hoist while it is lifting, T the time in seconds of a *complete operation*, and t the time of actual lifting; then the force-pump and its motor should have proportions sufficient to supply a quantity of water per second expressed by $\dfrac{V}{T}$, and the capacity of the accumulator must be sufficient to store the quantity $V\dfrac{T-t}{T}$ which is supplied to it in the interval $T-t$, during which the hoist is standing idle.

By the capacity of the accumulator we must not understand the whole volume of the cylinder B, but only the quantity of water Fl which is displaced by the lifting plunger in falling a distance l, for the space between the plunger and the wall of the cylinder always remains filled with water which can never be used as a driving force.

This determination of the capacity of the accumulator is the same when a greater number of hoists is to be supplied, for in this case V is the quantity of water which *all* the hoists would require for one operation were they all in action at the same time. For although in general the water supply is partly equalised in consequence of the strokes of the separate lifting plungers taking place at different times, still in order to ensure the working of the hoists we must provide for the possible case that all the hoists may be working at the same time.

In order that the pump may disengage itself when the accumulator is filled, it is customary to provide the accumulator plunger with a dog O, Fig. 72, which in its highest position acts upon a lever which closes the throttle valve of the motor, and thus by stopping the pump prevents the accumulator plunger from being completely lifted out of the cylinder.

The intensity of pressure in the accumulator due to the pressure column $H = \dfrac{G}{F\gamma}$ can be controlled at will for a given cross-section F of the plunger, by suitably choosing the weight G, this intensity being only limited by the strength of the cast-iron cylinder and supply pipes. Even in the ordinary accumulators very large heads are employed in many cases.

Armstrong has made use of a pressure of 50 atmospheres, which corresponds to a head of about 500 metres [1650 ft.]; such a head evidently would hardly be obtainable by the use of elevated reservoirs. The employment of great water pressure makes possible the use of small sectional areas for the working plungers, and by this means the hydraulic resistances in the supply pipes are considerably reduced. For example, let us suppose the water of the accumulator to be conducted to the separate hoists through water-mains of diameter d and length l, which latter quantity is sometimes very great, then for a velocity v the quantity of water under a head H which passes through the pipes in each second is $Q = \dfrac{\pi d^2}{4} v$, and when wasteful resistances are neglected, represents the work

$$A = QH\gamma = \frac{\pi d^2}{4} v H \gamma.$$

But on account of the friction of the water in the pipes there is a loss of head, which, according to vol. i. § 455, Weisb. *Mech.*, is given by

$$\zeta \frac{v^2}{2g} \frac{l}{d} = h_0 \,;$$

this shows that the loss is independent of the head H, *i.e.* for equal values of l, d, and v, but different values of H, the loss of head is a constant quantity. If therefore the ratio between the loss of work expressed by

$$A_0 = \frac{\pi d^2}{4} v h_0 \gamma,$$

and the energy expended $A = \dfrac{\pi d^2}{4} v H \gamma$, is determined, we shall obtain the relative loss of work due to the transmission of water in pipes; that is to say, the loss of work corresponding to each unit of energy expended, will be expressed by

$$\frac{A_0}{A} = \frac{h_0}{H} = \zeta \frac{l}{d} \frac{v^2}{2g} \frac{1}{H},$$

and therefore varies inversely as the head H. Similar remarks apply to the other hydraulic resistances; for example, to the loss of pressure due to the passage of the water through the contraction caused by regulators, bends, branch pipes, etc., for the loss of head at these places is independent of the absolute head H.   From this point of view the hydraulic transmission of power with a large head of water is very economical.

On the other hand, owing to the high pressure the construction of water pipes in accumulator plants requires the greatest care, as even slight leakage is attended with considerable loss of power.

In consequence of the large head under which the water of an accumulator works, there is another peculiarity belonging to the class of lifts under consideration.   When the load is to be lifted through a moderate height only, it is customary to form the upper part of the lifting plunger into a platform for its support, as illustrated in Figs. 70 and 71.   But for a great length of stroke this construction would lead to practical inconveniences, as the plunger would project out of the cylinder to a considerable distance, which would be a matter of great consequence, inasmuch as the diameter of the plunger would be comparatively small, especially when the load is light, the plunger for this condition assuming the character of a long slender pillar whose strength against fracture from bulging would be insufficient.   To meet this difficulty the common *Armstrong* mode of construction in such cases is to give the plunger a short stroke and to gain an increased range of motion by suitably arranged tackle.   The action of these lifts is the reverse of that mentioned in § 8, inasmuch as the load to be lifted acts at the free end of the chain, while the motion of the pulley or block is directly effected by the plunger.   The forward motion during the raising of the load therefore corresponds to the reverse motion of the ordinary tackle.   Of course it follows that the effort which the piston exerts on the lifting tackle is greater than the load to be lifted in the proportion of the velocity ratio.   The high pressure of the water renders it always easy to exert the required force by correspondingly increasing the sectional area of the plunger, the further advantage thus being gained that the resistance offered by the latter against compression is increased.

When an hydraulic plant is employed for both heavy and light work, then instead of using a single hoist it is more economical to use several so arranged as to work either separately for light loads, or together for heavier loads. If this arrangement were not followed it would be necessary to entirely fill the hoist cylinder in both cases, and a very small efficiency would result.

That it is possible to stop or retard the motion of a hoisting apparatus by simply closing the supply pipe is self-evident, and we have mentioned above that by adjusting the opening of the passage in the supply pipe we can control the descending load. When a platform is employed it is generally made sufficiently heavy to cause the descent by its own weight, and it is often necessary to balance a portion of this weight by the use of a counter-weight. It is also common practice to utilise the excess of platform weight over counter-weight to force the discharge water into an elevated tank which supplies the force-pumps of the accumulator.

When there is no platform, as in the case of cranes and many other lifts where the load is attached to a hook, it is generally necessary to load the latter with a special weight to ensure the backward motion of the lift. As this weight, besides keeping the chain tight and overcoming the resistances of the pulleys, must cause the descent of the lifting plunger, the latter is frequently connected with a *counter-plunger* having a smaller sectional area, but the same stroke as the lifting plunger, thus avoiding the use of a heavy weight on the hook. As the counter-plunger takes part in the direct motion of the working plunger, it is forced into its cylinder during the forward motion of the latter, and then driven out again by the water pressure, thus causing the return of the lifting plunger. The weight attached to the hook, therefore, only serves to give the chain the necessary degree of tension. Ropes are seldom employed in hydraulic hoisting apparatus. The class of hydraulic machinery under consideration, besides being used for the actual lifting of weights, is frequently employed to perform other kinds of work, for example, to lift sluice gates, and to swing crane-jibs. (See chapter on Cranes.) It is also employed in Bessemer works to rotate the converters, and in boiler works to form boiler heads, etc., but only in cases where intermittent power is required. For machinery which is to work continuously the accumulator is not an advantageous prime mover

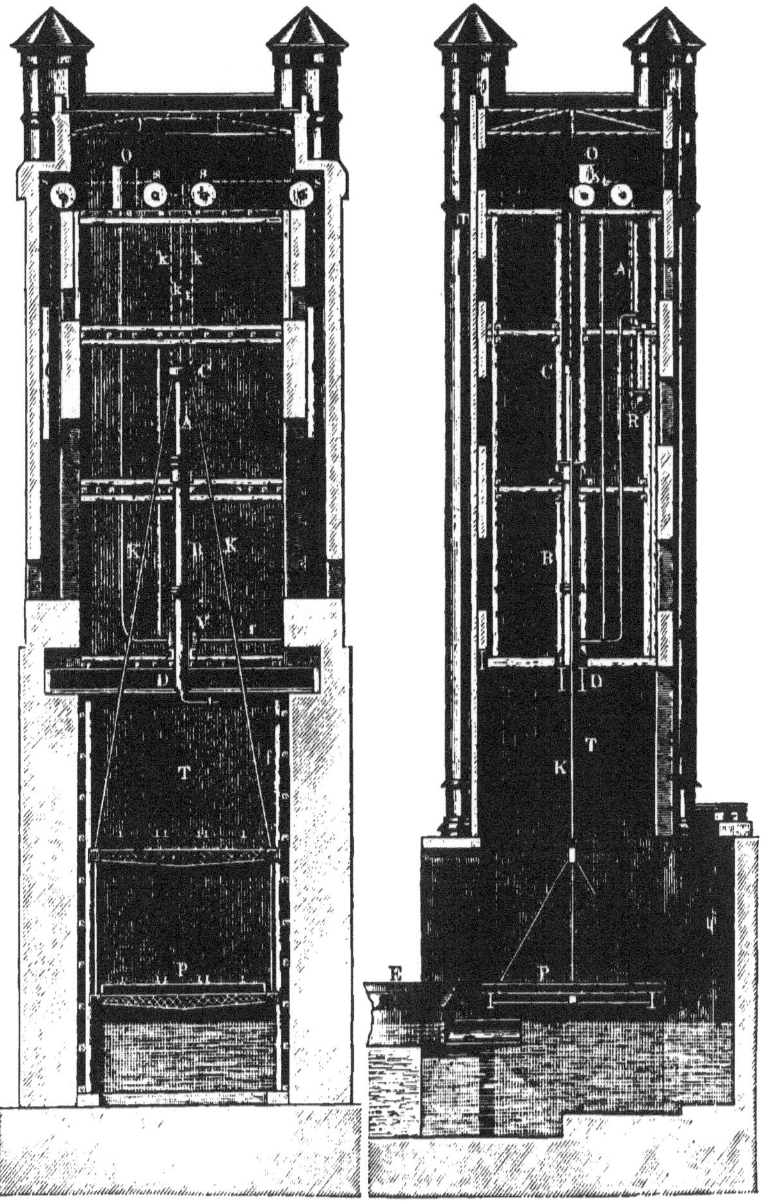

Fig. 73.

on account of its indirect mode of action.   In the following pages a few lifts operated by accumulators will be described.

§ 18. **Hydraulic Lifts.**—In Germany one of the first hydraulic plants constructed according to *Armstrong's* system was devised in 1856 for effecting railroad communication between Homberg and Ruhrort, being employed for lifting cars from the steam-ferry to the level of the rails.   To effect this each station is provided with the hoisting apparatus shown in Fig. 73.   The lifting cylinder is placed in the massive tower T, and fastened to two wrought-iron beams D.   The plunger A has a diameter of 0·314 m. [12·36 in.], and carries a cross-head C from which the chains K spread downward to the platform P, which is to be raised.   The latter is provided with tracks upon which are generally placed two loaded cars, each weighing about 300 cwt. [15,000 kg.]   These can be run directly from the ferry upon the platform when in its lowest position.   The vertical motion of the platform is secured by means of guides f attached to the walls of the tower, and is effected by admitting water through the pipe r leading from the accumulator.   The weight of the platform amounting to about 560 cwt. [28,000 kg.] is partly balanced by two counter-poises, together weighing 23,400 kg. [52,600 lbs.] which weights working in passages in the walls, are connected with the cross-head C of the lifting plunger by chains k, which pass over pulleys s.   The distance through which the platform is to be raised changes with the varying height of the water of the Rhine, the maximum lift being 8·72 metres [28·61 ft.] This is also the estimated length of stroke of the plunger A. There is still another plunger A_1 having a diameter of 0·196 metres [7·72 in.] and a length of stroke one-half that of plunger A.   By means of the movable pulley R attached to the cross-head, and the chain k_1, the smaller plunger can assist in lifting the platform when the weight of the load demands it.   Besides, it is employed alone in lifting the empty platform, preceding the lowering of the cars, no water being admitted into the large cylinder.   In order that the latter may still be supplied with the water needed as a brake during the next descent, there is located in the upper part of the tower an auxiliary reservoir into which the water used in the cylinder is forced as the platform descends.   By means of valves, water from the accumulator may be supplied to one or both cylinders as desired, the cylinder into which no water pressure is admitted being at the same time placed in communication with the auxiliary reservoir O. The valves are worked by hand, but a self-acting disengaging device is introduced, by which the platform, during the last two metres [6·5 ft.] of its up and down stroke acts upon a system of levers, causing a gradual closing of the admission or discharge port, in order to prevent the shocks which would occur if the motion of the masses were suddenly arrested.

The accumulator, whose pumps are driven by a steam-engine of 30 horse-power, has a diameter of 0·418 metres [16·46 in.] and a stroke of 5·33 metres [17·5 ft.], so that its capacity approximately corresponds to that of the two hoist cylinders for their maximum stroke of 8·72 metres [28·61 ft.] and 4·36 metres [14·3 ft.] respectively. The weight on the accumulator plunger is estimated to produce a pressure of 43 kg. per sq. centimetre [611 lbs. per sq. in.]

Fig. 74.

Fig. 74 shows an *Armstrong* lift for warehouses, Fig. 75 representing the hoist cylinder A on a larger scale. The piston K is driven downward by the water pressure causing the movable pulley carried by the cross-head C of the piston-

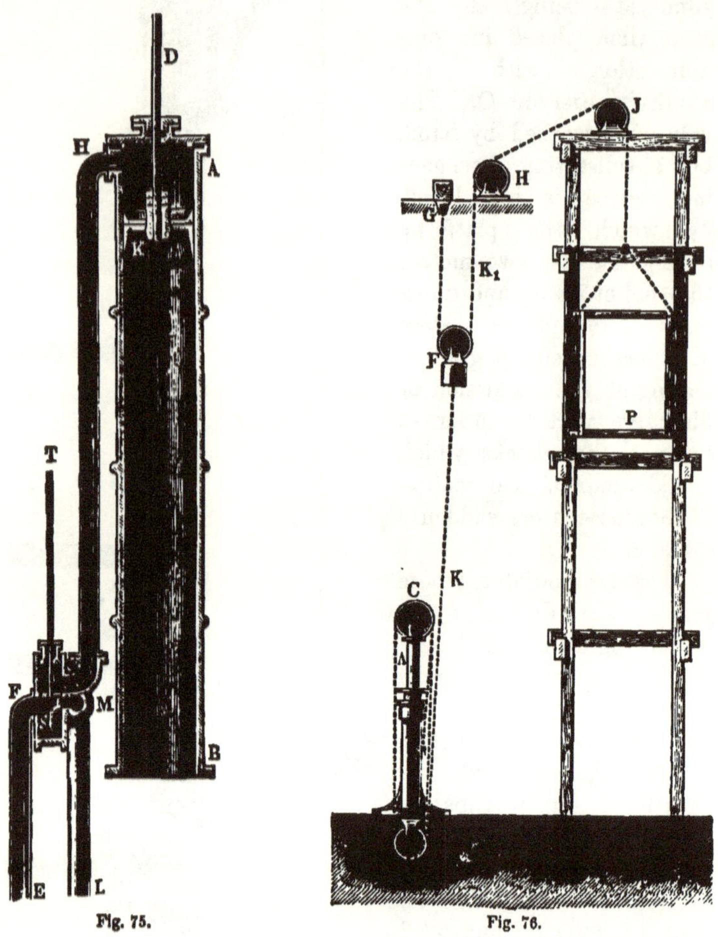

Fig. 75.Fig. 76.

rod to haul upon the rope that passes over the fixed pulley E. One end of this rope is made fast to the cross-head C, while the other end is fixed to the block of the movable pulley F. From the latter, in the manner shown in the figure, a second rope passes over the fixed pulleys G and H, whence it is carried horizontally between the guide-pulleys Z over the pulley K of the crane-jib, and then downwards to the ring L which carries the load. The jib turning about the axis MN allows the load to be taken into the warehouse through an opening in the wall. A detailed description of this part will be given under the head of cranes.

It is obvious from the figure that in the above arrangement of pulleys, the velocity of the load is nine times that of the piston rod.

Fig. 76, which represents the hydraulic lift for the custom-house at Harburg,[1] will be easily understood from what precedes. The lifting plunger which rises through the cylinder cover B, under the action of the water pressure carries two chain pulleys C on its cross-head, while the fixed pulley D is secured to the base of the hoist cylinder. The chain K is attached to the cylinder at E, and after passing over the four pulleys C, D, C, D, is secured to the block of the movable pulley F. One end of the chain K_1 is made fast at G, and the other after passing over the guide-pulleys H and J is carried downwards to the cage P. Here the velocity ratio is eight to one. In other respects the same considerations obtain as in the preceding case.

It is evident from Fig. 75 how the water is supplied to and discharged from the cylinders by means of the valves S, similar in construction to the ordinary slide-valves used in steam-engines. The water is supplied through the pipe EF and discharged from the cavity in the valve through the pipe LM. In the lowest valve position shown, it enters the cylinder through GH, effecting the lifting operation by forcing down the piston K, while in the highest position communication is opened between GH and the pipe LM, as required for lowering the load. When the valve is placed in its central position, no water enters nor leaves the cylinder, the load consequently remaining at a stand-still. By cocks or lift valves, additional means is provided for cutting off the communication between cylinders and reservoir.

The form of piston illustrated in the apparatus just described is seldom employed in hydraulic hoists, preference being given to the plunger type as more accessible and more easily made water-tight. The lift valve is also preferable to the slide-valve,

---

[1] See *Zeitschr. des hannoversch. Arch.- u. Ing.-Vereins* 1860, p. 443.

since the use of the latter is accompanied by considerable friction due to the great pressure of the water, which makes it more difficult to handle.

Since water is practically incompressible, it is necessary in all hydraulic hoisting machinery to use special precautions against shocks which occur when masses in motion are suddenly brought to rest. Let us, for example, suppose the platform of the lift, Fig. 73, to be descending, and let its motion be suddenly arrested by the closing of the discharge valve, then the actual energy $\dfrac{Mv^2}{2}$, stored in the moving mass M, would be expended in producing shocks, in consequence of which the pressure of the water in the hoist cylinder might become sufficiently great to cause fracture. In order to avoid this straining action, a relief valve is introduced into the pipe connecting the cylinder with the valve chamber; this valve, subjected on its upper surface to the water pressure in the accumulator, is generally kept closed, and is opened only when the pressure within the cylinder becomes excessive; when this happens, a certain quantity of water from the cylinder is forced back into the supply pipe communicating with the accumulator. A shock would also occur if the upward motion of the platform were suddenly arrested by cutting off the supply of water; this may be prevented by a valve—in this case, a suction valve. For example, let us suppose the platform to reach the end of its up-stroke with a velocity v; then after suddenly cutting off the supply of water, it would still be capable of rising to a height $h = \dfrac{v^2}{2g}$ due to this velocity, the effect of which would be to produce a "vacuum" or empty space, having a volume Fh. Now, immediately following the formation of such a vacuum, the platform would drop through this distance h, and, on striking the water, would cause a shock. To prevent the platform from thus falling, the pipe connecting the cylinder with the valve chamber is provided with a second valve. This is lifted during each up-stroke, and draws water from the discharge pipe into the cylinder, thus preventing the formation of a vacuum, and afterwards by its action as a brake, regulating the descent of the platform. The arrangement of these relief valves is best seen in Fig. 77, which represents the valve gear of the Ruhrort-Homberg lift.   The water from the accumulator is admitted through the pipe a, and, after it has been used in the cylinder, is conducted through the pipe b to the auxiliary reservoir O in the top of the tower.   By opening the stop valve c the water reaches the chamber d, from which it may pass either through valve g and pipe g_1 to the large hoist cylinder, or through k and k_1 to the small hoist cylinder, or it may be conducted at the same time through g and k to both cylinders, in which case the discharge valve f must be closed. If, after closing c, the discharge valve is opened, the plunger will complete its return stroke, since the water then escapes

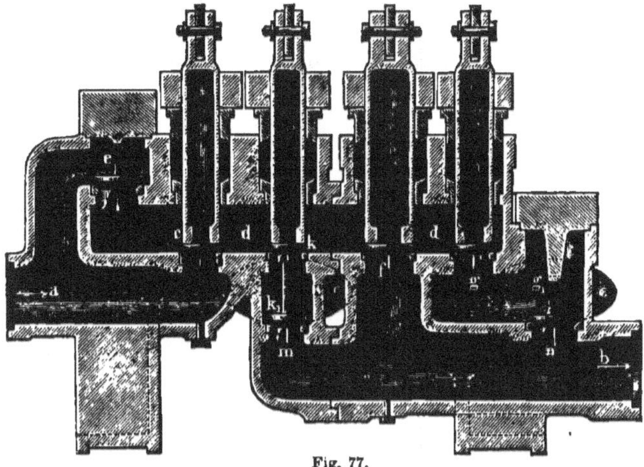

Fig. 77.

from g_1 and k_1 through f into the pipe b.   At e we have the relief valves already mentioned, whose lower side is acted on by the pressure of the water in the accumulator, while m and n represent the suction valves.   During the ascent of the platform these valves will be opened, if the pressure in the hoist cylinder falls below the pressure exerted on the lower surfaces of the valves by the water of the auxiliary reservoir O.   At the same time the valves m and n provide either cylinder—when not in communication with the accumulator—with water from the auxiliary reservoir, to act as a brake in the manner already described.

We see that the valves e, m, and n cannot extend their action to the balance weights connected by chains with the platform, and that these chains are exposed to unavoidable shocks by the sudden stopping of the machinery. To prevent these straining actions it is always advisable to gradually retard the motion, and it is with this object in view that the self-acting disengaging device mentioned above is made use of.

§ 19. **Action of Accumulators.**—It still remains for us to establish the principles which govern the working of hydraulic hoisting machinery by accumulators.[1] For this purpose let the fixed weight resting upon the accumulator plunger be replaced by a column of water of equal weight, having the same sectional area F_1 as the plunger, and a height represented in Fig. 78 by $BG = h$. Let us also suppose the part of the weight of the lifting plunger and platform, which is not balanced by counter-weights, to be replaced by a column of water having the same sectional area F as this plunger, and a height $AC = q_0$, and let the useful load Q to be lifted be represented by a similar column, having the height $CD = q$, so that $Q = Fq\gamma$, where γ represents the weight of a unit volume of water. F_2 may

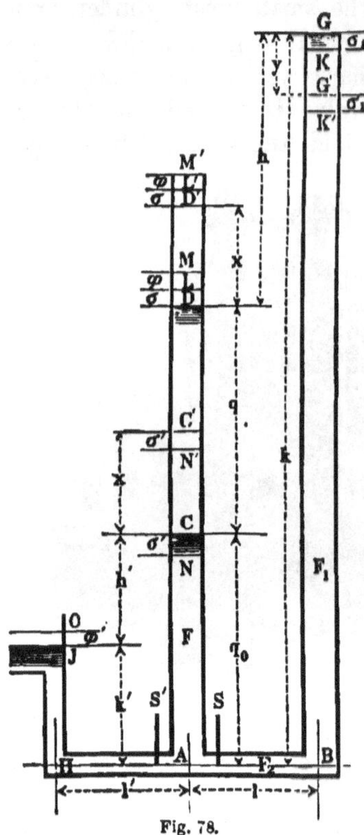

Fig. 78.

represent the sectional area of the horizontal supply pipe, l its length, BA and l' the length of the discharge pipe AH having the same cross-section F_2. Finally the height above the pipe BH of the water level J in the auxiliary reservoir —from which the pump replenishes the accumulator—will be denoted by $HJ = k'$. The water levels D and G in the figure correspond respectively to the *lowest* position of the lifting plunger, and the *highest* position of the accumulator plunger.

---

[1] See article by L. Putzrath, *Zeitschr. d. Ver. Deutsch. Ing.*, 1878, p. 505.

Suppose the admission valve S to be opened, then the lifting plunger will ascend, *i.e.* the surface of the water will rise from D to D′, while the level G in the accumulator will fall to G′, and we have

$$\mathrm{F}x = \mathrm{F}_1 y, \text{ hence } y = \frac{\mathrm{F}}{\mathrm{F}_1}x = \nu x,$$

where $x = \mathrm{DD}'$ represents the distance through which the lifting plunger has risen, and $y = \mathrm{GG}'$ the distance through which the accumulator plunger has fallen at a given instant, and ν denotes the ratio $\dfrac{\mathrm{F}}{\mathrm{F}_1}$ of the sectional areas of the two plungers.

During this motion certain wasteful resistances are to be overcome, the principal ones of which we will now state. First comes the friction of the lifting plunger in its stuffing box. This friction, which will be determined more exactly hereafter, is proportional to the pressure of the water on the lifting plunger, and is therefore proportional to the area F; hence we may suppose it to be replaced by a water column acting by its weight on the lifting plunger, and having a height $\sigma = \mathrm{DL} = \mathrm{D}'\mathrm{L}'$, where σ is a value that will be determined later. Likewise the friction of the accumulator piston in the stuffing box may be replaced by a column of section F_1 and height $\sigma_1 = \mathrm{GK} = \mathrm{G}'\mathrm{K}'$, which column must be subtracted from the column BG. Finally friction and certain other hydraulic resistances occur in the cylinders, and especially in the long and narrow supply pipe BA, which friction also corresponds to a column having a height $\phi = \mathrm{LM} = \mathrm{L}'\mathrm{M}'$, and acting by its weight on the lifting plunger.

Let us now suppose that the lifting plunger has been raised through a height $x = \mathrm{DD}'$, and that the valve S is closed, and the discharge valve S′ opened, as soon as the useful load Q has been removed from the platform; the latter will then descend from C′ until it has passed over the distance x, and reached its starting point C, meanwhile forcing the water through S′ into the reservoir J. During such a complete operation of the lift, consisting of an up and down stroke, the load $\mathrm{Q} = \mathrm{F}q\gamma$ is raised through a height x, and the useful work performed is consequently expressed by

$$\mathrm{A} = \mathrm{F}q\gamma \cdot x = \mathrm{F}x\gamma \cdot q = \mathrm{V}\gamma q \qquad \cdot \qquad \cdot \qquad (1)$$

where V denotes the quantity of water Fx that has passed from the accumulator into the cylinder of the hoist. In performing this work a volume of water $F_1 y = Fx = V$ has been conveyed from the space GG' of the accumulator to the reservoir J, and at the same time the centre of gravity of this water has, according to the figure, descended through the distance

$$h - \frac{y}{2} + q + h' = q + h + h' - v\frac{x}{2}.$$

Consequently the work performed by this mass of water in falling is expressed by

$$V\gamma\left(q + h + h' - v\frac{x}{2}\right).$$

We therefore see that the expression $V\gamma\left(h + h' - v\frac{x}{2}\right)$ represents the loss of work which attends the lifting of the load, and that the efficiency of the hydraulic lift is represented by

$$\eta = \frac{q}{q + h + h' - v\dfrac{x}{2}}.$$

The question now is under what conditions the above loss becomes a minimum and the efficiency a maximum. This is evidently the case when the value $h + h' - v\frac{x}{2}$ is made as small as possible. If the frictional resistances σ of the plungers in their stuffing boxes, and ϕ of the water in the pipes did not exist, we see that the height h, or the difference of level between G and D in the two cylinders at the beginning of the stroke, would be at least $h = x + y = x(1 + v)$, thus $h = 2x$ for equal cross-sections F and F_1 of the plungers, because when the lifting plunger is in its highest position the column of water in the cylinder A must be balanced by the column in B. Likewise during the descent of the plunger the height h' cannot be less than zero even in the absence of friction, for the level C in the hoist cylinder, corresponding to the *empty* platform, cannot fall below the level J in the auxiliary reservoir.

Under these ideal conditions of freedom from all frictional resistances, the loss of work for a complete operation becomes

$$V\gamma\left(h + h' - v\frac{x}{2}\right) = V\gamma\left[x(1 + v) + 0 - v\frac{x}{2}\right] = V\gamma x\left(1 + \frac{v}{2}\right) \qquad (2)$$

Hence for $F = F_1$ the loss is $V\gamma\frac{3}{2}x$. A portion $V\gamma\left(h - \frac{x + y}{2}\right)$ $= V\gamma x\frac{1 + v}{2}$ of this work is lost during the ascent, and the remainder $V\gamma\left(h' + \frac{x}{2}\right) = V\gamma\frac{x}{2}$ during the descent of the platform.

The explanation of this is found in the fact that the volume of water Fx which has descended does not, as would be the case with a rigid body, rise above the level D' or J in consequence of its velocity, the assumption being made that the velocity communicated to the water is destroyed by eddies or internal motion.

In reality, however, the loss of work is much greater owing to the friction σ and σ_1 of the plungers, and the friction ϕ in the pipes. Taking these frictional resistances during the ascent of the platform into account, the figure shows that the minimum value of h must satisfy the condition

$$h = x + \sigma + \phi + \sigma_1 + y = x(1 + v) + \sigma + \sigma_1 + \phi.$$

Further, let the height of the water column which corresponds to the friction of the stuffing box during the descent of the lifting plunger be denoted by $\sigma' = C'N' = CN$, which height is henceforth to be deducted from the water column $AC' = q_0 + x$, and let $\phi' = JO$ denote the height of the water column corresponding to the friction in the discharge pipe AH, then we likewise find the least value of h' to be

$$h' = \sigma' + \phi'.$$

Assuming these minimum, *i.e.* most favourable, values, the lost work during the raising of the load $Q = Fq\gamma$ to a height x becomes

$$V\gamma\left(h + h' - v\frac{x}{2}\right) = V\gamma\left[x\left(1 + \frac{v}{2}\right) + \sigma + \sigma_1 + \phi + \sigma' + \phi'\right].$$

If now the lift be arranged so as to fulfil these conditions, *i.e.* if the effective heads h and h' be so chosen as to satisfy the relations

$$h - (\sigma + \sigma_1 + \phi) = x(1 + \nu)$$

and

$$h' - (\sigma' + \phi') = 0,$$

then although the ascent and descent would be theoretically possible, yet for practical purposes the working of the machine would require too much time, inasmuch as the raising and lowering of the water levels at D and G during the up stroke, and at C′ and J during the down stroke, would take place in a manner similar to that observed when communication is opened between two vessels containing water to unequal heights, as in the case of a canal lock.

As the actual problem, however, is to lift the load to a certain height in a given time, and as the same conditions must be fulfilled for the down stroke, the solution requires that we assume the effective heads h and h' greater than the minimum heads determined above. The problem is thus reduced to finding the relation between the assumed effective heads h and h' and the time required for the ascent and descent. From what has preceded it follows that the loss of work during the up stroke, found to be $V\gamma\left(h + h' - \nu\dfrac{x}{2}\right)$, must increase with h and h', so that obviously the efficiency of the hydraulic lift diminishes in proportion as the velocity of the ascent and descent increases.

In order to determine the speed of the lifting plunger, it is necessary to consider the mass of the counter-weight W balancing the platform. Let this counter-weight be replaced by a water column of height $w = \dfrac{W}{F\gamma}$, whose sectional area is equal to the area F of the lifting plunger, while q_0 represents the height of the water column corresponding to the *total* weight Q_0 of the platform. Further, let η be the efficiency of the system of pulleys used in connection with the counter-weight; then during the up stroke the excess $Q_0 - \eta W$ of the weight of the platform must be overcome, and the descent of the platform must be effected by the excess of weight $Q_0 - \dfrac{W}{\eta}$.

During the up stroke the lifting plunger starts from its lowest position with a velocity zero, and under the action of the effective head the motion becomes one of acceleration. Now suppose the lifting piston to have moved through a distance $x = DD'$, and let the acceleration of the mass of water $\dfrac{F(q + q_0 + x)\gamma}{g} = m$, contained in the working cylinder at this instant, be denoted by c, then the counter-weight, whose mass is $\dfrac{W}{g} = m_3$, will have the same acceleration c. Further, let c_1 be the acceleration of the mass of water $m_1 = \dfrac{F_1(k - vx)\gamma}{g}$ contained in the accumulator, and c_2 that of the mass $m_2 = \dfrac{F_2 l\gamma}{g}$, contained in the supply pipe AB at the same instant. If now P denotes the effort which causes this acceleration, then according to the fundamental law that

$$\text{Acceleration} = \frac{\text{Force}}{\text{Mass}},$$

we have

$$(m + m_3)c + m_1 c_1 + m_2 c_2 = P \qquad . \qquad . \qquad (3)$$

Remembering that $c_1 = c\dfrac{F}{F_1}$ and $c_2 = c\dfrac{F}{F_2}$, we find

$$(m + m_3)c + m_1 c_1 + m_2 c_2$$

$$= F(q + q_0 + x + w)\frac{\gamma}{g}c + F_1(k - vx)\frac{\gamma}{g}\frac{F}{F_1}c + F_2 l\frac{\gamma}{g}\frac{F}{F_2}c$$

$$= F\frac{\gamma}{g}[q + q_0 + x(1 - v) + w + k + l]c = Mc \quad . \quad (4)$$

where M denotes the coefficient of c. Hence for the acceleration c of the platform we have

$$c = \frac{P}{M},$$

in which the quantity

$$M = F\frac{\gamma}{g}(q + q_0 + w + k + l)$$

may be considered a constant, as the value $x(1 - v)$ is comparatively small.

Now the driving force P, and hence the acceleration c, diminish with every instant, this diminution being due not only to the reduction in effective head h by $(1 + \nu)$ times the increase in x, but also to the expenditure of a portion of this head towards giving motion to the water. A clear idea of this fact may be gathered by noting the difference between the *hydrostatic* pressure of water when at rest, and the *hydraulic* pressure when it is in motion. According to vol. i. § 427, Weisb. *Mech.*, the *hydraulic* head, at a point where the water flows with a velocity v_2, is less than the *hydrostatic* head at the same point by $\dfrac{v_2^{\,2} - v_1^{\,2}}{2g}$, where v_1 denotes the velocity of the water at the entrance. In accordance with this principle the pressure which causes the acceleration of the mass M is determined as follows. Suppose the lifting plunger to be in its lowest position, and the valve S just opened to the water, then for the effective head $h - \sigma - \sigma_1 - \phi$ at this point, the *initial* hydrostatic pressure is $f(h - \sigma - \sigma_1 - \phi)\gamma$, where $h = k - q - q_0 + \eta w$, and f is the area of the passage offered by the valve to the water. But the pressure diminishes after the motion of the plunger has begun. For example, assuming that the lifting plunger has reached a height $DD' = x$, at which instant its velocity may be taken equal to v; then from what precedes a diminution of the hydrostatic head h, equal to $x + y = x(1 + \nu)$, occurs; besides, the hydraulic head of the flowing water in S is less than the hydrostatic head by $\dfrac{v_2^{\,2} - v_1^{\,2}}{2g}$ where v_2 and v_1 denote the velocities in S and G'.

Now $v_1 F_1 = vF$, hence $v_1 = \dfrac{F}{F_1}v = \nu v$, and $v_2 f = vF$, hence $v_2 = \dfrac{F}{f}v = \mu v$, where μ denotes the ratio $\dfrac{F}{f}$ of the sectional area of the lifting cylinder to that of the inlet. Consequently the hydraulic head at S at this instant is

$$h - \sigma - \sigma_1 - \phi - (1 + \nu)x - (\mu^2 - \nu^2)\frac{v^2}{2g},$$

in which expression evidently ν^2 can be disregarded in comparison with μ^2, as ν is always less than 1, while μ generally lies between 20 and 40, thus making ν^2 equal to hardly $\frac{1}{10}$

per cent of μ^2. We thus obtain the acceleration of the lifting plunger at the position referred to, that is at a distance x from its lowest position

$$c = \frac{P}{M} = \frac{f\gamma\left[h - \sigma - \sigma_1 - \phi - (1 + v)x - \mu^2\dfrac{v^2}{2g}\right]}{F\gamma(q + q_0 + w + k + l)}g$$

$$= \frac{a - (1 + v)x - \mu^2\dfrac{v^2}{2g}}{\mu b}g,$$

where for the sake of brevity we put

$$h - \sigma - \sigma_1 - \phi = a$$
$$\text{and } q + q_0 + w + k + l = b.$$

Substituting the general value $c = \dfrac{\delta^2 x}{\delta t^2}$ for the acceleration, and placing $v = \dfrac{\delta x}{\delta t}$, we finally obtain the equation applicable to all hydraulic hoisting machines :

$$\frac{\delta^2 x}{\delta t^2} = \frac{a - (1 + v)x - \mu^2\left(\dfrac{\delta x}{\delta t}\right)^2\dfrac{1}{2g}}{\mu b}g \qquad . \qquad . \qquad (5)$$

The integration of this differential equation offers great difficulties, but it is easy to obtain a solution sufficiently accurate for our purpose. According to the above investigation, the height $\mu^2\dfrac{v^2}{2g}$ due to the velocity increases from zero at the beginning to a value depending upon the speed v of the lifting plunger. The velocity v is usually small in all hoisting machines, owing to the large masses put in motion, seldom exceeding 0·3 metres [0·98 ft.] per second; consequently, though the corresponding velocity of the water through the supply valve is greater in the proportion of $\dfrac{F}{f}$, the head due to the latter velocity is nevertheless small compared with the large hydrostatic heads employed in the accumulator plant.

The error will therefore be trifling if, when determining the accelerating force, instead of simply subtracting the above head at the instant corresponding to the maximum speed v of the lifting plunger, we subtract it throughout from the beginning of the motion. Conducting the calculation in this manner, we shall find that the time t of the up stroke as obtained from the formula will be slightly in excess of the actual time required, as in reality the acceleration is somewhat greater than we assume at the commencement of the motion, in view of the fact that the head which properly should have been deducted from h at this point of the stroke, would fall short of the maximum value actually deducted. Thus the principal equation (5) developed above, changes to

$$c = \frac{f\gamma(h - \sigma - \sigma_1 - \phi) - (1 + \nu)x}{F\gamma(q + q_0 + w + k + l)}g = \frac{a - (1 + \nu)x}{\mu b}g \qquad . \qquad (6)$$

where h no longer represents the actual excess of head of the water level G above D, Fig. 78, but a quantity which has been diminished by $\mu^2\dfrac{v^2}{2g}$.

In order to deduce an expression for this time t from equation (6), let us substitute $\dfrac{\delta v}{\delta t}$ for c, then

$$\frac{\delta v}{\delta t} = \frac{a - (1 + \nu)x}{\mu b}g.$$

Multiplying this formula by

$$v = \frac{\delta x}{\delta t},$$

we have

$$v\delta v = g\frac{a - (1 + \nu)x}{\mu b}\delta x,$$

from which by integration we get

$$\frac{v^2}{2} = g\left(\frac{a}{\mu b}x - \frac{1 + \nu}{2\mu b}x^2\right) + \text{const.}$$

Since $v = 0$ when $x = 0$, we find const. $= 0$. Hence

$$v^2 = \left(\frac{\delta x}{\delta t}\right)^2 = g\left[\frac{2a}{\mu b}x - \frac{1 + \nu}{\mu b}x^2\right] \qquad . \qquad . \qquad . \qquad (7)$$

or

$$\delta t = \sqrt{\frac{\mu b}{g}}\frac{\delta x}{\sqrt{2ax - (1 + \nu)x^2}} \qquad . \qquad . \qquad . \qquad (8)$$

Now let s denote the length of stroke of the lifting plunger; then the time of the lifting operation is

$$t = \sqrt{\frac{\mu b}{g}} \int_0^s \frac{\delta x}{\sqrt{2ax - (1+\nu)x^2}}.$$

But according to a well-known rule for integration

$$\int_0^s \frac{\delta x}{\sqrt{2ax - (1+\nu)x^2}} = \frac{-1}{\sqrt{1+\nu}} \left(\text{arc sin } \frac{a - (1+\nu)s}{a} - \text{arc sin } 1 \right)$$

$$= \frac{1}{\sqrt{1+\nu}} \left(\frac{\pi}{2} - \text{arc sin } \frac{a - (1+\nu)s}{a} \right)$$

$$= \frac{1}{\sqrt{1+\nu}} \text{ arc cos } \frac{a - (1+\nu)s}{a};$$

hence

$$t = \sqrt{\frac{\mu b}{g(1+\nu)}} \text{ arc cos } \left(1 - \frac{1+\nu}{a}s \right) \qquad . \qquad . \quad (9)$$

or after substituting the values

$$a = h - \sigma - \sigma_1 - \phi \text{ and } b = q + q_0 + w + k + l:$$

$$t = \sqrt{\frac{\mu}{g} \frac{q + q_0 + w + k + l}{1+\nu}} \text{ arc cos } \left(1 - \frac{1+\nu}{h - \sigma - \sigma_1 - \phi}s \right) \quad (10)$$

When the time t corresponding to the up stroke is given, the above equation may be written as follows:

$$1 - \frac{1+\nu}{a}s = \cos \frac{180°}{\pi} t \sqrt{\frac{g}{\mu} \frac{1+\nu}{b}},$$

from which is obtained

$$a = \frac{(1+\nu)s}{1 - \cos \dfrac{180°}{\pi} t \sqrt{\dfrac{g}{\mu} \dfrac{1+\nu}{b}}} = h - \sigma - \sigma_1 - \phi . \qquad . \quad (11)$$

Hence we get

$$h = \sigma + \sigma_1 + \phi + \frac{(1+\nu)s}{1 - \cos \dfrac{180°}{\pi} t \sqrt{\dfrac{g}{\mu} \dfrac{(1+\nu)}{q + q_0 + w + k + l}}} \qquad (12)$$

The velocity v, with which the lifting plunger reaches the end of its stroke s, is likewise determined from formula (7):

$$v = \sqrt{\frac{g}{\mu} \frac{2as - (1+\nu)s^2}{b}} = \sqrt{\frac{g}{\mu} \frac{2(h - \sigma - \sigma_1 - \phi)s - (1+\nu)s^2}{q + q_0 + w + k + l}} \quad (13)$$

Corresponding to this velocity of the lifting piston we have for the velocity of the water through the orifice of the supply valve $v_2 = \mu v$, and if we add to h the head

$$\frac{v_2^2}{2g} = \mu \frac{2as - (1+\nu)s^2}{2b} = h_1,$$

due to this velocity v_2, we shall obtain the effective head by which the water level in the accumulator must exceed that of the lifting plunger in order to accomplish the desired object; that is, in order to pass through the length of stroke s in the given time t.

The formulæ for the down stroke are determined in the same manner. For this case the acceleration c' of the moving masses, when the lifting plunger is at a distance x' from its highest position, and the valve opening is f', is determined by

$$c' = \frac{\delta v'}{\delta t'} = \frac{f'(h' - \sigma' - \phi' - x')}{F(q_0 + w + k' + l')} g = \frac{a' - x'}{\mu' b'} g,$$

when we note that here q_0 represents the driving water column and $k' + w$ the driven. Hence we must assume for the effective head

$$h' = q_0 - \frac{w}{\eta} - k' \text{ and } \nu = 0,$$

as the area of the free surface of the auxiliary reservoir is many times greater than the area of the plunger. Analogous to the calculation for the up stroke, we have placed

$$\frac{F}{f'} = \mu'; \ h' - \sigma' - \phi' = a' \text{ and } q_0 + w + k' + l' = b'.$$

Likewise, multiplying by $v' = \dfrac{\delta x'}{\delta t'}$, will give

$$v' \delta v' = g \frac{a' - x'}{\mu' b'} \delta x',$$

and by integration:

$$v'^2 = \left(\frac{\delta x'}{\delta t'}\right)^2 = g \left(\frac{2a'}{\mu' b'} x' - \frac{x'^2}{\mu' b'}\right),$$

hence the velocity of the platform when it reaches its lowest position is

$$v' = \sqrt{\frac{g}{\mu'} \frac{2a's - s^2}{b'}} = \sqrt{\frac{g}{\mu'} \frac{2(h' - \sigma' - \phi')s - s^2}{q_0 + w + k' + l'}} \qquad . \quad (14)$$

For the time t' of the descent, we have from the above equation

$$\delta t' = \sqrt{\frac{\mu' b'}{g}} \; \frac{\delta x'}{\sqrt{2a'x' - x'^2}},$$

hence

$$t' = \sqrt{\frac{\mu' b'}{g}} \int_0^s \frac{\delta x'}{\sqrt{2a'x' - x'^2}} = \sqrt{\frac{\mu' b'}{g}} \; \text{arc cos} \left(1 - \frac{s}{a'}\right) \; . \quad (15)$$

from which we deduce

$$a' = \frac{s}{1 - \cos \dfrac{180°}{\pi} t' \sqrt{\dfrac{g}{\mu' b'}}} \qquad . \qquad . \qquad . \quad (16)$$

Therefore we have

$$h' = \sigma' + \phi' + \frac{s}{1 + \cos \dfrac{180°}{\pi} t' \sqrt{\dfrac{g}{\mu'(q_0 + w + k' + l')}}},$$

to which value we must again add the head

$$h'_1 = \frac{\mu'^2 v'^2}{2g} = \mu' \frac{2a's - s^2}{2b'},$$

due to the velocity of the water through the discharge orifice, in order to obtain the head of the water level q_0 relatively to the discharge tank, it being assumed that the time of descent is limited to t' seconds.

From careful experiments made by *John Hick*,[1] the friction of a plunger working through a leather collar varies directly as the water pressure per unit of the surface of the plunger, and as the diameter of the latter, but is independent of the depth of the wearing surface of the collar.

According to these experiments, the *total* friction of the plunger is $R = \kappa \dfrac{K}{D}$, where K is the *total* pressure on the plunger expressed in kilograms, D the diameter, and κ a coefficient depending upon the state of lubrication, which ranges from 1·009 to 2·48, when D is expressed in millimetres [κ ranges from 0·0398 to 0·0977, when D is expressed in inches and K and R in pounds]. Let this friction be expressed in the form

<hr>

[1] *Engineer*, June 1 1866, and in abstract *Verhandlg. d. Ver. z. Bef. d. Gew.* 1866.

of a pressure column having a head δ, and an area of cross-section $F = \dfrac{\pi D^2}{4}$, equal to the area of the plunger ; then this head in metres can be determined from

$$\frac{\pi D^2}{4}\frac{\sigma}{1000} = \kappa\frac{K}{D} = \frac{\kappa}{D}\frac{\pi D^2}{4}\frac{k}{1000},$$

where k is the head which measures the plunger pressure K.

$$\left[\frac{\pi D^2}{4}\frac{G}{144}\sigma = \kappa\frac{K}{D} = \frac{\kappa}{D}\frac{\pi D^2}{4}\frac{G}{144}k,\right.$$

where D is expressed in inches, σ in feet, and G is the weight of a cubic foot of water.]

Hence we have

$$\sigma = \frac{\kappa}{D}k,$$

which varies between

$$\frac{1\cdot009}{D\,\mathrm{mm.}}k \text{ and } \frac{2\cdot48}{D\,\mathrm{mm.}}k.$$

$$\left[\frac{\cdot0398}{D\,\mathrm{inches}}k \text{ and } \frac{\cdot0977}{D\,\mathrm{inches}}k\right]$$

Therefore, for a plunger diameter of 100 mm., the loss of head is, roughly, from 1 to 2·48 per cent. According to the older estimates of *Rankine*, the friction of the plunger in hydrostatic presses is much greater, and amounts to about 10 per cent of the load in ordinary cases. In an article on Accumulators, *Zeitschr. deutsch. Ing.* 1867, page 65, *Horner* assumes 5 per cent of the load as a mean value for this friction. Hence, corresponding to each set of dimensions, we must assume a different value for σ. The frictional resistances ϕ and ϕ^1 of the water in the pipes are to be computed according to the rules laid down in vol. i. sec. vii. ch. 4, Weisb. *Mech.* If we suppose that the load Q, instead of resting upon the lifting piston—as assumed in the preceding pages—is indirectly connected with the ram, say by interposing pulleys, as illustrated in Figs. 74 and 76, the calculation remains the same, except that q will then represent not the load itself, but a pressure column equivalent to the resistance $\dfrac{nQ}{\eta}$, where n denotes the velocity ratio of the mechanism and η its efficiency.

If, in place of an accumulator, an elevated reservoir is employed, then, since the area of the free surface of the reservoir is generally much greater than that of the plunger, the ratio $\nu = \dfrac{F}{F_1}$ may be placed equal to zero in the preceding formulæ.

EXAMPLE.—Let us choose the hydraulic lift shown in Fig. 73. Here $F = 0.0774$ sq. metres [120 sq. in.], $F_1 = \dfrac{0.418^2}{4}\pi = 0.1372$ sq. metres [212.65 sq. in.]; hence $\nu = \dfrac{F}{F_1} = 0.565$. The weight of the platform is $Q_0 = 28,500$ kg. [62,840 lbs.], hence $q_0 = \dfrac{28.5}{0.0774} = 368$ m. [1207 ft.], and as the total weight of the counter-weights is 23,500 kg. [41,820 lbs.], we must place $w = \dfrac{23,500}{0.0774} = 302$ m. [991 ft.] Now, suppose the platform to be loaded by a car weighing about 400 cwt. or 20,000 kg. [44,100 lbs.], which gives $q = \dfrac{20}{0.0774} = 257$ m. [843 ft.], and that this load is to be lifted to the maximum height $s = 8.7$ m. [28.5 ft.] in 45 seconds, then we will proceed to determine the minimum effective head h, or to investigate whether the load on the accumulator plunger corresponding to a pressure head $k = 431$ m. [1414 ft.] will be sufficient to accomplish the desired result. For this purpose we make use of the equation

$$a = h - (\sigma + \sigma_1 + \phi) = \frac{(1+\nu)s}{1 - \cos \dfrac{180^\circ}{\pi} t \sqrt{\dfrac{g}{\mu}\dfrac{1+\nu}{q + q_0 + w + k + l}}} .$$

The diameter of the opening in the supply valve being 52.3 mm. [2.06 in.], we have $\mu = \left(\dfrac{314}{52.3}\right)^2 = 36$, and the length l of the supply pipe being about 50 m. [164 ft.], we have

$$a = h - (\sigma + \sigma_1 + \phi) = \frac{1.565 \times 8.7}{1 - \cos \dfrac{180^\circ}{3.14}45 \sqrt{\dfrac{9.81}{36}\dfrac{1.565}{257 + 368 + 302 + 431 + 50}}}$$

$$= \frac{13.61}{1 - \cos 44^\circ 40'} = \frac{13.61}{0.2888} = 47.1 \text{ m. } [154.5 \text{ ft.}]$$

Assuming the efficiency of the chain-pulleys to be $\eta = 0.95$, and the friction of the leather packing to be 3 per cent of the pressure on the plunger (according to *Hick* we should have only $\dfrac{2\cdot48}{314} =$ 0·008 for the lifting plunger, and $\dfrac{2\cdot48}{418} = 0\cdot006$ for the accumulator), we must place

$$\sigma = 0\cdot03(q + q_0 - \eta w) = 0\cdot03(257 + 368 - 0\cdot95 \times 302) = 10\cdot0 \text{ m. [32·8 ft.]}$$

and

$$\sigma = 0\cdot03 \times 431 = 13 \text{ m. [42·6 ft.]}$$

In order to determine ϕ we may take the mean velocity of the lifting plunger at $\dfrac{8\cdot7}{45} = 0\cdot194$ m. [0·64 ft.], so that the mean velocity of flow in the supply pipe of diameter $d = 104\cdot6$ mm. [4·12 in.], is $\left(\dfrac{314}{104\cdot6}\right)^2 \times 0\cdot194 = 1\cdot75$ m. [5·74 ft.]   Hence the loss of head according to vol. i. § 456, Weisb. *Mech.*, is given by the formula

$$\phi = \zeta\frac{l}{d}\frac{v^2}{2g} = 0\cdot02\frac{50}{0\cdot104}\frac{1\cdot75^2}{2\times9\cdot81} = 1\cdot5 \text{ m. [4·9 ft.]},$$

and we have roughly $\sigma + \sigma_1 + \phi = 10 + 13 + 1\cdot5 = 25$ m. [82 ft.]

Further, the velocity with which the lifting plunger reaches the end of its up stroke is found from (13) to be

$$v = \sqrt{\frac{g}{\mu}\frac{2ns - (1+\nu)s^2}{q + q_0 + w + k + l}}$$

$$= \sqrt{\frac{9\cdot81}{36}\frac{2\times47\cdot1\times8\cdot7 - 1\cdot565\times8\cdot7^2}{257 + 368 + 302 + 431 + 50}} = 0\cdot370 \text{ m. [1·21 ft.]}$$

The corresponding velocity of the water through the opening of the supply valve is therefore $\mu v = 36 \times 0\cdot370 = 13\cdot3$ m. [43·6 ft.], the producing of which requires an additional head

$$h_1 = \frac{13\cdot3^2}{2\times9\cdot81} = 9\cdot04 \text{ m. [29·66 ft.]}$$

Consequently the minimum effective head required to work the machine is

$$h = a + \sigma + \sigma_1 + \phi + h_1 = 47\cdot1 + 25 + 9\cdot04 = 81\cdot14 \text{ m. [266·2 ft.]}$$

In the present case the available balance actually amounts to

$$k - q - (q_0 - \eta w) = 431 - 257 - 368 + 287 = 93 \text{ m. [305 ft.]},$$

which proves that the working of the lift is sufficiently ensured, and that if we desire to prevent the platform from moving at a greater speed, the excess of head must be reduced by throttling the opening of the supply valve.

For the descent of the platform we have $k' = 18$ m. [59 ft.], $l' = 0$, $\phi' = 0$, $\mu' = 19$; hence, again assuming the time for lowering to be 45 seconds, we find according to (16)

$$a' = \cfrac{s}{1 - \cos\dfrac{180°}{\pi}\, t\; \sqrt{\dfrac{g}{\mu'(q_0 + w + k')}}}$$

$$= \cfrac{8 \cdot 7}{1 - \cos\dfrac{180°}{3 \cdot 14}\, 45 \;\sqrt{\dfrac{9 \cdot 81}{19(368 + 302 + 18)}}} = \frac{8 \cdot 7}{1 - \cos 70°30'} = 13 \cdot 0 \text{ m. } [42 \cdot 65 \text{ ft.}]$$

The final velocity of the platform may then be determined from (14), which gives

$$v' = \sqrt{\frac{9 \cdot 81}{19}\;\frac{2 \times 13 \times 8 \cdot 7 - 8 \cdot 7^2}{368 + 302 + 18}} = 0 \cdot 335 \text{ m. } [1 \cdot 16 \text{ ft.}]$$

From this we obtain the velocity through the opening of the discharge valve to be $19 \times 0 \cdot 335 = 6 \cdot 36$ m. [20·87 ft.], which represents a head of 2·06 metres [6·76 ft.] If we assume a value of

$$\sigma' = 0 \cdot 03\left(q_0 - \frac{w}{\eta}\right) = 0 \cdot 03 \times (368 - 318) = 1 \cdot 5 \text{ m. } [4 \cdot 9 \text{ ft.}]$$

for the friction of the leather packing, the effective head required will be

$$h' = 13 + 1 \cdot 5 + 2 \cdot 06 = 16 \cdot 56 \text{ m. } [54 \cdot 33 \text{ ft.}]$$

As, however, the available effective head has the greater value, represented by

$$q_0 - \frac{w}{\eta} - k' = 368 - 318 - 18 = 32 \text{ m. } [105 \text{ ft.}],$$

the time for the descent of the empty platform may be reduced. Deducting, for instance, from these 32 m. a head of say 5 m. [16·41 ft.] for overcoming the friction of the packing, and producing the required velocity through the discharge valve, there still remains a head of $a' = 27$ m. [88·59 ft.], which gives the time t' required for lowering from (15),

$$t' = \sqrt{\frac{\mu' b'}{g}}\; \text{arc cos}\left(1 - \frac{s}{a'}\right)$$

$$= \sqrt{\frac{19 \times (368 + 302 + 18)}{9 \cdot 81}}\; \text{arc cos}\left(1 - \frac{8 \cdot 7}{27}\right) = 30 \cdot 2 \text{ seconds.}$$

It is evident that when the platform is loaded, too rapid motion must be prevented by throttling the discharge valve.

§ 20. **Pneumatic Hoists.**—At the present day pneumatic

hoists have gained an extended application. Two furnace-lifts of this class are illustrated in Figs. 79 and 80. That shown in Fig. 79 was designed by *Gibbons* for four blast furnaces near Dudley, and has been in operation for a number of years. It consists of a wrought-iron tube AB 1·75 m. [5·74 ft.] wide, and 16 m. [52·49 ft.] long, which is filled from below with

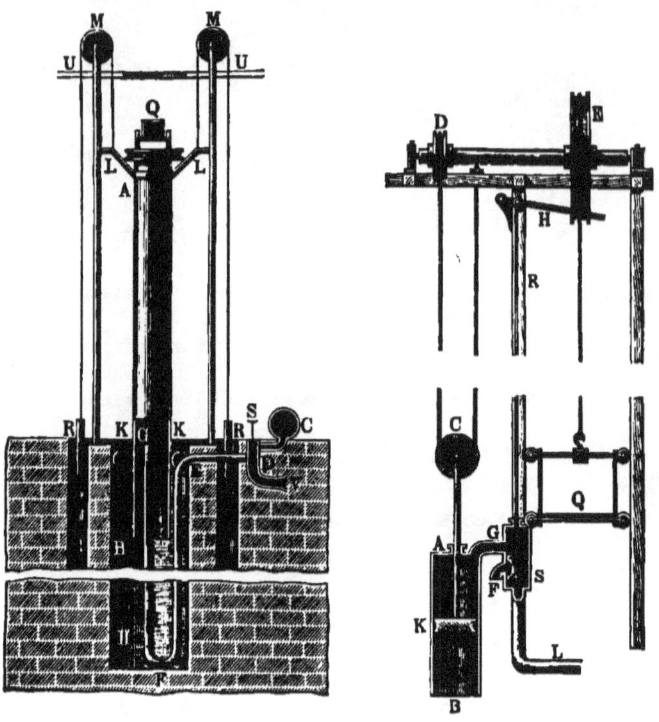

Fig. 79.   Fig. 80.

compressed air. The load Q is placed on a platform at the top A, and is lifted together with the tube by the action of the air. The latter is supplied through the pipes CDEFG from the blast chamber which provides the furnace with air, and the lower end of the tube AB is closed by the water which nearly fills the pit BEF, that is enclosed by brick work. In its lowest position, the tube AB rests on a support at the bottom of the pit, and it is guided during its upward motion both by rolls KK inside the pit, and by four columns above the latter, which are reached by the four arms LL at the top of the tube.

For controlling the up-and-down motion of the tube AB, the supply pipe is provided with a valve cylinder DS, in which a *piston valve* D may be moved up or down. When the tube is in its lowest position, and the load Q has been placed on the platform, the piston is pushed down in the position shown by the figure. This operation brings the inside of the tube AB in communication with the blast chamber, thus causing it to ascend with the load. When the latter has arrived nearly on a level with the charging platform UU, the tube, through a suitable lever motion, pulls up the piston valve, thus placing the inside of AB in communication with the open air, through the exhaust pipe V. If the tube is nearly balanced by the counter-weights R which are suspended from ropes passing over the pulleys M, and attached to the arms L, it will descend slowly, after Q has been unloaded, and in its descent force out the enclosed air through the pipe V. Besides this outlet V, the top of the tube is also provided with a valve by which the motion may be controlled. [For further information in regard to this lift see the *Civil-Eng. and Arch. Journal*, 1849; and *Polytechn. Centralblatt*, Jahrgang 1850.]

In place of the long tube, a common cylinder AB, Fig. 80, may be employed, together with a piston and rod. In this case the load is not applied directly to the piston rod, the connection between the two being accomplished by means of ropes and pulleys; in the lift shown the stroke s of the piston is first doubled by the movable pulley C, and then further increased by the shaft DE, with the pulleys D and E.

Assuming, for instance, the diameter of the pulley D equal to four times that of E, then to every foot of the stroke of the piston corresponds a lift of the load Q equal to 8 feet. Consequently the force acting on the piston must be equal to 8Q, if the wasteful resistances be neglected. The air is admitted and released through the valve S which is operated by the lever H. When the valve is in the position shown, the air enters the cylinder *via* LSG, and when it is in its highest position the release takes place *via* GF.

A more modern pneumatic furnace-lift, which is in extensive use, is that shown in Fig. 81 and constructed by *Gjers*.[1] Here a heavy piston K, provided with close-fitting packing

---

[1] See *Engineering*, 1872, p. 343.

rings, travels in a vertical tube A which is accurately bored out. A square platform B is connected to this piston by four ropes D, one in each corner, these ropes being carried over four guide-pulleys C placed diagonally. By this arrangement any movement of the piston K causes an equal and opposite movement of the platform. The piston K is made of sufficient weight to balance the platform, together with the empty ore and coke waggons, as well as part of the load.

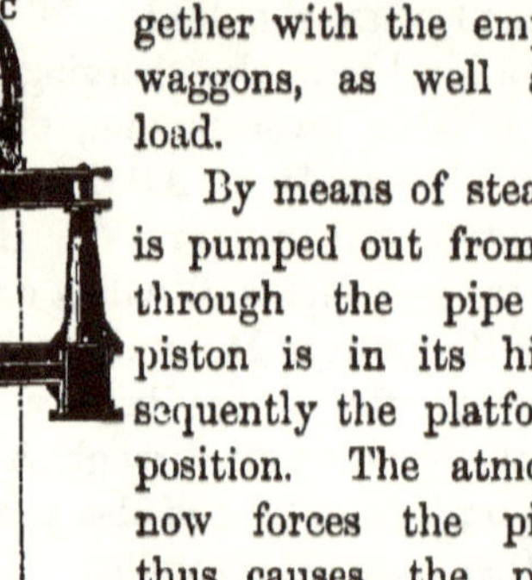
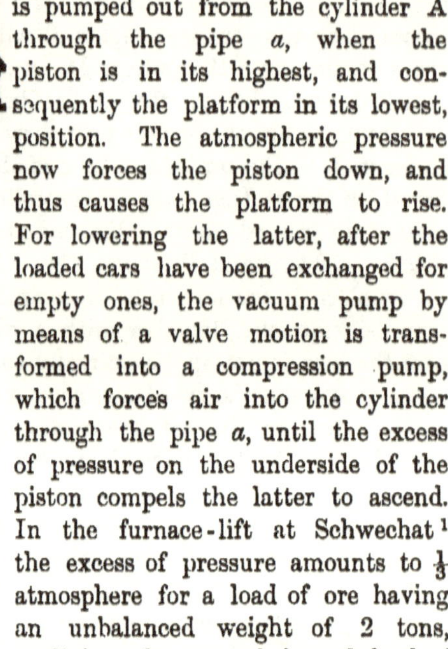
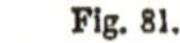

Fig. 81.

By means of steam power the air is pumped out from the cylinder A through the pipe a, when the piston is in its highest, and consequently the platform in its lowest, position. The atmospheric pressure now forces the piston down, and thus causes the platform to rise. For lowering the latter, after the loaded cars have been exchanged for empty ones, the vacuum pump by means of a valve motion is transformed into a compression pump, which forces air into the cylinder through the pipe a, until the excess of pressure on the underside of the piston compels the latter to ascend. In the furnace-lift at Schwechat[1] the excess of pressure amounts to $\frac{1}{3}$ atmosphere for a load of ore having an unbalanced weight of 2 tons, while $\frac{1}{12}$ atmosphere is sufficient for an unbalanced load of coke weighing $\frac{1}{2}$ ton.

The calculations for a pneumatic hoist may be made in the following manner:—Let W denote the resistance to be overcome by the piston, including wasteful resistances, and with due attention paid to the counter-weights; further, let F represent the area of cross-section of the piston, and p_0 the

---

[1] *Excursionsbericht*, von Riedler, 1876, Sketch 74.

outside, and p the inside pressure of the air per unit of area, then

$$W = F(p - p_0).$$

When the piston has travelled through a distance s, it has performed an amount of work

$$A_0 = F(p - p_0)\,s = V(p - p_0),$$

if the volume Fs passed through by the piston is denoted by V.

In order to determine the power which must be developed by the engines which drive the air pumps, let f be the area and l the length of stroke of the pump piston B, Fig. 82. While the latter travels from B to B_1, the air in the cylinder is compressed from the original pressure p_0 to p.   Beyond this point no further compression takes place, the air in the pump cylinder during the re-

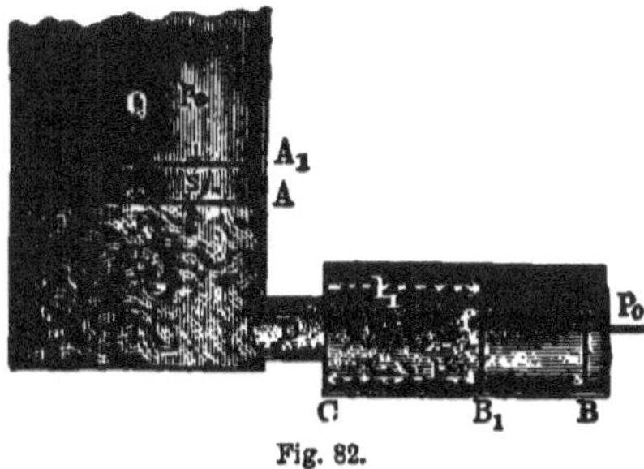

Fig. 82.

mainder of the stroke $B_1C = l_1$ being instead forced into the lifting cylinder through the pipe D, thus causing the piston to move through the distance s from A to A_1, so that $Fs = fl_1 = V$.   According to vol. i. § 415, the work done by the engine during the period of compression is given by

$$A_1 = fl_1 p \, \log_e \frac{p}{p_0} - fp_0(l - l_1)$$

$$= Vp \, \log_e \frac{p}{p_0} - V\frac{p_0}{l_1}(l - l_1),$$

as the external atmosphere acts with a force fp_0 through the distance $(l - l_1)$.

The work done during the second period $B_1C = l_1$ is

$$A_2 = f(p - p_0)l_1 = V(p - p_0),$$

and consequently the total work performed by the engine for one stroke of the piston is

$$A = A_1 + A_2 = Vp \, \log_e \frac{p}{p_0} - V\frac{p_0}{l_1}(l - l_1) + V(p - p_0) = Vp \, \log_e \frac{p}{p_0},$$

as

$$V\frac{p_0}{l_1}(l - l_1) = V(p - p_0).$$

The efficiency η of the pneumatic lifting cylinder is thus found to be

$$\eta = \frac{A_0}{A} = \frac{p - p_0}{p \, \log_e \dfrac{p}{p_0}} = \frac{\nu - 1}{\nu \, \log_e \nu},$$

when the ratio of compression $\dfrac{p}{p_0}$ is denoted by ν.

For the vacuum-lift, Fig. 81, in which $W = F(p_0 - p)$, we obtain in a similar manner the useful work

$$A_0 = Ws = V(p_0 - p) = Vp_0 \frac{\nu - 1}{\nu},$$

when the ratio of density $\dfrac{p_0}{p}$ is denoted by ν.

The work actually expended will be

$$A = flp_0 - Vp_0 \log_e \frac{p_0}{p} - fl_1 p,$$

or, as $fl_1 = Fs = V$, and $lp = l_1 p_0$,

$$A = Vp_0 \left(\frac{p_0}{p} - \log_e \frac{p_0}{p} - \frac{p}{p_0} \right) = Vp_0 \left(\nu - \log_e \nu - \frac{1}{\nu} \right),$$

which gives for the vacuum-lift an efficiency

$$\eta' = \frac{A_0}{A} = \frac{\nu - 1}{\nu^2 - \log_e \nu - 1}.$$

By further investigation we find that the efficiency of pneumatic hoists decreases as the ratio of densities ν increases, and that the result is more unfavourable in vacuum-lifts than when compressed air is employed.   This may be seen from the following table, which gives the efficiencies η and η' deduced from the above formulæ for different values of ν.

TABLE GIVING THE EFFICIENCY OF PNEUMATIC HOISTS.

$\nu =$	1·1	1·2	1·3	1·5	2	3	5
$\eta = \dfrac{\nu - 1}{\nu \, \log_e \nu}$	0·953	0·914	0·878	0·822	0·721	0·607	0·496
$\eta' = \dfrac{\nu - 1}{\nu^2 - \log_e \nu - 1}$	0·950	0·905	0·860	0·781	0·619	0·425	0·251

The losses connected with the use of pneumatic hoisting apparatus may be explained from the fact that the lifting cylinder for every operation must be filled with air of the required density, which necessitates an expenditure of a corresponding amount of work that is entirely lost.

EXAMPLE.—In a lift like that shown in Fig. 80, the load together with the platform weighs 600 kg. [1323 lbs.], and the ratio between the rope-pulleys D and E is $\frac{a}{b} = \frac{1}{4}$; accordingly the resistance which opposes the motion of the piston is

$$W = \frac{1}{\eta_1} Q \times 2 \times 4,$$

where η_1 represents the efficiency of the driving gear, consisting of the two rope drums and the movable pulley C, the friction of the piston, and that in the stuffing boxes being also included. If, in accordance with previous deductions, the value $\eta_1 = 0\cdot85$ be given to it, we obtain

$$W = \frac{600 \times 8}{0\cdot85} = 5647 \text{ kg. } [12,434 \text{ lbs.}]$$

Assuming an air pressure of $p_1 = 1\frac{1}{3}$ atmospheres, we shall require an area of the piston

$$F = \frac{5647}{\frac{1}{3} \times 10336} = 1\cdot639 \text{ sq. m. } [17\cdot65 \text{ sq. ft.}],$$

which corresponds to a diameter of the latter equal to $1\cdot445$ m. [$4\cdot74$ ft.]

The efficiency of the pneumatic cylinder will be

$$\eta_2 = \frac{\frac{4}{3} - 1}{\frac{4}{3} \log_e \frac{4}{3}} = \frac{0\cdot25}{0\cdot288} = 0\cdot867,$$

and thus the efficiency of the whole apparatus

$$\eta = \eta_1 \eta_2 = 0\cdot85 \times 0\cdot867 = 0\cdot737,$$

if no attention is paid to the losses due to the air pump.

§ 21. **Brakes for Lowering.**—These devices are used for retarding the motion in lowering a load, so as to obtain a smooth and uniform descent, and prevent shocks when the load reaches its point of destination.

A simple brake is sometimes employed as a life-saving apparatus at fires, for letting down persons from the upper stories of burning buildings. Here the coil friction acting between a rope and a cylinder is utilised, as may be seen from Fig. 83. The apparatus consists of a spool-shaped object C, around which the rope AB is wound in several coils. The upper end of the rope is secured in a suitable manner, while the lower end hangs down freely. Now, if the person is suspended from the hook by means of a belt, the coil friction is brought into play by the sliding of the cylinder along the rope. The friction may be estimated in the following manner. Let the tension in the upper end of the rope, which is equal to the suspended load, be denoted by S_1, and that in the free end by S_2, then, according to vol. i. § 199, $S_1 = S_2 e^{\phi a}$, when ϕ is the coefficient of friction and a the arcs surrounded by the rope. Thus the friction is $F = S_1 - S_2 = S_2(e^{\phi a} - 1)$, which is independent of the diameter of the cylinder, and it is evident that it can be greatly increased by increasing the number of

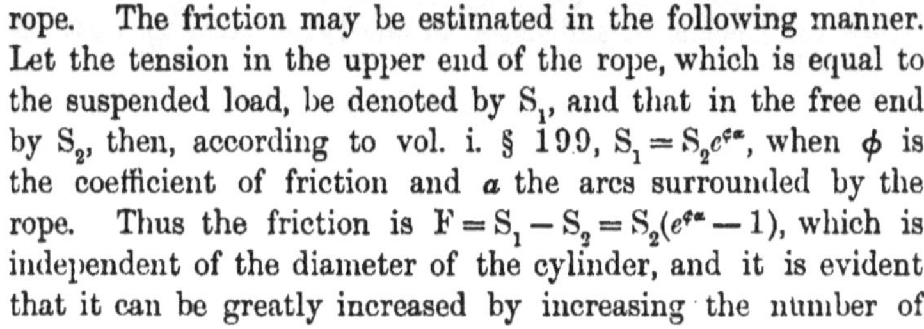

Fig. 83.

coils of the rope.   To check the sliding motion, it is only
necessary to apply a slight force at the free end of the latter,
which may be done at will by the occupant of the apparatus.
Assuming $\phi = 0{\cdot}3$, and that the rope makes two full coils
around the cylinder, then

$$e^{\phi a} = 2{\cdot}7182^{0{\cdot}3 \times 2 \times 2\pi} = 43{\cdot}3,$$

and consequently $F = 42{\cdot}3S_2$, which shows that for arresting

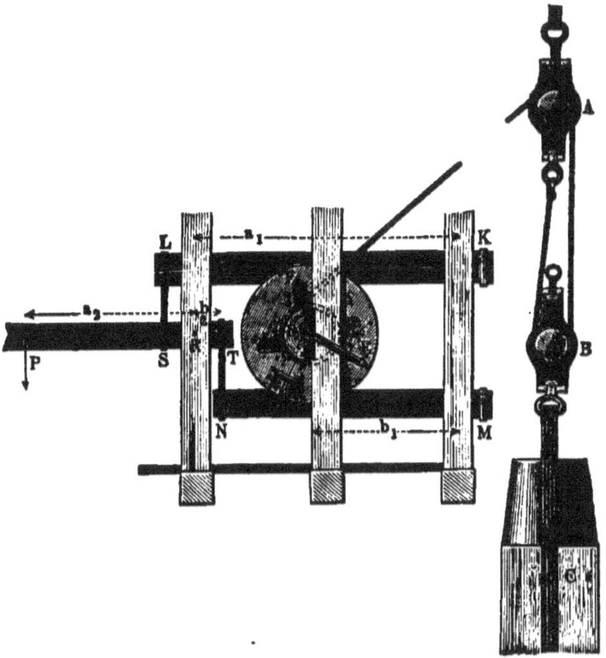

Fig. 84.

the motion, the application of a force equal to $\frac{1}{42}$ of the weight
of the occupant would be sufficient.   The lower end of the
rope may be secured to the knobs D, when it is desired to stop
the descent for any length of time.

In machinery, the mechanism employed for lowering loads
consists chiefly in a horizontal shaft or barrel provided with a
brake wheel or disc.   Around this barrel the rope which holds
the load is wound, and gradually unwinds during the descent,
while the brake lever or strap is pressing against the wheel.

M

A brake winch, as used for lowering machines and materials into shafts of mines, is shown in Fig. 84. AB is a common tackle, and C is the load attached to it; D is the winch drum, DE the crank, and FG the brake disc, which is fast on the drum. The brake beams, which are movable about K and M, are forced against the disc by means of the lever PR which turns on the fixed stud R. While one workman applies the brake lever, and thus counteracts the weight of the load C, another slowly turns the crank DE, thus causing the rope to unwind from the drum, and allowing the load to descend.

If Q is the load, and n is the number of ropes AB in the tackle, then, neglecting the wasteful resistances, the force at the circumference of the drum is $Q_1 = \dfrac{Q}{n}$; letting b denote the radius of the drum, including half the thickness of the rope, and a the radius of the brake disc, we now obtain the force required at the circumference of the latter $R = \dfrac{b}{a} Q_1 = \dfrac{bQ}{na}$. Placing the braking force at the end of the brake lever $= P$ and its lever arms $\dfrac{KL + MN}{2} = a_1$, and $RP = a_2$, while the lever arms of the load are $KF = MG = b_1$, and $RS = RT = b_2$, the coefficient of friction at the brake disc being $= \phi$, then $R = \phi \dfrac{a_1}{b_1} \dfrac{a_2}{b_2} P$; hence we get

$$\phi \dfrac{a_1}{b_1} \dfrac{a_2}{b_2} P = \dfrac{bQ}{na},$$

which gives

$$P = \dfrac{bb_1b_2}{aa_1a_2} \dfrac{Q}{\phi n}.$$

As the frictional resistances assist the action of the effort applied in all apparatus for lowering, it follows that the force P actually required falls below the value just deduced.

EXAMPLE.—A load Q weighing 1000 kg. [2205 lbs.] is to be lowered by means of the brake winch in Fig. 84; the ratios of lever arms are:

$$\frac{b}{a} = \tfrac{2}{3}; \quad \frac{b_1}{a_1} = \tfrac{1}{2}; \quad \text{and} \quad \frac{b_2}{a_2} = \tfrac{1}{10};$$

the number of ropes in the tackle AB is $n = 6$, and the coefficient of friction at the brake pulley is assumed equal to $\phi = 0\cdot3$; then,

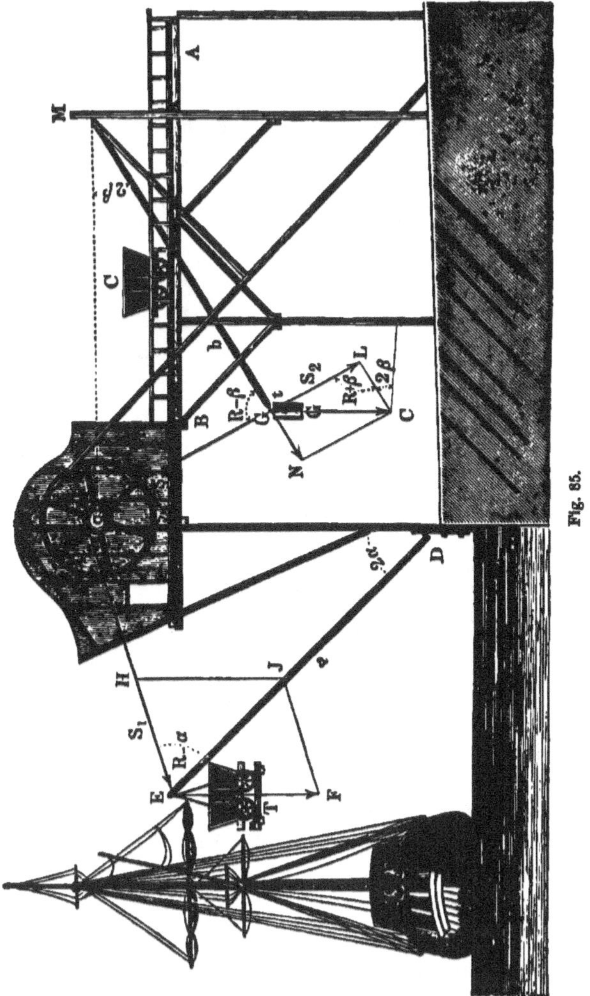

neglecting wasteful resistances as well as the force applied at the
crank, the effort at the brake lever will be

$$P = \frac{b}{a}\,\frac{b_1}{a_1}\,\frac{b_2}{a_2}\,\frac{Q}{\phi n} = \tfrac{2}{3} \times \tfrac{1}{2} \times \tfrac{1}{10} \times \frac{1000}{0\cdot3 \times 6} = 11\cdot1 \text{ kg. } [24\cdot48 \text{ lbs.}]$$

To machinery for lowering loads must be counted so-called
Drops, which name is applied to the arrangements employed in

England for lowering coal cars to vessels, as shown in Fig. 85. AB is a track on which the cars C arrive; DE is an arm which swings about the point D, and carries a platform T; in its highest position this platform forms a continuation of the track AB, and is made to receive a coal-car. A rope EK attached to the end E of the lever, unwinds from the drum K, when the loaded car is lowered to the vessel, and causes the empty car to return when wound on to the drum. In order to accomplish the latter operation without the aid of special motive power, a counter-weight G is made use of, which is suspended partly by the arm GM pivoted at M, and partly by a rope GK, which is wound on to the drum K while the car is being lowered, thus causing G to rise and storing up sufficient power for the return of the empty car.

A large brake wheel RS operated by a strap S, is fixed on the drum K with a view to securing a uniform motion and a moderate rate of speed during both ascent and descent.

In designing a drop of this kind, the counter-weight should be so chosen that the braking force required to counteract the accelerated up and down motions of the car will be as small as possible, and also the same in both cases.

For the calculation let us assume that, when the car is in its lowest position, both arms DE and MG are horizontal, and that it is sufficiently correct to place

$$DK = DE = a \text{ and } MK = MG = b.$$

Now let for any position 2α denote the angle EDK which the load arm makes with the vertical, and 2β the angle GMK which the counter-weight arm makes with the horizon; further, let Q be the load in the car, W the weight of the latter empty, together with the platform and one-half of the arm DE, and G the counter-weight, including one-half of the arm MG. By the use of the parallelogram of forces, we then find the tension in the rope EK during the lowering of the loaded car amounts to

$$S_1 = (Q + W)\frac{\sin 2\alpha}{\cos \alpha} = (Q + W)2 \sin \alpha,$$

and while the empty car is being hoisted

$$S'_1 = W2 \sin \alpha.$$

The tension in the rope KG is in both cases

$$S_2 = G \frac{\cos 2\beta}{\cos \beta}.$$

The force P acting in the circumference of the rope drum, and which must be overcome through the brake action, is in lowering

$$P = S_1 - S_2 = (Q + W)2 \sin a - G \frac{\cos 2\beta}{\cos \beta},$$

and in hoisting

$$P' = S_2 - S_1 = G \frac{\cos 2\beta}{\cos \beta} - W2 \sin a.$$

If these forces are to be equal, as required, we obtain

$$G = (Q + 2W) \frac{\sin a \cos \beta}{\cos 2\beta}.$$

Evidently this requirement cannot be fulfilled for all values of a and β; assuming, however, that this condition holds for the lowest position, when $2a = 90°$ and $\beta = 0$, then we get

$$G = (Q + 2W) \sin 45° = (Q + 2W) \sqrt{\tfrac{1}{2}}.$$

Besides, it is possible to make the two brake pressures equal in a second position if the relation between the angles a and β is suitably chosen. If this occurs in the highest position of the car, in which the angles may be denoted by $2a_1$ and $2\beta_1$, we have the equation:

$$G = (Q + 2W) \sqrt{\tfrac{1}{2}} = (Q + 2W) \frac{\sin a_1 \cos \beta_1}{\cos 2\beta_1},$$

or

$$\sin a_1 \cos \beta_1 = \cos 2\beta_1 \sqrt{\tfrac{1}{2}},$$

from which when a_1 is given we obtain

$$\cos \beta_1 = \frac{\sin a_1 + \sqrt{4 + \sin^2 a_1}}{2 \sqrt{2}}.$$

When the arrangement is contrived in the above manner, it is then evident that the braking force required for uniform lowering, either in the highest or lowest position, equals that needed for hoisting. In order to establish the desired relation between the angles a_1 and β_1, it is necessary to give the proper lengths to the arms a and b.

These will be determined with reference to the fact that the same length of rope is wound on to the drum K on one side as is unwound on the other, so that in general

$$KG = ED \sqrt{2} - EK,$$

or for the highest position

$$2b \sin \beta_1 = a \sqrt{2} - 2a \sin a_1.$$

If now the lever arm a of the load is known as well as its angle $2a_1$ with the vertical in the highest position, we may easily compute β_1 and b from the equations given. The condition regarding equal braking force for lifting and lowering is only approximated to in the intermediate positions.

EXAMPLE.—A drop is to be constructed like that in Fig. 85 for a load $Q = 1000$ kg. [2205 lbs.], the car together with platform, etc., having a weight $W = 400$ kg. [882 lbs.] In this case the counter-weight must be made equal to

$$G = (1000 + 2 \times 400)0 \cdot 7071 = 1273 \text{ kg. [2807 lbs.]}$$

The force required at the circumference of the drum will be

$$P = (1000 + 400)1 \cdot 414 - 1273 = 1273 - 400 \times 1 \cdot 414 = 707 \text{ kg. [1558 lbs.]}$$

For a diameter of brake wheel of 6 times that of the drum, the friction at the strap must equal 118 kg. [260 lbs.] The calculations for ascertaining the required pressure at the lever are treated of in vol. iii. 1, Weisb. *Mech.*, in the chapter on brakes.

If the load is to be let down from a height $h = 12$ m. [39·37 ft.] and the arm from which it is suspended makes an angle of 20° with the vertical in its highest position, so that $a_1 = 10°$, then the length of this arm will be

$$a = \frac{h}{\cos 2a_1} = \frac{12}{0 \cdot 9397} = 12 \cdot 77 \text{ m. [41 \cdot 9 ft.]}$$

The angle β_1 is obtained from

$$\cos \beta_1 = \frac{\sin 10° + \sqrt{4 + \sin^2 10°}}{2 \sqrt{2}} = 0 \cdot 7713,$$

which gives $\beta_1 = 39°32'$, and accordingly the length b of the counter-weight arm

$$b = \frac{a\sqrt{2} - 2a \sin a_1}{2 \sin \beta_1} = \frac{12\cdot77 \times 1\cdot414 - 25\cdot54 \times 0\cdot1736}{2 \times 0\cdot6365} = 10\cdot69 \text{ m. } [35\cdot07 \text{ ft.}]$$

To the class of contrivances under consideration we must also count the *Inclined planes*, the principal feature of which is a sloping track for lowering loaded cars by the action of their own weight. Such inclined planes [1] are constructed with either *single* or *double* track. In the latter case the loaded car, by reason of its greater weight, is always utilised for returning the

Fig. 86.

empty car to the top of the plane, a rope for this purpose being carried from the cars around a pulley at this point.

This arrangement resembles to a great extent the inclined furnace-lift in Fig. 63, the principal difference being that in place of a driving mechanism a simple brake pulley is made use of for controlling the motion of the cars. The power required may be ascertained in a similar manner to that employed in the case of the furnace-lift just mentioned.

When but a single track is used, a special *counter-weight* is employed which is suitably guided and sufficiently heavy to cause the return of the empty car. The guides for the counter-weight are arranged either below or on one side of the main

[1] See Serlo, *Leitfaden zur Bergbaukunde*.

track. In Fig. 86 an arrangement of this kind is illustrated, as carried out in practice at the *Saarbrücker* mines.[1] The car W travels on the rails *a*, between which a second and lower track *b* is placed for the counter-weight G, which is made of cast iron and provided with small rollers. The rope S leading from the car unwinds from the drum B on the shaft A, while the rope S_1 secured to the counter-weight is wound on to the smaller drum C on the same shaft. It is evident from the figure how the motion may be controlled by the brake wheel D and the lever E.

Such inclined railways have been constructed for a variety of grades, ranging as low as $1° 50'$; where the inclination is so slight, however, it would be impossible to attain the desired result without the exercise of great care in the construction so as to reduce the hurtful resistances to a minimum.

If for the inclined plane in Fig. 86 we denote by Q the weight of the load in the car, by W that of the empty car, by G the counter-weight, and by *a* the inclination of the railway to the horizon, then we obtain for the turning moment exerted on the shaft A when the car is descending

$$[(Q + W)b - Gc] \sin a,$$

where *b* and *c* are the radii of the rope drums B and C, and when it is ascending, the moment is

$$(Gc - Wb) \sin a,$$

neglecting wasteful resistances. By placing these two expressions equal, we find that it would be necessary to make

$$G = \frac{(Q + 2W)b}{2c}.$$

In order to determine the minimum inclination of the plane, we must take account of the frictional resistances. Let ϕ be the coefficient of journal friction, ν the ratio of the radius of the car journals to that of the wheels, and ν_1 the same ratio for the counter-weight, *r* the radius of the drum shaft A, and σ and σ_1 the coefficients due to the stiffness of the ropes S and

<hr>

[1] *Zeitschr. f. d. Berg-, Hütten-, und Sal-Wesen*, 1856.

S_1, then the turning moment acting on the drum in lowering will be

$$\left(1 - \sigma - \phi\frac{r}{b}\right)(Q + W)(\sin a - \nu\phi \cos a)b$$

$$- \left(1 + \sigma_1 + \phi\frac{r}{c}\right)G(\sin a + \nu_1\phi \cos a)c,$$

and in lifting

$$\left(1 - \sigma_1 - \phi\frac{r}{c}\right)G(\sin a - \nu_1\phi \cos a)c$$

$$- \left(1 + \sigma + \phi\frac{r}{c}\right)W(\sin a + \nu\phi \cos a)b.$$

These expressions should always give values considerably greater than zero, in order to make the motion possible.

§ 22. **Mine Hoists.**—These are used, as the name implies, for lifting ore, coal, etc., in the shafts of mines. The direction in which shafts are sunk may approach to the *perpendicular*, when the angle varies from $75°$ to $90°$ to the horizon, or it may be nearly *horizontal*, in which case the angle is $15°$ or less. The essential part of a mine hoist is a drum with two ropes so attached that while one is wound on to the drum the other is being unwound. Thus if a loaded receptacle hangs at the end of one of the ropes, and an empty one at the other, the former may be lifted while the latter is lowered. After an exchange of receptacles at each end the operation may be repeated by turning the drum in the opposite direction. It is chiefly by this arrangement, which is similar to the foundry hoists already described, that the mine hoists differ from common winches and ordinary lifts, which, as a rule, are provided with only one rope carrying a hook or a cage.

In cases where small loads only are to be hoisted from moderate depths, the common *winch* is made use of, which may be operated by two or more workmen. For greater loads and considerable depths the *whin*, or upright drum, is employed, which is either driven by hand power or revolved by horses or oxen. All hoisting of any great consequence, however, is nowadays done by water or steam as the motive power, the hoisting machine being provided with a drum which, in most cases, is horizontal. Evidently the arrangement differs according as a *water-wheel, turbine, water-motor,* or a *steam-engine* is made use of as the prime mover.   More recently *compressed air* [1] has been introduced for the same purpose, the air being compressed by steam or water power, and afterwards used in a motor coupled to the hoisting machine, in a manner similar to the mode of using steam in a steam-engine.

The arrangement of a mine winch is shown by Fig. 87. A is the drum around which a rope B is wound; at the ends C and D of the latter the iron-bound buckets are suspended. Only the arriving or departing bucket E at the top of the shaft is shown in the cut, the other bucket at the same moment being near the point where the loading is being done.

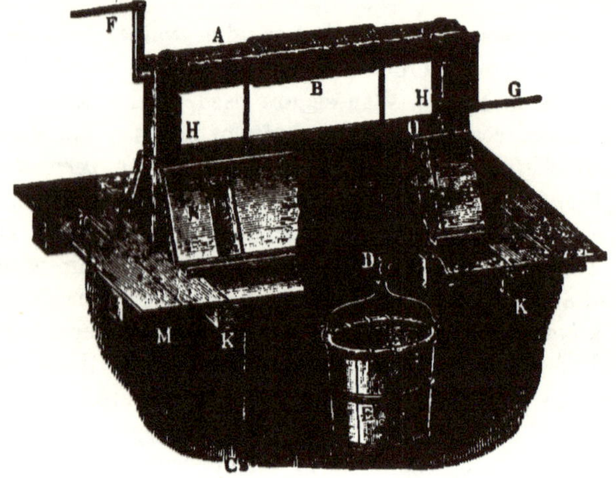

Fig. 87.

The journals of the drum rest on the posts HH, secured to the timbers KK, placed horizontally over the mouth of the shaft.   At their upper ends the posts are provided with iron-lined slots for the reception of the journals.   The workmen stand on a platform LL, supported by the timber framework KM placed over the shaft.   The rod *a* running parallel with the drum, and fastened to the posts at each end by means of iron brackets, serves as a hand-rail for the workmen in removing the loaded bucket and replacing an empty one.   Finally, in order to prevent the falling into the shaft of objects which might create obstructions, the doors N and O are made use

[1] And *electric motors.*   Translator's remark.

of, leaning against the braces at the lower ends of the winch posts. When the shaft is perpendicular the buckets are freely suspended; when it is inclined the latter travel in a trough formed by boards or slabs, covering the sides and bottom of the shaft, a board partition besides being placed in the centre to prevent the buckets from colliding with each other.

The buckets are made either of wooden staves or sheet-iron, the cross-section usually being of elliptic form. At the bottom the axis of the ellipse measures 0·235 and 0·390 metres [9·25 and 15·35 in.], and at the upper end 0·288 and 0·470 metres [11·73 and 18·5 in.], so that with a depth of 0·390 metres [15·35 in.] the cubic contents amount to 42 litres [1·48 cub. ft.] Making the assumption that only two-fifths of the total cubic contents of the bucket is occupied by the ore, and that the average specific gravity of the latter is 2·5, we obtain a total weight of 42 kilograms [92·61 lbs.] In the mines at Freiberg it is estimated that two men in an eight-hour shift can hoist 120 buckets of ore from a perpendicular shaft 40 m. [131·24 ft.] in depth. Hence we obtain the useful work done by each workman

$$\frac{40 \times 42 \times 120}{2} = 100,800 \text{ m. kg. } [729,126 \text{ foot-pounds}],$$

or per second

$$\frac{100,800}{8 \times 60 \times 60} = 3\cdot5 \text{ m. kg. } [25\cdot3 \text{ foot-pounds}].$$

This low figure for the useful work done in comparison with the average effect of 8 m. kg. per second [57·87 ft.-pounds], developed by a workman turning a crank, may be explained not only from the influence of the wasteful resistances, but chiefly from the fact that owing to the numerous pauses in the work of hoisting while the buckets are being exchanged, the actual working time is reduced to considerably less than eight hours.

For greater depths it is impossible to wind the whole length of the rope in one layer on the drum, as this would require the latter to be of inconvenient length; it is therefore necessary to wind the rope in several layers. The lever arm of the load is then no longer constant, which fact may be taken into account in the following manner. Let s be the length of the rope which is to be wound on to the drum, r the radius, and l the length of the latter, then for a thickness d of the rope the number of coils in each layer is represented by $n = \dfrac{l}{d}$. For m layers of rope, that next to the drum has a radius of $r + \dfrac{d}{2}$, and the radius of the outside layer is $r + \left(m - \dfrac{1}{2}\right)d$, hence the mean radius is to be taken equal to $r + \dfrac{m}{2}d$. Consequently the total length of the rope will be

$$s = mn\,2\pi\left(r + \frac{m}{2}d\right),$$

which equation gives

$$m^2 + \frac{2r}{d}m = \frac{s}{\pi nd},$$

and hence

$$m = -\frac{r}{d} + \sqrt{\frac{s}{\pi l} + \left(\frac{r}{d}\right)^2} = \frac{r}{d}\left(\sqrt{\frac{sd^2}{\pi r^2 l} + 1} - 1\right).$$

The mean radius is therefore

$$r + \frac{m}{2}d = r + \frac{r}{2}\left(\sqrt{\frac{sd^2}{\pi l r^2} + 1} - 1\right) = \frac{r}{2}\left(1 + \sqrt{1 + \frac{sd^2}{\pi l r^2}}\right),$$

or approximately

$$r\left(1 + \frac{sd^2}{4\pi l r^2}\right).$$

The methods employed for equalising the influence of the weight of the rope by varying the radii of the drum, will be explained in the following.

In order to make possible the use of a winch for heavy loads, a gear and pinion are occasionally employed for driving the drum. This is necessary, owing to the fact that the diameter of the drum cannot be reduced beyond a certain size consistent with proper strength, and also depending on the stiffness of the rope, and besides it is not practicable to make the cranks longer than the usual measure of from 0·36 to 0·42 m.

(from 14 to $16\frac{1}{2}$ inches). Mine winches with two sets of gears are rarely employed.

A geared winch is shown in Fig. 88. Here the drum A is about 0·3 or 0·4 m. [12 or 15 inches] in diameter, and carries a large gear BD with 40 to 60 teeth, while the pinion E is keyed to the crank shaft EF (not shown in the fig.) As a rule one workman is stationed at each crank, though it is evident that provided the crank handles are of sufficient length,

Fig. 88.

three or four men may profitably engage in the hoisting operation.

The distribution of labour may be so contrived that, while two men are regularly employed at the cranks, a third one attends to the emptying of the buckets, besides assisting at the cranks part of the time. The wire ropes used for winches of this kind are made from four strands, each consisting of four wires twisted together. The thickness of wire is from 1 to 1·5 mm. [0·04 to 0·06 in.] The drum and crank shaft are supported by two pairs of posts K, K . . . which are let into the timbers L, L placed across the opening of the shaft, the drum being carried by the beams M, M, and the crank shaft by the cross-pieces N, N. The floor O, O serves to give a firm footing to the workman.

§ 23. **Hand- and Horse-Gins.**—The general arrangement of vertical drums operated either by hand or horse power is explained in vol. ii., Weisb. *Mech.*, and it is here only intended to treat of their application for mining purposes. The essential feature of the arrangement is the rope drum, which may be either cylindrical or conical. It is usually made in two parts, one of which can be disconnected from the shaft in order to allow of a change being made in the hoisting depth whenever required. The disengagement takes place when the bucket, operated by this part of the drum, has reached the top, motion on the shaft being prevented by means of a brake arrangement. In the meantime, the lower empty bucket is transferred from the previous point of loading to its new location. When this has been accomplished, the loose part of the drum is again rigidly connected with the shaft, and the hoisting may go on as before. The radius of the drum is generally equal to one-fourth of the length of the sweep, and the length of the drum is from 30 to 60 centimetres [1 to 2 feet]. By the use of flanges at each end of the drum the rope may be wound on to a depth of 0·3 to 0·6 metres [1 to 2 feet]. One of the flanges also serves as a brake wheel, the brake being arranged essentially as described in vol. iii. 1, Weisb. *Mech.*

The horizontal direction in which the rope leaves the drum is changed into a direction parallel with the shaft, by the use of guide-pulleys or idlers, placed about 6 m. [19·68 feet] above the mouth of the shaft. These pulleys are about 2 or 3 m. [6·56 to 9·84 feet] in diameter, and provided with a groove for the rope. It is advisable to make the distance from the guide-pulleys to the drum equal to at least twenty times the height of the coils of rope, in order to cause the rope to wind uniformly on to the drum. The portion of the rope between the guide-pulley and the drum should also, for this purpose, be supported by *counter-weights*. The plane in which each guide-pulley is to be placed, is determined by the direction of the rope in the shaft, and that of the part leading from the drum to the pulley. When the shaft is vertical, both guide-pulleys will be placed vertically; when it is inclined and the horizontal distance between the two ropes in the shaft is less than the diameter of the drum (which is usually the case), the guide-pulleys as well as their journals will be inclined to the horizon.

At the present day wire ropes are generally used for this kind of hoisting. They are twisted as nearly round as possible, and consist of from three to six strands of four or six wires each. The connection between the rope and the bucket or cage is accomplished by means of chains. For hoisting in vertical shafts, bucket-shaped receptacles are generally used; for inclined shafts, on the other hand, where suitable guides are necessary for the receptacles, the box shape is more commonly adopted. In order to prevent the buckets from revolving, in vertical shafts, flat driving-ropes made of several round ropes joined together are used in many places. It is far better, however, to provide guides for the buckets; in this case the latter are of rectangular section and fitted with four rollers, two on each side, movable between two pairs of vertical guide-beams. When the shaft is inclined, the receptacles are prismatic boxes with trapezoidal sides, and in addition to the side rollers, are provided with four wheels travelling on the stringers, to which nowadays iron rails are always fastened. To avoid unnecessary delay in loading and unloading, especially in vertical shafts, a cage is often suspended from the rope in place of the bucket, and hoisted together with the car in which the ore is brought to the shaft.

In inclined shafts carrying pulleys are located at suitable intervals to sustain the weight of the rope and prevent it from wearing against the walls of the shaft.

Finally, for discharging the load at the top of the shaft, a special tilting motion must be arranged, consisting of two hooks and two studs. The former are placed on the stringers above the shaft, the latter protrude from the sides of the bucket somewhat below the middle. When the load is to be dumped, the hooks are let down by means of a lever and allowed to engage with the studs on the bucket. When a cage with a car is employed in place of the bucket, it is necessary to provide safety catches so as to prevent the cage from going down at the wrong moment; the unloading in this case is accomplished simply by removing the full car from the cage and substituting an empty one.

The arrangement of a vertical drum operated by *hand power* as used for an inclined shaft is shown in Fig. 89. A is the upright spindle, and B, B, B the three sweeps, each about 2 m.

[6·56 ft.] in length, and secured to the former, the ends being

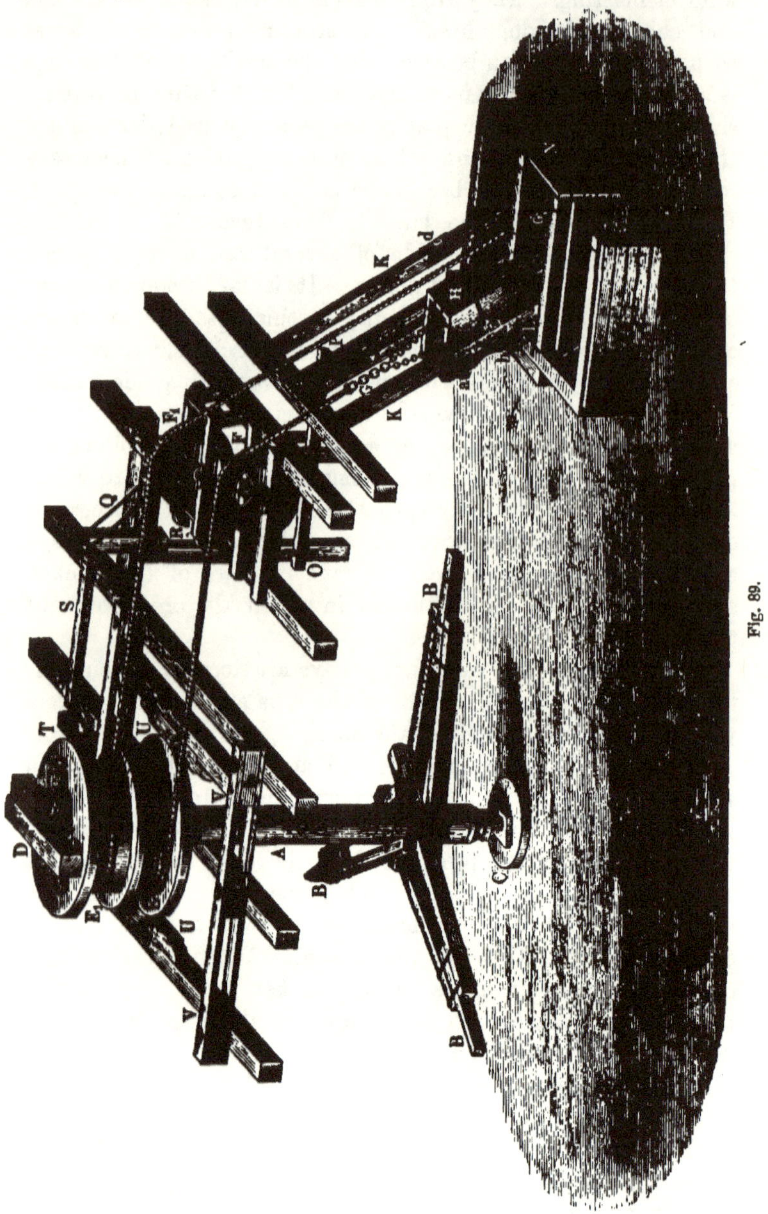

grasped and pushed forward by the workmen.   The pivot of

the spindle rests in a step which is let into the block C, and
the upper end is held by a bearing secured to one of the overhead beams in the hoisting shed.   The two drums E and E_1
are fast on the vertical shaft.   The ropes EFG and $E_1F_1G_1$ pass
over the rope-pulleys F and F_1 placed side by side, one above
the other, and are thus given the proper inclination corresponding to the slope of the shaft.   One of the buckets H has just
arrived to the surface, and has been grasped by the dumping-
hooks; the other has at the same moment reached the place of
loading, and for that reason is not visible in the figure.   The
guide-beams K, K, etc., may be seen as well as the upper ends
of the stringers at L, on which the bucket travels by means of
the wheels $a, b, \ldots$ Two of the four dumping-hooks which
are attached to the guide-beams are shown at c and d.   MN is
the so-called *collar*, or the horizontal timbering which surrounds
the ladder shaft at M, and the two divisions of the working
shaft at N.   The lower flange of the drum also serves as a
brake wheel when a brake is required.   The brake is operated
by pulling down the lever OP, which is movable about O; by
means of a vertical rod (scarcely noticeable in the figure), this
lever acts on a bell-crank QR, movable around R, and connected
through the bar S with the well-known block brake TUV.
The labourer may be assumed to move at a velocity of one
metre (3·28 ft.) per second.   Hand power is nowadays seldom
used in this kind of hoisting, as the efficiency of a workman
operating a vertical drum is less than when he is turning a
crank.[1]

In Fig. 90 is shown a *Saxon horse-gin*.   As before, A is the
upright shaft, B the sweep, and C the supporting block for the
former, provided with a step for the steel pivot.   Of the two
drums D and D_1 the lower one is fast to the shaft, while the
upper one is movable.   For regular working the upper drum
rests on the lower one, the two being connected by means of
pins protruding from the arms of the lower drum and locked
into corresponding recesses in the upper one.   When it is
desired to move the position of the empty bucket in the shaft,
the upper drum may be lifted and thus disconnected from the
lower one by means of the winch a, which operates the horizontal shaft d by the action of the chain b and an arm c, and

[1] See Serlo, *Leitfaden der Bergbaukunde*, vol. ii.

ultimately lifts the upper drum by the carrying chains *e.* It

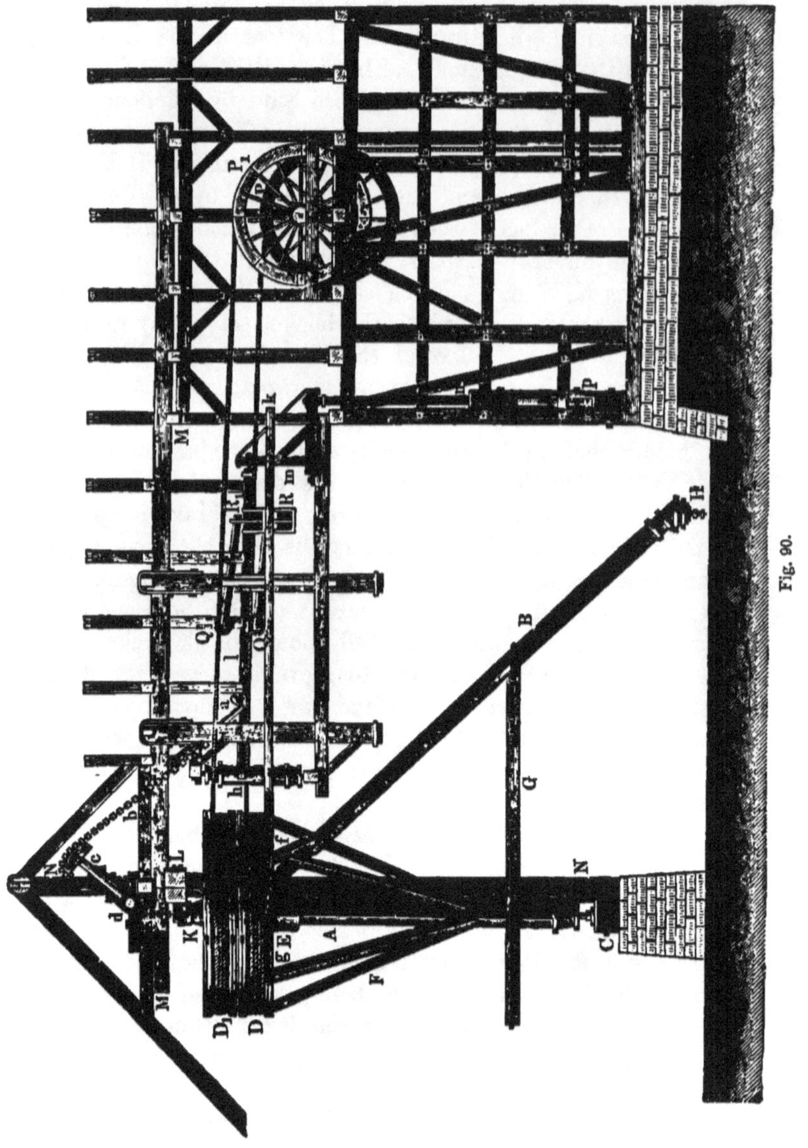

Fig. 90.

is now possible to rotate the upright shaft together with the
lower drum, and thus move the bucket to a higher or lower point

without disturbing the loaded bucket which is suspended from the upper drum.

The lower drum D is supported by lugs E and braces F, and the sweep B is let into and secured to the upright shaft at the top, besides being fastened to the lower drum by wedges, and braced to the latter and the upright shaft through the strut f and the straining-beam G. The pole by which the horses pull is about 3 metres [9·84 ft.] long, and is carried by a pin H at the bottom end of the sweep, which thus allows the drum to be revolved alternately in opposite directions. Two pointed sticks are dragged along by the sweep, which enter the ground of track when the pulling force relaxes, thus preventing backward motion. The upper pivot K of the vertical shaft is held by a bearing bolted to the double cross-beam L, which carries two horizontal beams M, and is joined to the rafters NN, which extend over the whole track. These rafters are supported at the bottom end by a wall O shown in Fig. 91, and at the top

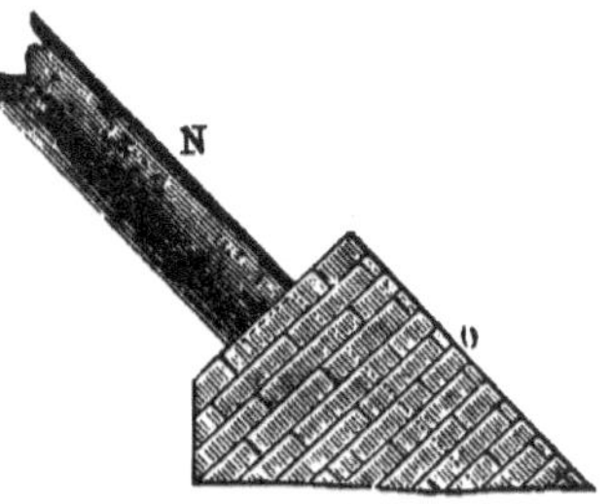

Fig. 91.

either joined to each other or to the ridge beam. The guide-pulleys P and P_1 give the proper direction to the ropes in the shaft, the ropes being also supported by the pulleys Q and Q_1, which are pressed upwards by the weights R and R_1 attached to levers. Finally, the machine is provided with a brake arrangement, which has already been described. In Fig. 90 gk shows one of the two levers to which the brake shoes are attached, which are applied to the rim of the upper or the lower drum. In order to allow of the brake being operated without effort from the hoisting shed, the turning-roller h is introduced, which by means of pull rods (not visible in the figure) is connected to the brake lever, and by the wooden beam l is operated through the bell-crank m, from which a third connection n is suspended, which finally may be pulled down by the action of a brake lever p. For securing the brake in any position a toothed rack is applied at one side of this lever, which is provided with a catch for engaging the rack.

The apparatus is generally driven by one or two horses (seldom four), for which may be assumed a velocity of 0·9 m. [3 ft.], and a pulling force of 45 kg. [100 lbs.] for each horse. The length of sweep is usually taken to be from 6 to 10 m. [19·68 to 32·81 ft.], and the radius of the drum, which is determined by the load to be hoisted, is generally from 1·8 to 2 m. [5·9 to 6·56 ft.] when one horse is used, and from 3·5 to 4 m. [11·48 to 13·12 ft.] when two horses are employed. The average velocity of the load may be assumed to be about 0·3 m. [1 ft.], and rarely reaches 0·5 m. [1·64 ft.] per second.

§ 24. **Hoisting Machines operated by Water Power.**— Water-wheels are frequently employed for driving hoisting machinery in mines. In most cases no gears are made use of, the drums, one movable and one fixed, being secured directly to the shaft of the water-wheel. It is then evidently impossible to carry the ropes to the drums horizontally from the guide-pulleys above the shaft of the mine. A special shaft must be constructed when the water-wheel is located at some depth below the surface, and the ropes conducted through it to the drums after passing over two pairs of guide-pulleys at the top. It is thus necessary to increase the length of the ropes by an amount equal to the depth of the special shaft, as compared with the case where the drum is located above ground. Although the journal friction is increased by this mode of driving, the loss of power ensuing is not so great as would be the case were the drum located above ground and a rod connection employed for driving it from the water-wheel below. The arrangement of a rod connection is based on the various

Fig. 92.

types of parallel cranks described in vol. iii. 1, § 137, Weisb. *Mech.* Each end of the water-wheel shaft, as well as the drum shaft, is provided with a double crank with pins diametrally opposite, as in Fig. 92, the cranks on one side being set at 90° to the cranks on the other side. Four connecting rods of equal length connect the corresponding crank pins. Evidently dead centres are avoided by this arrangement; and besides, the advantage is gained that the rods will be submitted to a pulling strain only, which is a matter of great moment, inasmuch as it would be impossible to arrange suitable guides for them owing to their swing motion.

An essential feature in this kind of hoisting apparatus is the *reversing wheel.* In order to hoist the two buckets alternately the drum must revolve alternately in opposite directions. As a simple water-wheel turns in one direction only, it would be necessary to employ a reversing motion were such a wheel used to drive the hoist. With a view to greater safety, however, a water-wheel with two sets of buckets and two gates placed in opposite directions is made use of. According as the water is let into one or the other of the two sections of the wheel, the latter, together with the drum, will be made to rotate in one direction or the other.

The gates are operated by a double lever, which is movable about a point between the two gate rods, and passes through longitudinal slots in the latter. The double lever is operated by a single lever placed above ground, from which a rod leads down to the former. When the machine is to be stopped after the loaded bucket has reached the surface, the operation consists in closing the gate, and at the same time applying the brake, which acts on the partition between the buckets in the rim of the water-wheel. The arrangement of this brake is essentially the same as that employed with drums driven by horses. The necessary brake lever is located beside the gate lever, and near the levers for pulling up and down the dumping-hooks.

As a rule the buckets are larger than when horse power is employed, and move at a greater velocity. While in the latter case they are made to hold 8 or 10 baskets (*kibbles*), and move at speed of 0·3 to 0·5 m. [0·98 to 1·64 ft.], they are made of a capacity of from 12 to 15 baskets when water power is used, and allowed to move at a velocity of from 0·5 to 1 m. [1·64 to 3·28 ft.]

The method of filling the buckets quickly and with the least possible travel is shown by Fig. 93. A is the bucket to be filled, which travels along the guide-beams BC on the rollers *aa*, and during the filling rests on the timbers DE placed on the traverse beams. FH is a car or truck (or buggy) containing the ore or coal, and travelling on rails to the point of loading. The load is dumped into the bucket by tipping the box-shaped top around a shaft in the front part of the truck ; when the car holds the same quantity as the bucket the least possible time is lost.

Still less time is required for filling when the method shown in Fig. 94 is made use of.   Here the car A in which the ore arrives on the railway is placed on a cage BCD, which hangs on the rope in place of the bucket, and travels between the guide-beams EE, FF by the side rollers aa.   When the machine is at rest the cage is supported by the struts G and H, which are movable around a horizontal axis, and must be

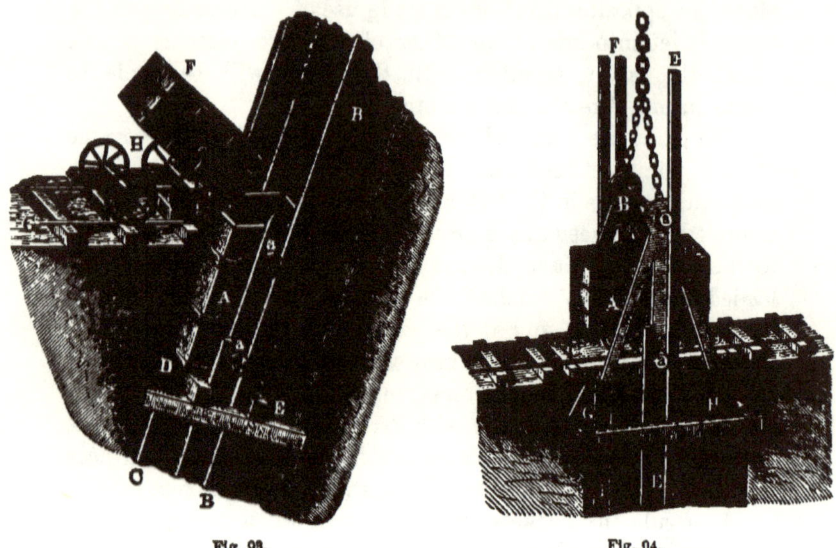

Fig. 93.Fig. 94.

thrown out of the way when hoisting at greater depth is to be done.

The general arrangement of a hoisting machine operated by a water-wheel *directly* is shown by two views in Figs. 95 and 96.   A is the reversing wheel, B, B_1 are the two drums which are fast on their own shafts, and connected to the water-wheel shaft by means of releasing couplings $a, a,$ which consist of two discs and a pin passing through diametrally.

The rope which passes around the drum B is carried around the pulleys C in the rope shaft, and then over the idlers D and

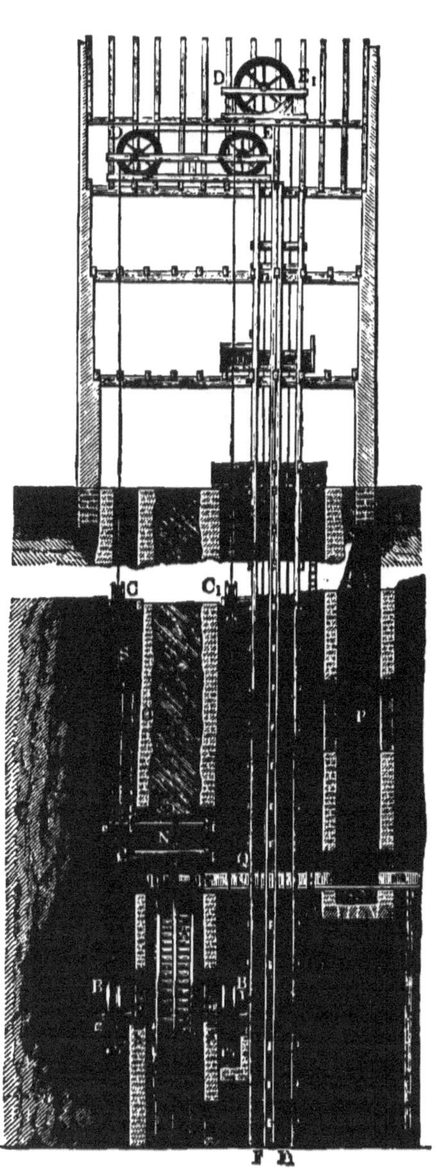

Fig. 95.

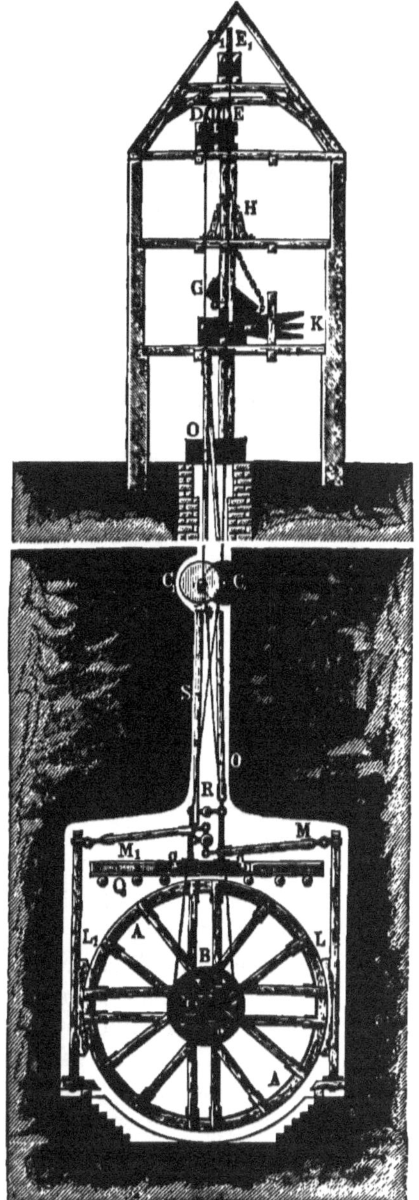

Fig. 96.

E located under the roof of the hoisting shed, wherefrom it hangs down in the working shaft in the direction E, F. The rope leading from the drum B_1 passes around the guide-pulley C_1, and over the idler D_1E_1, finally having the direction E_1F_1 in the working shaft. Fig. 96 shows at G the dumped bucket and at H the safety catches, which allow the bucket to travel clear up to the idlers, in case the machine should be stopped too late, but prevent it from falling back into the shaft. At K the levers are shown which serve for operating the gates, brakes, and dumping-hooks. In the latter figure the double brake is represented at L, L_1, which acts on the middle rim of the wheel, and is connected with the brake lever above ground by means of the pull rods M, M_1, the double bell-crank N, and the rod O. Finally, at P, Fig. 95 shows the pen-stock for the pump wheel, and at Q the flume in which the water is conducted from this pen-stock to the reversing wheel A; both figures also illustrate at q, q_1 the gates for both sections of the wheel, at R the bell-crank, and at S the rod for operating the gates.

A side view of a hoist driven by a water-wheel *through rod connections* is shown in Fig. 97. Here A is the reversing wheel, and at BC and DE are seen the two rods on one side of the wheel, which connect the double return crank BD of the latter with that of the hoisting drum CE. At G is shown the guide-pulley over which the rope is carried from the drum to the working shaft, and at H the bucket suspended from the rope. The brake KLK, and the arrangement MNM for operating the gates, are identically the same as in the direct acting hoist already described.

Hoisting machines operated by *turbines* always require the use of one or more pairs of gears in order to reduce the large number of revolutions of the turbine shaft to the small number of four to eight turns per minute required of the hoisting drum, which is usually $2\frac{1}{2}$ to 3 m. [8·20 to 9·84 ft.] in diameter, and moving at a surface velocity of 0·5 to 1 m. [1·64 to 3·28 ft.] Fig. 98 gives an idea of how a machine of this class may be arranged. Here a turbine hoist is illustrated, as constructed by *Braunsdorf* at Freiberg. A is the turbine, on the shaft of which is placed the cast-iron brake wheel B and the small bevel-pinion C with 20 teeth, which engages the large bevel

Fig. 97.

L
L₁
Q
P
S₁
O
M
G
F
F₁
T'
D
N
R₁
H
E
E
R
S
V
K₁
R
C
X
D
B
B
U
U
W
T₁
Q₁
n n₁
m
M
M
K
M
C
W
L
S
A
A
M
M
n n₁

Fig. 99.

Fig. 98.

D with 108 teeth. In spite of the fact that this turbine works under a head of 4·4 m. [14·4 ft.] only, the reduction $\frac{108}{20} = 5·4$ is insufficient for advantageous running of the machine, and it is therefore necessary to introduce a second set of bevels, consisting of the pinions F and F_1 with 13 teeth each, and the larger bevel G with 56 teeth. The pinions are not fast on the shaft E, however, but run loosely on a conical bushing keyed to the shaft. To accomplish a rigid connection between the shaft EE and the pinions F, F_1, the sleeves HH provided with clutches, and arranged to slide on the feathers a in the shaft E, are thrown into gear with the clutch-shaped hubs of the pinions by means of the forked levers KL, K_1L_1. Now, according as the sleeve H is made to engage the hub of F, or the sleeve H_1 that of F_1, the bevel G together with the drum M secured to it is made to revolve in one or the other direction. This coupling makes unnecessary the use of a double water-wheel with opposite buckets. The coupling is operated by the lever N, which turns on the shaft O, and is connected to the forked levers by the rod P. The brake Q is forced against the fixed drum MM by pressing down the lever RS with the rod T. The brake Q_1 for the movable drum is operated by the lever R_1O and the rod S_1T_1; and the action of the brake UU for the upright shaft is controlled by the lever V attached to the horizontal shaft W, and provided with a latch which locks into a horizontal rack.

In Fig. 99 a side view and section of a movable drum is shown, the arrangement being also applicable when a vertical reversing wheel is employed. The drum consists of two discs MM bolted together, the cylindrical part m, m, and the wooden lining nn, n_1n_1 on the inside of the drum flanges, which holds the staves forming the cylindrical part in position, and together with the latter forms the space for the rope. The drum is loose on the shaft W; for securing it to the latter a disc S, provided with four claws, is keyed to the shaft, and a latch-lever KCL movable around C is made to engage one of the claws by means of the hook-shaped end L. The latch-lever is firmly locked in position by the catch K.

NOTE.—By numerous experiments on water-wheel hoists in

neighbouring mining districts, the author has found that under the most favourable circumstances, that is when no gears or connections are used, and the hoisting is done in vertical shafts of an average depth of 300 m. [984 ft.], this kind of hoist gives a mean efficiency of $\eta = 0.75$, and that under unfavourable conditions, that is, when long rod connections are employed, and the hoisting depth is very great, an efficiency of $\eta = 0.30$ only can be expected.

§ 25. **Hoisting Machines operated by Water-Pressure Engines** are rarely employed at the present day. In order to obtain the smoothest possible running, machines of this type are made with two double-acting cylinders, and besides provided with a large fly-wheel. An excellent example of this kind of hoisting engine is that constructed by *Adriany* at *Schemnitz*. The arrangement and working are clearly shown by Fig. 100. A is a four-way cock, to which the inlet and discharge pipes are connected at B and C respectively, and at D and O the pipes leading to the driving cylinders. The latter pipes DE and NO branch off at E and N, and lead directly into the valve cylinders LMH and $L_1M_1H_1$ at M and M_1, and then at F and F_1 into other pipe connections which lead to the valve cylinders at G and Q and at G_1 and Q_1. One of the two valve cylinders HLM, and one of the two driving cylinders LKH, are shown cut open lengthwise. Short pipes H, L, H_1, and L_1 connect the driving cylinders with the valve cylinders. Each valve rod is provided with two piston valves R and S, and is operated by an eccentric T. Each driving piston K transmits its force through the piston rods KU and the connecting rod UV to the crank V attached to the drum shaft. The cross-heads are fitted with friction rollers U, U_1 and W, W_1, which travel in horse-shoe shaped guides. The construction of the drums X and X_1, as well as the mode of connecting the cast-iron rim YY of the fly-wheel to the shaft by means of wooden arms Z, are plainly indicated on the figure.

The operation of this engine is as follows. The stream of water which is admitted to the regulating valve at B, and then flows into the pipes DE, is divided at E, and from this point runs partly to F and partly to F_1. The portion of the water which reaches F is conducted through the pipes FG into the valve cylinder at G, and hence arrives to the driving cylinder through the short pipe H, and is there utilised to propel the

Fig. 100.

driving piston K. When this piston has reached the end of

its stroke, the water flows back into the valve cylinder through the short pipe L, and from here returns to the regulating valve or four-way cock A by the pipe MNO, and is here finally discharged through the pipe CP. At the latter part of the stroke the eccentric T, through the action of the eccentric rod TW, and the valve rod WRS, pushes the piston valves R and S far enough over to the opposite side of the short pipes L and H to cut off the admission of water from G, and allow the latter to enter at the other end from Q. Consequently the driving water now flows by the path FQLK, and forces the piston K towards the outer end, while the discharge water reaches the outlet by the path KHMNOP. Shortly before the piston arrives at the end of this stroke, the valves are pulled back by the eccentric T, which reopens the communication between G and H, as well as between L and M, thus allowing the piston to start on a new stroke.

The second engine $L_1M_1H_1$ is driven in identically the same manner as LMH, the two being similarly constructed, and having the inlet and outlet pipes BDE and NOP in common. In order to make the resultant tangential effort as nearly uniform as possible, the cranks and eccentrics of the two engines are set quartering, so that one engine is half a stroke in advance of the other. If the four-way cock A is turned through an angle of 45° by means of the lever Aa, both admission and discharge of water are stopped, and if it is turned at a right angle to the original position, the inlet is made into a discharge and *vice versa*. When the hoisting machine is to be brought to rest, after the loaded bucket has reached the surface, it is then only necessary to turn the valve lever Aa at an angle of 45°, and when the load has been dumped and the empty bucket has been refilled, the motion will be reversed if the lever be turned through another 45°.

It is a matter of importance in water-pressure engines, which produce a rotating motion, as in the case at hand, that the length and height given to the piston valves R and S should not exceed the diameter of the pipes L and H, so as to prevent the water from being entirely shut off from the driving cylinder. This would cause very injurious shocks, especially at the points where the piston valves pass the inlets to these pipes, owing to the great resistance of water to expansion and compression. Cylindrical inlet pipes to the driving cylinder are employed in these engines, as a rectangular section would be apt to give too short or low piston valves. A slight loss of driving water is occasioned by the incomplete closing of the piston valves in their middle position, when they, for a moment, place the admission pipe in communication with the discharge pipe.

NOTE.—The water-pressure engine in the "Andreas shaft" at Schemnitz utilises a fall of 111 m. [364 ft.]; it has a piston diameter of 0·16 m. [·52 ft.], and a stroke of 1 m. [3·28 ft.] When in operation the engine on an average makes $4\frac{1}{4}$ turns per minute, giving to the buckets an average velocity of 0·5 m. [1·34 ft.] per second. More recently water-pressure engines, driven by accumulators according to *Armstrong's* method, have gained application for hoisting in mines.

§ 26. **Steam Hoisting Engines.**—Hoisting by steam power is nowadays the most common method employed in mines, especially when the depth is great and the masses to be removed are considerable, which is the case in coal-mines, for instance. The advantages of steam power are to be found in the ease with which it can be procured at any place, and the greater hoisting velocity which it makes possible, factors of great importance, especially where fuel is cheap as in coal-mines. The arrangement of the hoisting machine proper is materially the same as when water power is employed, embodying the drum shaft with the two drums which are turned alternately in each direction.

The engines are always provided with a link reversing motion, that originally invented by *Stephenson* for locomotives being almost exclusively in use. Only in small hoisting engines, which are often made oscillating and given the name *steam winches*, is the reversing accomplished by a valve as in the steam hoist illustrated in Fig. 47. *Gear wheels* are hardly ever used for the purpose of reversing the motion of the drum.

In Germany the engine *cylinders* are generally made *horizontal*, while in England *vertical* beam engines, or engines of the *inverted* type, are largely employed, the latter type chiefly with a view to giving the drum as high a location as possible; by this arrangement, the bend of the hoisting rope around the guide-pulleys is reduced to a minimum.

It is evidently necessary to build the vertical engines in a more substantial manner, and provide a firmer foundation than is needed with horizontal engines, whereas the latter require more floor space, which cannot always be spared. High-pressure steam is generally used together with expansion, while condensers are nowadays rarely employed, although they were formerly quite common in this class of engine. As absolute safety is required, however, in the operation, simplicity of construction is of greater importance, and, besides, in many places the necessary condensing water is wanting. To facilitate the reversing of the engine, the latter is generally arranged with two cylinders acting on cranks set at right angles. By this method the objections to a single-cylinder engine, which can be reversed only when the crank is at a sufficient distance from the dead centres, are obviated. Single-cylinder engines also call for a heavier fly-wheel to ensure uniform motion, and are therefore more difficult to reverse than the double-cylinder engines, in which the inertia of the hoisting drum alone is often sufficient for this purpose.

In older constructions the crank shaft of the engine was arranged to drive the hoisting drum, with reduced velocity, through the medium of a pair of gears, but nowadays the cranks are generally placed directly on the ends of the drum shaft; in this manner greater hoisting velocity is obtained, provided that the pistons are made sufficiently powerful. In good constructions a velocity of 6 to 8 m. [19·68 to 26·25 ft.][1] per second may thus be allowed, and below[2] an instance is cited, where, in an English mine, the hoisting from a depth of 737 m. [2418 ft.] is done in 55 seconds, which involves a velocity of 13·4 m. [43·96 ft.]

Such great speed evidently necessitates that the whole hoisting plant should be constructed in the most substantial and satisfactory manner, and that special safety appliances should be provided to meet possible accidents, such as the breaking of a rope, etc. The drum shaft, more particularly, must be fitted with a reliable and powerful brake. A *steam*

---

[1] See Serlo, *Leitfaden der Bergbaukunde*, vol. ii.

[2] *Berg- und Hüttenmännische Zeitung*, by Kerl and Wimmer, 1876, p. 126.

brake is commonly made use of for this purpose, the pressure
of the steam on a piston working in a separate cylinder being
utilised for applying the brake blocks or strap (see vol. iii. 1,
fig. 721, Weisb. *Mech.*)   The brake wheel is usually secured
to the drum shaft between the two drums; when a fly-wheel
is used the rim of the latter also serves as brake wheel.

When only a small amount of power is needed, *portable*

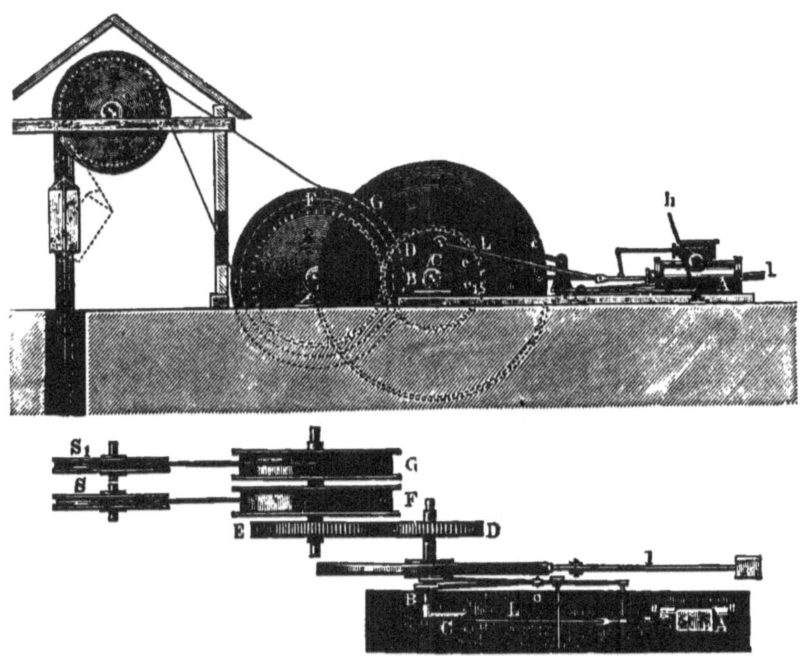

Fig. 101.

hoisting engines are occasionally brought into service.   It is
often required to locate the hoisting engine in a shaft below
ground, and it is then necessary to conduct the steam to it
from the boilers which are placed above ground.   As this
is accompanied with considerable loss from condensation, and
besides it is difficult in such cases to get rid of the exhaust
steam, *compressed-air engines*[1] have been introduced, operating
in a manner similar to that of steam-engines.   The air

---

[1] See Hasslacher, *Zeitschr. f. Berg-, Hütten-, und Salinenwesen,* 1869.

compressor is then placed above ground and driven by a steam-engine. The exhausting air may be used for ventilating purposes. By this indirect method of driving a very small fraction only of the driving force is utilised, however, and for this reason the arrangement is justified only in special cases.

A single-cylinder geared hoisting engine is illustrated in Fig. 101. The cylinder A is horizontal, and the piston rod transmits motion through the connecting rod L to the crank

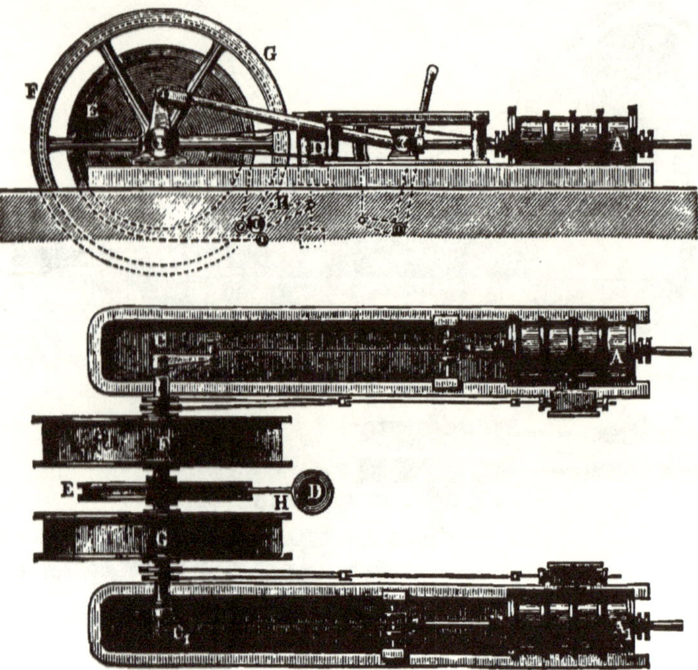

Fig. 102.

C on the fly-wheel shaft B, which carries the smaller gear D, and by this drives the larger gear E on the drum shaft. The ropes lead from the two drums F and G to the pulleys S and S_1 located above the shaft, and from them vertically downwards. The reversing is done by means of the two eccentrics e and e_1 operating the link c, which may be raised or lowered at will by the hand lever h. The under side of the fly-wheel rim is fitted with a brake strap operated by placing the foot on the end of the lever l.

Fig. 102 represents the arrangement of a horizontal double-cylinder hoisting engine without gears, as constructed at *Salms'* machine works [1] at *Blansko*. Here the two connecting rods rotate the cranks C and C_1 set at right angles on the drum shaft, which carries a brake wheel E between the two hoisting drums F and G. The brake band is operated by the steam cylinder D, the steam being admitted on the under side of the piston by the operation of a valve lever, and allowed to act on a lever H connected to the piston rod. The valve lever is turned by hand, and in addition an arrangement is made use of by which the admission valve to the brake cylinder may be opened automatically when the cage has reached the top, in case the party in charge should neglect to stop the engine in time.

Besides *automatic brakes* of this kind, the machines are generally provided with automatic *signal apparatus* by which a bell is sounded after a certain number of revolutions of the drum, to warn the operator. The essential feature of this attachment is a screw driven in either direction by gears from the drum shaft. A nut on this screw is thus moved back and forth longitudinally, and allowed to strike the bell when the drum has made the number of revolutions required to bring the bucket to the surface.

The arrangement of *vertical hoisting engines*,[2] Fig. 103, is evident from the foregoing. Also in this case the force is transmitted from the two cylinders A and A_1 through the connecting rods directly to the cranks on the drum shaft. The eccentrics e and e_1 which operate the links c and c_1 are attached to the wrist-pins of the return cranks K and K_1. As flat ropes or bands are used, pulleys with deep flanges F and F_1 are substituted for the ordinary hoisting drums, and the bands allowed to coil on these in layers on the top of each other. In regard to the equalisation of the weight of the rope obtained by this method, see the following paragraph. One of the pulleys F is firmly secured to the shaft while the other F_1 is loose on the latter, and may be brought into rigid

---

[1] See *Excursionsbericht d. Maschinenbauschule zu Wien*, under direction of Riedler, 1876, Sketch 17.

[2] *Portfeuille*, John Cockerill, vol. iii. pl. 19 and 20 ; and Rühlmann, *Allgem. Maschinenlehre*, vol. iv.

connection with it by the aid of the clutch coupling G.   As
before, this arrangement admits of a change in the hoisting
depth.   The mode in which the piston of the brake cylinder
D acts on the brake blocks H by means of the lever combination h is plainly indicated in the cut.

Fig. 104 shows a small hoisting engine (*winch*)[1] with two
oscillating cylinders A and A_1, as constructed at *Salms'* works.

Fig. 103.

Here the steam distribution is effected by the oscillation of
the cylinders, the steam being admitted and discharged through
the central portion of the journal.   The reverse motion is
brought about by turning the valve V, which is so contrived
as to admit of the admission ports being turned into exhaust
ports, and *vice versa.*

Finally, in Fig. 105 a *two-cylinder geared steam winch*[2] is
represented.   The gear Z is here rigidly connected with the
two drums F and F_1, and the reversing is accomplished by a

[1] See *Excursionsbericht d. Maschinenbauschule zu Wien*, under direction of
Riedler, 1876, Sketch 15.

[2] See note above.

peculiar valve s, which likewise changes the function of the ports.   (For further particulars we refer to the paper mentioned in the note below.)

§ 27. **Balancing the Rope Weight.**—When hoisting is

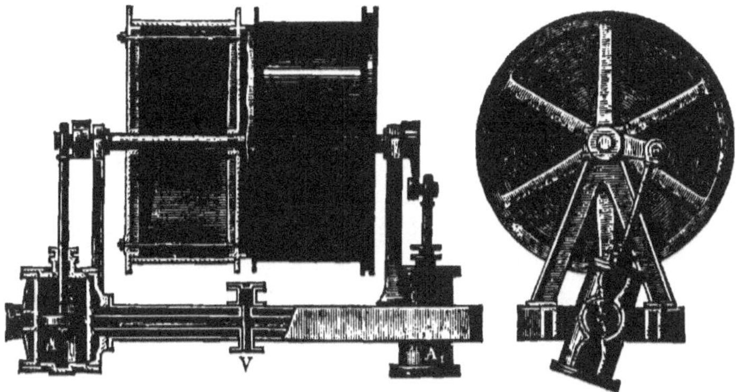

Fig. 104.

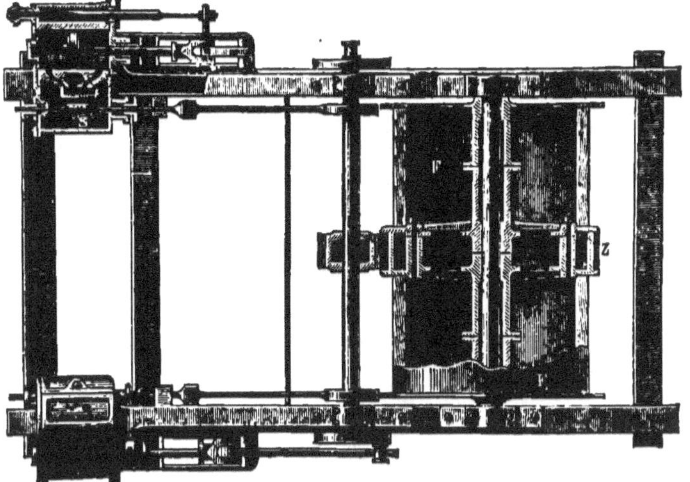

Fig 105.

being done from great depths, the weight of the rope alone forms quite an essential factor in the total resistance to be overcome by the motive power.   It is evident that the resistance is continually growing smaller as the rope which carries the load is gradually wound on to the drum.   At the same time the rope from which the empty bucket is suspended is uncoiled by the same amount, and thus furnishes a steadily increasing extra driving force. Denoting by Q_1 the resistance offered by the loaded bucket during the hoisting, and by Q_2 the addition to the driving force furnished by the empty bucket, then, if S is the weight of a length of rope equal to the hoisting depth s, we obtain for the resistance to be overcome at the beginning of the hoisting operation the value $Q_1 + S - Q_2$, and at the end of the operation $Q_1 - (S + Q_2)$; while at the point during the motion of both buckets where the two ropes exactly balance each other, the resistance is expressed by $Q_1 - Q_2$. The limits of the variable resistance are thus found to be $Q_1 - Q_2 + S$ in the lowest, and $Q_1 - Q_2 - S$ in the highest position, the two values differing by double the weight of the hoisting rope. Consequently, the greater the length s and the weight γ per unit of length, the greater the variation in the resistance, the range being especially very considerable when hemp ropes are used, which for the same strength must be made twice as heavy as wire ropes. In case the weight S of the rope should exceed that of the load proper $Q_1 - Q_2$, which might happen when the shaft is very deep, it is apparent that at the latter portion of the hoisting the resistance would change into a negative one which must be counteracted by the brake. This condition of things would cause great waste of power, and also make the driving arrangement very unsatisfactory, inasmuch as the motor must be made powerful enough to overcome the maximum resistance at the beginning of the hoisting, and then merely be required to develop a gradually decreasing amount of power, which would finally become negative.

Several methods have been employed with a view to overcoming this difficulty by balancing the weight of the ropes.

The remedy nearest at hand is to apply counter-weights in such a manner that they will assist the drum shaft by descending during the first part of the hoisting operation, and oppose its motion during the latter part by ascending again to their original position. To this end the two cages are connected from underneath by a second rope which passes around a guide-pulley at the bottom of the shaft, the effect being practically the same as if an endless rope were made use of, the two branches being always in perfect balance. The only objection to this simple arrangement is to be found in the fact that the additional weight introduced by the balance-ropes materially increases the pressure on the journals of the rope-pulleys, thus creating greater frictional resistances.

In place of this method a *balance-carriage* has been tried, that is a counter-weight, which during the earlier half of the hoisting operation allows a bucket to descend on a curved track, and then pulls it up during the latter half. Also *link-chains* have been proposed, which are wound around a special drum on the hoisting shaft, and serve the purpose of balancing by winding on or off in the same manner as balance-chains for the leaves of drawbridges. None of these methods, however, have gained an extended application.

Another mode of balancing the rope, which was introduced at a comparatively early date, and has been retained up to the present time, consists in the employment of rope drums with *varying radii* for the different coils. Drums of this type are commonly termed *conical drums* or *spiral drums*, as the coils of the rope form spirals on the cone-shaped drum.

In order to equalise the influence of the weight of the rope by this method, the drum must be so designed that the lever arm of the resistance W gradually increases in the same ratio as the latter decreases, thus making the product of the two, or the moment of the load, a constant quantity. In the lowest position of the bucket the rope will then act at the small radius r_1, while in the highest position it will run on to the drum at a point where the radius is a larger one, r_2.

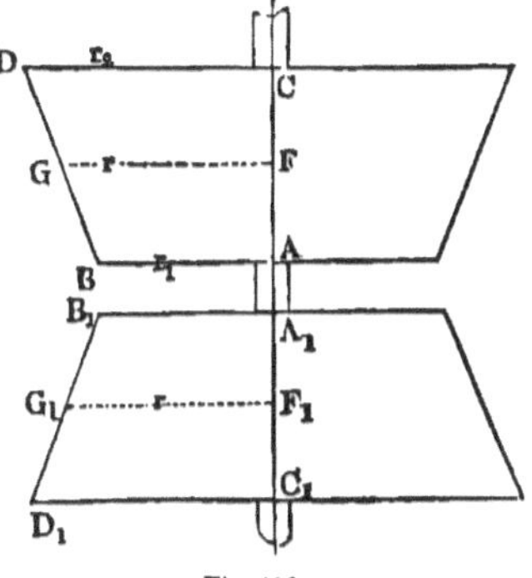

Fig. 106.

Let us assume a conical drum BD, Fig. 106, with a minimum radius $AB = r_1$ for the lowest position of the bucket, and a maximum radius r_2 for the highest position, it will then be necessary to determine the ratio of r_1 to r_2 in accordance with the above-mentioned provision. In the first place, it is evident that the two drums $\dot{B}D$ and B_1D_1 must be made of the same dimensions, as the same conditions pertain to the winding of both ropes ; also it is clear that the part which holds the empty bucket leaves the drum at the largest radius, in the highest position, at the same moment as the loaded rope is wound on at the smallest radius, when the loaded bucket is on the point of starting from the bottom of the shaft. Further, the number of coils n must be the same for both drums, and it is easily seen that after half the total number of revolutions the two ropes touch the drums at the same distances from the centre

$FG = F_1G_1 = r = \dfrac{r_1 + r_2}{2}$. In this position the two buckets

must be at the same height in the shaft, as the ascending bucket has risen through a distance l_1 equal to the length of the coils on the smaller half of the drum BD, while the empty bucket has descended a distance l_2, which is equal to the length of the coils on the larger half of the drum DG, and the two lengths l_1 and l_2 together equal the total hoisting depth l. Hence it follows that the point where the two buckets meet will fall *below the middle* of the shaft, because l_1 is smaller than l_2. Only in the case of cylindrical drums will the point of meeting fall at the middle of the shaft.

To determine the proportions of the conical drums, let us as above denote by Q_1 the resistance due to the ascending bucket, and by Q_2 the auxiliary driving force derived from the empty bucket. Further, let γ be the weight per unit of length of the rope in case of *vertical* shafts, or its component in the direction of the rope for the event that the shaft makes an angle a to the horizon, that is for the latter condition $\gamma = q \sin a$, when q signifies the weight per unit of length of the rope.

The moment of the forces acting on the drum shaft through the ropes, for the lowest position of the bucket about to be hoisted, will then be

$$M_1 = (Q_1 + l\gamma)r_1 - Q_2 r_2 \quad . \qquad . \qquad . \qquad (1)$$

and for its highest position

$$M_2 = Q_1 r_2 - (Q_2 + l\gamma)r_1 \quad . \qquad . \qquad . \qquad (2)$$

If these two moments are to be of the same magnitude,

and each equal to M, we obtain by adding together the two equations (1) and (2)

$$2\,M = (Q_1 - Q_2)(r_1 + r_2)$$

or

$$r_1 + r_2 = 2r = \frac{2M}{Q_1 - Q_2} \qquad . \qquad . \qquad . \qquad (3)$$

At the point above referred to where the buckets meet, and where both ropes have the same lever arm r, the moment $M_0 = Q_1 r - Q_2 r = (Q_1 - Q_2)r$, as the ropes here balance each other. Hence we find, by combining this equation with (3), that the moment of resistance M_0 at this point is also equal to M. By using the assumed conical drum we thus obtain equal moments of resistance for three different positions of the buckets, the *highest, lowest*, and that where they meet in the shaft.

To determine r_1 and r_2, when r is given, we subtract (2) from (1), thus getting

$$(Q_1 + Q_2 + 2l\gamma)r_1 - (Q_1 + Q_2)r_2 = 0,$$

or

$$\frac{r_1}{r_2} = \frac{Q_1 + Q_2}{Q_1 + Q_2 + 2l\gamma} \qquad . \qquad .. \qquad . \qquad (4)$$

By combining this equation with

$$r_1 + r_2 = 2r \qquad . \qquad . \qquad . \qquad . \qquad (5)$$

we easily obtain from

$$2r = r_1 + r_1 \frac{Q_1 + Q_2 + 2l\gamma}{Q_1 + Q_2} = r_1 \left(2 + \frac{2l\gamma}{Q_1 + Q_2} \right)$$

$$r_1 = \frac{r}{1 + \dfrac{l\gamma}{Q_1 + Q_2}} = r \frac{Q_1 + Q_2}{Q_1 + Q_2 + l\gamma} = r \left(1 - \frac{l\gamma}{Q_1 + Q_2 + l\gamma} \right) \qquad (6)$$

and

$$r_2 = 2r - r_1 = r \left(1 + \frac{l\gamma}{Q_1 + Q_2 + l\gamma} \right) \qquad . \qquad . \qquad (7)$$

The mean radius r is determined by the hoisting depth l and the total number of coils n on each drum. From $2\pi r n = l$ we have

$$r = \frac{l}{2\pi n} .$$

The point where the buckets meet is given by its distance from the bottom of the shaft (measured in the direction of the latter)

$$l_1 = \pi n \frac{r + r_1}{2} = \pi n \left[r - \frac{l\gamma}{2(Q_1 + Q_2 + l\gamma)} \right],$$

and its distance from the surface (*grass*) above

$$l_2 = \pi n \frac{r + r_2}{2} = \pi n \left[r + \frac{l\gamma}{2(Q_1 + Q_2 + l\gamma)} \right].$$

The length of each drum will be nb, when b denotes the distance from centre to centre of two adjoining rope grooves, which distance may be taken to $b = 1\cdot5\delta$, approximately, if δ is the thickness of the rope. It is evident that the equality of moments just obtained for three different positions of the buckets will only exist under the supposition that the hoisting depth is equal to l.

The formulæ above deduced are also applicable to the case when flat ropes are used, wound spirally in coils on the top of each other, around spool-shaped pulleys (see Fig. 103). The mean radius must then not be chosen arbitrarily, however, inasmuch as the difference in radii of two successive coils must be equal to the thickness δ of the rope. Thus, besides the equations (4) and (2) we obtain, for n coils

$$r_2 - r_1 = n\delta = \frac{l}{2\pi r}\delta.$$

After introducing for r_2 and r_1 the values from (7) and (6), we get

$$r_2 - r_1 = r\frac{2l\gamma}{Q_1 + Q_2 + l\gamma} = \frac{l}{2\pi r}\delta,$$

from which equation follows

$$r = \sqrt{\frac{Q_1 + Q_2 + l\gamma}{4\pi\gamma}\delta} \qquad . \qquad . \qquad . \qquad (8)$$

In practice it is generally sufficient to balance the ropes in the three positions mentioned, and for this reason the drums are made in the shape of conic frusta. In all other positions during the hoisting the moments of the resistance are variable. If it is desired that the moment shall be constant in *any* position, it is necessary to design the drums in such a manner that the radii will vary according-

ing to a different law from that governing the shape of the conical drum. Assuming as before that the distance between two adjoining coils (measured in the direction of the axis) is always the same, the outline of the drum will be found as follows.

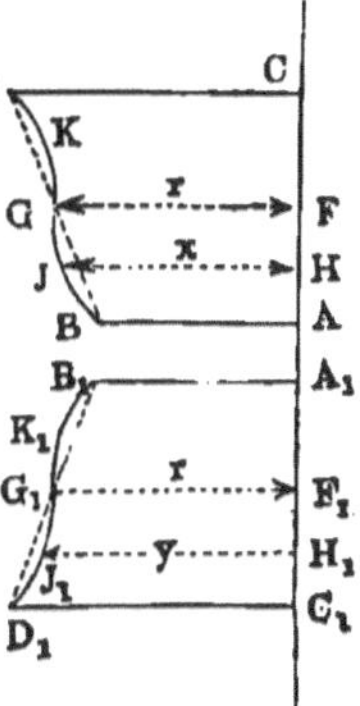

Fig 107.

Let us suppose that the drum shaft CC_1, Fig. 107, has turned through any angle w, measured from the beginning of a hoisting operation, then the rope from which the loaded bucket is suspended will have travelled a certain. distance AH in the direction of the axis of the drum, the radius HJ for this point being x, and the length of rope wound on to the part BJ being denoted by u. At the same time the rope carrying the empty bucket has been unwound by a length r, and has advanced a distance C_1H_1 in the direction of the axis of the drum; and as the distance between adjoining coils is everywhere the same, we have $C_1H_1 = AH$. As the two drums must be identical in every respect, the two ropes will therefore in every moment be at the same distance from the middle sections FG and F_1G_1. Denoting by y the radius H_1J_1 at the point where the descending rope uncoils, then the moment of resistance at the position under consideration will be

$$[Q_1 + (l - u)\gamma]x - (Q_2 + v\gamma)y \quad . \quad . \quad . \quad . \quad (9)$$

and for the reversed condition, when the empty bucket is suspended from J and the loaded one at J_1, it will be

$$(Q_1 + v\gamma)y - [Q_2 + (l - u)\gamma]x \quad . \quad . \quad . \quad . \quad (10)$$

Each of these values must be equal to

$$M = (Q_1 - Q_2)r,$$

and consequently we obtain by addition

$$(Q_1 - Q_2)(x + y) = 2M = 2(Q_1 - Q_2)r,$$

from which equation follows

$$x + y = 2r ;$$

that is, the lever arms of the two ropes at the same moment will here, as was also the case with conical drums, differ by equal amounts from the mean radius r. For determining the fractions of the rope u and v wound on and off while the turning has taken place through the angle w, we apparently have, for an infinitely small angle δw

$$\delta u = x\delta w; \text{ or } x = \frac{\delta u}{\delta w},$$

and

$$\delta v = y\delta w; \text{ or } y = \frac{\delta v}{\delta w};$$

and hence

$$\delta u + \delta v = (x + y)\delta w = 2r \cdot \delta w.$$

By integration we then obtain

$$u + v = 2rw, \text{ or } v = 2rw - u \quad . \quad . \quad . \quad . \quad (11)$$

The constant of integration is in this case zero, as for $w = 0$ also $u = v = 0$. By substituting the value $\frac{\delta u}{\delta w}$ for x and $\frac{\delta v}{\delta w}$ for y in the equation (9), we get for the moment of resistance $M = (Q_1 - Q_2)r$ the expression

$$(Q_1 - Q_2)r = Q_1\frac{\delta u}{\delta w} + (l - u)\gamma\frac{\delta u}{\delta w} - Q_2\frac{\delta v}{\delta w} - v\gamma\frac{\delta v}{\delta w},$$

and thus by integration

$$(Q_1 - Q_2)rw = (Q_1 + l\gamma)u - \gamma\frac{u^2}{2} - Q_2v - \gamma\frac{v^2}{2}.$$

After substituting for v its value $(2rw - u)$, we obtain

$$(Q_1 - Q_2)rw = (Q_1 + l\gamma)u - \gamma\frac{u^2}{2} - Q_2(2rw - u) - \gamma\frac{(2rw - u)^2}{2},$$

which equation after some transformation gives

$$u^2 - u\left(\frac{Q_1 + Q_2}{\gamma} + l + 2rw\right) = -\frac{(Q_1 + Q_2)r}{\gamma}w - 2r^2w^2.$$

If for the sake of brevity we place

$$\frac{Q_1 + Q_2}{\gamma} + l = A \text{ and } \frac{(Q_1 + Q_2)r}{\gamma} = B,$$

and then solve the equation, we find

$$u = \frac{A}{2} + rw \pm \sqrt{\left(\frac{A}{2} + rw\right)^2 - Bw - 2r^2w^2},$$

and hence finally by differentiation :

$$x = \frac{\delta u}{\delta w} = r \pm \frac{1}{2}\frac{Ar - B - 2r^2w}{\sqrt{\frac{A^2}{4} + Arw - Bw - r^2w^2}} = r\left(1 \pm \frac{l - 2rw}{2\sqrt{\frac{A^2}{4} + lrw - r^2w^2}}\right)$$

$$= r\left(1 \pm \frac{l - 2rw}{\sqrt{A^2 + 4(l - rw)rw}}\right) \quad . \quad . \quad . \quad . \quad (12)$$

The two signs always give the two values of x and y, which belong together, and both differ from the mean radius r by the fraction

$$\frac{l - 2rw}{\sqrt{A^2 + 4(l - rw)rw}}$$

For $w = \dfrac{l}{2r}$ we have $x = y = r$, and for $w = 0$, the extreme radii r_1 and r_2 are found to be

$$r\left(1 \pm \frac{l}{A}\right) = r\left(1 \pm \frac{l\gamma}{Q_1 + Q_2 + l\gamma}\right)$$

as in the case of the conical drum.

EXAMPLE.—Let the mean radius of a spiral drum be $r = 2$ m. [6·56 ft.], the weight of the empty bucket 200 kg. [441 lbs.], and its load 500 kg. [1102 lbs.]; let further the weight of the rope per metre be $\gamma = 0·5$ kg. [0·29 lb. per foot], and the hoisting depth 400 m. [1312 ft.], then, neglecting frictional resistances, we have $Q_1 = 700$ kg. ; $Q_2 = 200$ kg. ; $l\gamma = 200$ kg., and hence we obtain for the minimum radius

$$r_1 = 2\left(1 - \frac{200}{700 + 200 + 200}\right) = \frac{2·9}{11} = 1·636 \text{ m. [5·366 ft.]},$$

and for the maximum radius

$$r_2 = 2\left(1 + \frac{2}{11}\right) = 2·363 \text{ m. [7·75 ft.]}$$

The number of coils will be found to be $\dfrac{400}{2\pi \times 2} = 31·8$, and thus the axial length of the drum, if the distance from centre to centre of coils is taken about 40 mm. [1·57 in.], will be $0·040 \times 31·8 = 1·272$ m. [4·17 ft.] If the drum is to be shaped like a *conoid*, with a view to getting a constant moment of resistance for all positions, it would be necessary to calculate the radii x and y corresponding to a number of angles of revolution w, taken at intervals of every other coil of 2π, for instance. After eight revolutions, that is for $w = 16\pi = 50·26$, the corresponding radii would thus be

$$r\left(1 \pm \frac{400 - 2 \times 2 \times 50·26}{\sqrt{\left(\frac{700 + 200}{0·5}\right)^2 + 4(400 - 2 \times 50·26)2 \times 50·26}}\right) = 2(1 \pm 0·110),$$

hence $x = 1·780$ m. [5·84 ft.] and $y = 2·220$ m. [7·28 ft.]

For a *cone*-shaped drum the radii corresponding to the same number of eight revolutions would be

$$x' = r_1 + \frac{8}{31·8}(r_2 - r_1) = 1·636 + \frac{8}{31·8}0·727 = 1·820 \text{ m. [5·97 ft.]},$$

and

$$y' = r_2 - \frac{8}{31·8}(r_2 - r_1) = 2·363 - \frac{8}{31·8}0·727 = 2·180 \text{ m. [7·15 ft.]}$$

The radii of the two styles of drum, at the assumed position, thus differ by the amounts

$$x - x' = 1\cdot780 - 1\cdot820 = -0\cdot040 \text{ m. } [-0\cdot13 \text{ ft.}],$$

and

$$y - y' = 2\cdot220 - 2\cdot180 = +0\cdot040 \text{ m. } [+0\cdot13 \text{ ft.}]$$

By calculating in this manner a number of radii, and laying them off as ordinates in Fig. 107, we obtain as profile the wavy line BJGKD, which changes its curvature in the central point G. Evidently this profile is correct only under the supposition that the axial distance between the rope grooves is everywhere the same. Were this assumption to be ignored, it would be possible to so locate the grooves on an ordinary cone, Fig. 106, as to accomplish a perfect balancing of the ropes. Such an arrangement, however, would cause an irregular side motion of the ropes, and make the guiding of the latter more difficult.

If, in place of a round rope, a flat band of a thickness $\delta = 15$ mm. [0·59 in.] is to be used, the mean radius r of the coils will be found to be

$$r = \sqrt{\frac{Q_1 + Q_2 + l\gamma}{4\pi\gamma}}\delta = \sqrt{\frac{1100}{12\cdot56 \times 0\cdot5}}0\cdot015 = 1\cdot620 \text{ m. } [5\cdot31 \text{ ft.}],$$

and the radius of the flanged pulleys

$$r_1 = r\left(1 - \frac{200}{700 + 200 + 200}\right) = \frac{9}{11}1\cdot620 = 1\cdot326 \text{ m. } [4\cdot35 \text{ ft.}],$$

while that of the extreme coil will be

$$r_2 = \frac{13}{11}r = 1\cdot915 \text{ m. } [6\cdot28 \text{ ft.}]$$

The number of coils will in this case be

$$n = \frac{400}{2 \times 3\cdot14 \times 1\cdot620} = 39\cdot32,$$

and as a natural consequence we have

$$39\cdot32 \times 0\cdot015 = 0\cdot589 = 1\cdot915 - 1\cdot326 \text{ m.}$$

$$\left(39\cdot32 \times \frac{0\cdot59}{12} = 1\cdot93 = 6\cdot28 - 4\cdot35 \text{ ft.}\right)$$

The construction of a conical spiral drum, designed by *v. Gerstner*, is shown in Fig. 108. Here A represents the lower and B the upper drum of a hoisting machine operated by horse power, the upright shaft being indicated by CD. The two nearly horizontal ropes S_1 and S_2, leading to the rope-pulleys above the shaft, are guided automatically by the sheaves R_1 and R_2 mounted in the frame J, which is suspended from a lever L. The frame is balanced by the counter-weight K, and is given a rising and falling movement by the worm Z arranged in the circumference of the upper drum, and engaging the pins P, which serve the purpose of rack-teeth. The pitch of the worm, as well as the distance between adjoining pins, is equal to the distance b between the coils axially, and for this reason it is apparent that, for every revolution of the drum, the frame together with the sheaves is carried up or down a sufficient distance to guide the ropes properly to the grooves in the circumference of the drums. Such a guiding arrangement, in common to both drums, necessitates that the rope should always be kept equally far apart—that is, the drums must

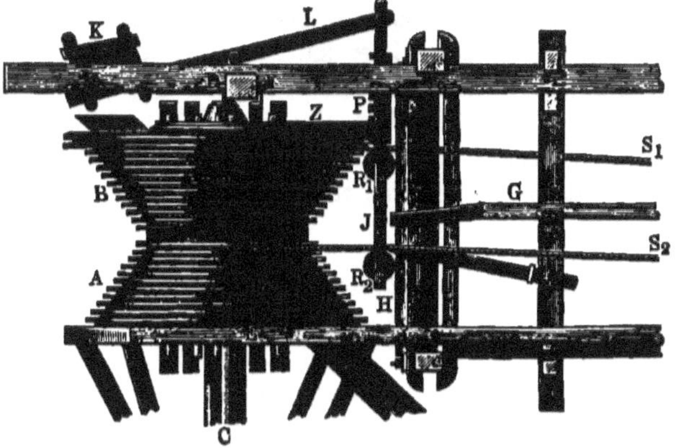

Fig. 108.

always be placed with either both small ends or both large ends together. The construction of the brake is evident from the figure, where E represents the brake blocks suspended at F, and which may be forced against the lower drum rim by iron rods operated by the bar G and the rocking shaft H.

§ 28. **Efficiency of Mine Hoists.**—In order to investigate the conditions of equilibrium of this class of hoisting machines, let Q denote the load proper, G the weight of a bucket, or a cage including the empty receptacle placed thereon, and S the weight of a rope of the length l. Further, let a be the angle of inclination of the shaft to the horizon, and assume for the rollers on the buckets, or cage, a radius r_1 and for their pins $\mathfrak{r}_1$, while the same quantities for the rollers supporting the rope are r_2 and $\mathfrak{r}_2$. At the point where the buckets meet, which, in the case of cylindrical drums, is located in the middle of the shaft, we then have for the tensions in the ropes hanging down from the idler-pulleys

$$Z_1 = (Q+G)\left(\sin a + \phi \frac{r_1}{r} \cos a\right) + \frac{S}{2}\left(\sin a + \phi \frac{r_2}{r_2} \cos a\right)$$

for the driving rope, and

$$Z'_1 = G\left(\sin a - \phi \frac{r_1}{r_1} \cos a\right) + \frac{S}{2}\left(\sin a - \phi \frac{r_2}{r_2} \cos a\right)$$

for the rope descending with the empty bucket.

If now s_1 and s'_1 denote the resistances due to stiffness of the rope in passing over the idler-pulleys overhead, whose radii may be taken at r_3, and whose journals have a radius r_3; and if R_3 and R'_3 represent the pressures of these journals against their bearings, resulting from the weight G_3 of the pulleys, together with the respective tensions in the ropes; then the tensions in the ropes between the idlers and the hoisting drum are found to be

$$Z = Z_1 + s_1 + \phi R_3 \frac{r_3}{r_3},$$

for the driving rope, and

$$Z' = Z'_1 - s'_1 - \phi R'_3 \frac{r_3}{r_3}$$

for the descending rope.

Denoting by R the pressure against the bearings of the drum shaft, caused by the weight of the drum and the tensions Z and Z', and by s and s' the resistances due to stiffness of the rope in passing around the drum, which has a radius r and journals of radii r, it will then be necessary to apply at the circumference of the drum a moment of force

$$M = Pr = (Z+s)r - (Z'-s')r + \phi Rr,$$

which equation gives for the force required in the circumference

$$P = Z - Z' + s + s' + \phi R \frac{r}{r}.$$

Without hurtful resistances, the same force for a useful load Q would be $P_0 = Q \sin a$, and consequently we have, exclusive of the prime-mover, an efficiency for the hoisting machine of

$$\eta = \frac{P_0}{P} = \frac{Q \sin a}{Z - Z' + s + s' + \phi R_3 \dfrac{r}{r}}$$

The wasteful resistances which appear in the shape of friction between the teeth and at the journals, when gears are employed, must be treated in accordance with § 3 ; that is, the above value η should be multiplied by the efficiency of the gearing, in order to find the efficiency of the machine including the gearing. For determining the frictional resistances connected with the operation of the motor, we refer to the special deductions pertaining to prime-movers in vol. ii.

To determine the value s for the resistances due to stiffness of the ropes, the formulæ given in vol. i. § 199, Weisb. *Mech.*, may be made use of; in accordance with these, we obtain in the case of wire ropes and a tension Z

$$s = 0{\cdot}57 + 0{\cdot}000694\frac{Z}{r} \left[s = 1{\cdot}26 + 0{\cdot}002276 \frac{Z}{r} \right].$$

In determining the pressure R against the journals of the rope-pulleys and the drum shaft, the weight of these parts must not be neglected; on the contrary, the resultant pressure must be calculated from these weights and the corresponding tensions. For this purpose we can proceed in accordance with *Poncelet's* Theorem (see vol. i. § 186, Weisb. *Mech.*), and resolve all the forces in their vertical and horizontal components. If the respective sums of these are found to be V and H, we shall have for the friction of the journals, approximately

$$R = \sqrt{V^2 + H^2} = V\sqrt{1 + \left(\frac{H}{V}\right)^2} = 0{\cdot}96V + 0{\cdot}4H.$$

EXAMPLE.—The efficiency of a hoisting apparatus operating in a shaft at 70° inclination, and 500 m. [1640 ft.] vertical depth, is to be determined, when the load to be hoisted weighs 800 kg. [1764 lbs.], and the empty cage G = 300 kg. [661·5 lbs.]

The length of each rope is given by $\dfrac{500}{\sin 70°} = 532$ m. [1745 ft.], and its weight, from the weight of 1 kg. per metre of wire rope

18 mm. [0·70 in.] in thickness [0·67 lb. per foot], is found to be
$S = 532$ kg. [1173 lbs.]   Taking the ratio $\dfrac{r_1}{r_1} = \dfrac{r_2}{r_2} = \dfrac{12 \text{ mm.}}{60 \text{ mm.}} = 0·2$
$\left[= \dfrac{0·47 \text{ in.}}{2·36 \text{ in.}} \right]$, both for the bucket rollers and supporting rollers
for the rope, we shall then obtain the tensions in the ropes hanging
down from the idler-pulleys from

$$Z_1 = (800 + 300)(0·940 + 0·1 \times 0·2 \times 0·342) + 266(0·940 + 0·1 \times 0·2 \times 0·342)$$
$$= 1041·5 + 251·8 = 1293·3 \text{ kg. } [2851·7 \text{ lbs.}]$$

and

$$Z'_1 = 300(0·940 - 0·02 \times 0·342) + 266(0·940 - 0·02 \times 0·342) = 280·0 + 248·2$$
$$= 528·2 \text{ kg. } [1164·7 \text{ lbs.}]$$

The loss due to stiffness at the idlers of radii $r_3 = 1$ m. [3·28 ft.]
will be

$$s_1 = 0·57 + 0·000694 \times 1293·3 = 1·5 \text{ kg. } [3·3 \text{ lbs.}],$$

and

$$s'_1 = 0·57 + 0·000694 \times 528·2 = 0·95 \text{ kg. } [2·09 \text{ lbs.}]$$

If each idler-pulley weighs 600 kg. [1323 lbs.], and the rear ends
of the ropes form an angle of $\beta = 15°$ with the horizon, we obtain
for the idler of the driving rope

$$V = 600 + 1293 \,(\sin 15° + \sin 70°) = 2150 \text{ kg. } [4443 \text{ lbs.}],$$

and

$$H = 1293(\cos 15° - \cos 70°) = 807 \text{ kg. } [1779 \text{ lbs.}],$$

and hence the pressure on the journals

$$R_3 = 0·96 \times 2150 + 0·4 \times 807 = 2387 \text{ kg. } [5264 \text{ lbs.}]$$

In like manner we get for the idler of the descending rope

$$V' = 600 + 528(\sin 15° + \sin 70°) = 1233 \text{ kg. } [2719 \text{ lbs.}],$$

and

$$H' = 528(\cos 15° - \cos 70°) = 330 \text{ kg. } [728 \text{ lbs.}],$$

and thus the pressure on the journals

$$R'_3 = 0·96 \times 1233 + 0·4 \times 330 = 1316 \text{ kg. } [2902 \text{ lbs.}]$$

Assuming a radius $r_3 = 40$ mm. for the journals of the idler-pulleys,
we then obtain the tensions in the ropes at the drum

$$Z = 1293 + 1·5 + 0·1 \times 2387 \frac{40}{1000} = 1304 \text{ kg. } [2876 \text{ lbs.}],$$

and

$$Z' = 528 + 0·95 + 0·1 \times 1316 \frac{40}{1000} = 534 \text{ kg. } [1177 \text{ lbs.}]$$

Hence follow the resistances due to stiffness at the drum of radius $r = 2$ m. [6·56 ft.]

$$s = 0·57 + 0·000694\frac{1304}{2} = 1·0 \text{ kg. } [s = 1·26 + 0·002276\frac{2876}{6·56} = 2·2 \text{ lbs.}],$$

and

$$s' = 0·57 + 0·000694\frac{534}{2} = 0·76 \text{ kg. } [s = 1·26 + 0·002276\frac{1177}{6·56} = 1·67 \text{ lbs.}]$$

Finally we determine the pressure R against the bearings of the drum shaft, which weighs about 5000 kg., from

$$V = 5000 - (1304 + 534) \sin 15° = 4225 \text{ kg. } [9316 \text{ lbs.}],$$

and

$$H = (1304 + 534) \cos 15° = 1772 \text{ kg. } [3907 \text{ lbs.}],$$

to be

$$R = 0·96 \times 4225 + 0·4 \times 1772 = 4765 \text{ kg. } [10506 \text{ lbs.}]$$

If the radius of the journal of the drum shaft is then assumed to be 0·10 m. [0·328 ft.], the driving force required at the circumference of the drum will be found from

$$P = 1304 - 534 + 1·0 + 0·8 + 0·1 \times 4765\frac{0·1}{2} = 794·6 \text{ kg. } [1751 \text{ lbs.}]$$

For a hoisting velocity of 3 m. [10·14 ft.], the *net* power which the motor must develop at the drum shaft will be

$$\frac{794·6 \times 3}{75} = 31·8 \text{ horse-power} \left[\frac{1751 \times 10·14}{550} = 32·28 \text{ h.-p.}\right]$$

The efficiency of the hoisting apparatus, apart from the motor, is found to be

$$\eta = \frac{800 \times \sin 70°}{794·6} = 0·946,$$

which high figure may be explained from the fact that the resistances due to the stiffness of the wire rope are very slight, on account of the large radii of rope-pulleys and drum.

§ 29. **Safety Apparatus.**—In order to prevent the falling of a cage into the shaft in case a rope should break, various designs of *safety-catches* and *safety-brakes* have been brought out during the past thirty or forty years, so arranged as to be thrown into action when a rope breaks, and thus check the motion of the cage. Such apparatus have proved the more indispensable the more the hoisting velocity has been increased, and besides the greatest possible safety is a matter of prime importance where the shafts are of great depth, and it is required to transport large gangs of men into and out of the mine. Since so-called *man-engines* (see the following paragraph) are not in use everywhere, the hoisting machines are frequently employed for the latter purpose. The lives and safety of the miners are then in a large measure dependent on the prompt action of the safety appliances, inasmuch as absolute security against breakages can never be attained in spite of daily inspection and the utmost care in the maintenance of the ropes.

These safety apparatus may be divided in *safety-catches*, which cause a sudden stop at the cage by their locking action at the instant a rope breaks, and *safety-brakes*, which produce a frictional resistance, sufficient not only to destroy the momentum of the cage, but also to counteract the acceleration due to the force of gravity, and finally bring the sinking load to rest. The latter kind can, therefore, never bring about a sudden stop, but must be in action for some length of time. It must be admitted, however, that the brakes possess advantages in the point of safety over the catches, in that the latter, when they are instantaneously thrown into action in the manner outlined, are apt to generate such enormous shocks as to cause the breaking of the locking parts.

It makes considerable difference whether the break occurs in the ascending or descending rope. In the former case, the moving mass, which travels upwards at a velocity v, is after the breakage carried a certain distance higher, like a body thrown into the air ; for a hoisting velocity of 6 m. [19·68 ft.] this height will be $\dfrac{v^2}{2g} = 1\cdot83$ m. [6 ft.], if hurtful resistances are neglected. After covering this distance, the mass is for a moment at rest, and when the catch is thrown into action, it will only have to resist the weight of the cage, which load it may easily be designed to withstand. On the other hand, if the descending rope should break, it is evident that, even for the most favourable event, that is, when the catch is immediately thrown into gear, a shock will be caused by the living force $G\dfrac{v^2}{2g}$ of the cage. As it is never possible, however, to make the action of the catches perfectly instantaneous, a certain length of time t_1 being always needed for effecting the engagement, the cage will have time to fall through a distance $h_1 = \frac{1}{2}gt_1^2$, so that the shock which must be sustained by the catches will be expressed by $G\left(\dfrac{v^2}{2g} + h_1\right)$. It is clear from the above how much the danger from a breakage is increased by *tardy* action of the safety-catches, since even for a length of time t_1 of only one second, we should have

$$h_1 = \tfrac{1}{2}g = 4\text{·}905 \text{ m. } [16\text{·}09 \text{ ft.}]$$

Besides, the majority of known apparatus of this class are very unreliable, from the fact that the catches are thrown into action by rubber or steel springs, which may easily lose their elasticity, thus causing the whole apparatus, which is usually inactive, to fail or operate too late in critical moments. It must be noted that weights cannot be used for operating the catches, as they would take part in the motion of the cage, and therefore would be unable to exert a pressure against it; the force of gravity would be wholly spent in accelerating these weights, or, as it is sometimes expressed, falling bodies lose their weight in their fall. The elastic action of compressed air has also been successfully used, in place of springs, for engaging the catches.

In braking apparatus, shocks from the above cause need not be apprehended, at least not when the friction is generated to such a degree only that the retardation of the moving masses will cause no damage to the vital parts of the mechanism. In reality all good *safety-catches* are so arranged that the motion will not be arrested instantaneously, irons provided with sharp edges or teeth being mostly employed and allowed to enter into the wooden guides. By this method a certain amount of slip against the guides is not excluded at the beginning of the fall, a fact which as a rule may be ascertained from the traces left in the wooden guides. Safety-catches made of iron, and engaging an iron rack, are only applicable for slow speeds and slight inclinations, as in the case of furnace-lifts (see Fig. 63). A few of the more important safety appliances,[1] which have been devised and carried out into practice, will be described below.

---

[1] *Zeitschr. deutsch. Ing.* 1868, page 353; v. Hauer, *Die Fördermaschinen;* Serlo, *Bergbaukunde.*

In *Büttenbach's* arrangement, Fig. 109, the rope carries the cage by means of the eye-bolt A, the double elliptic spring being kept in tension by the weight of the cage. In case the rope should break, the action of the spring on the links C would force the latches R against the wooden guides F, and allow them to enter into pockets made in the timbers. This apparatus can only be recommended for very small hoisting velocities.

Fontaine's construction has gained a more extensive application. In place of latches which enter into pockets, two levers are here used which are moved on their centres by means of springs when a rope happens to break, their sharp, claw-shaped ends being at the same time forced into the wooden guides. A safety-catch of this kind, as built by *Borgsmüller*, is shown in Fig. 110. The two chisel-pointed catch-levers A, movable about the fixed point B in the framework of the cage, are ordinarily held at a distance away from the guide-beams L by the pull in the chains K, which connect the cage with the rope S, the springs F being at the same time kept in a state of tension. In case of a breakage of the rope, the springs cause the lever A to turn on B, thus allowing the claws H to enter into the guide-beams L. As soon as the claws have taken hold, the weight of the cage, which will then be thrown on the pins B, causes them to penetrate still farther into the woodwork. The lever G serves as a means of throwing the apparatus into action by hand, when the machine is used for transport of people. Many modifications of this arrangement have been brought into practice, as, for instance, shaping the lever ends like forks so as to make them enter the guide-beams at the sides in place of in the centre ; rubber springs have also been tried instead of steel springs, but have not proved quite reliable.

The apparatus shown in Fig. 111, of *White* and *Grant's* design, has been largely used in practice. The locking action

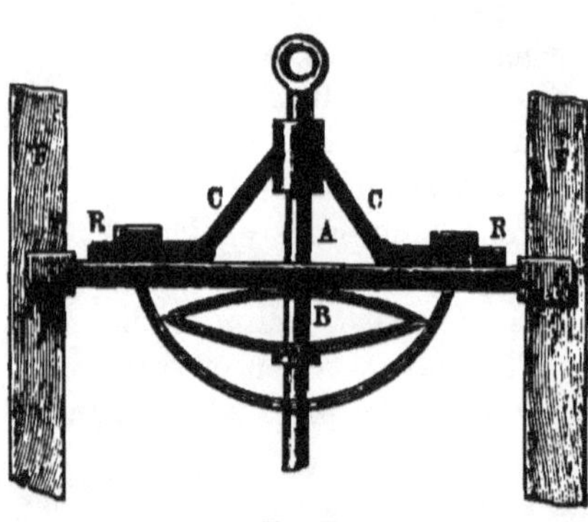

Fig. 109.

is here accomplished by means of two eccentrics A, provided
with pointed teeth. These eccentrics are secured to the ends
of two shafts B, which in the centre carry the pulleys R to
which the chains K are fastened. By the pull acting in these
chains under ordinary conditions, such a position is given to
the shafts B, that the eccentrics A do not touch the guide-

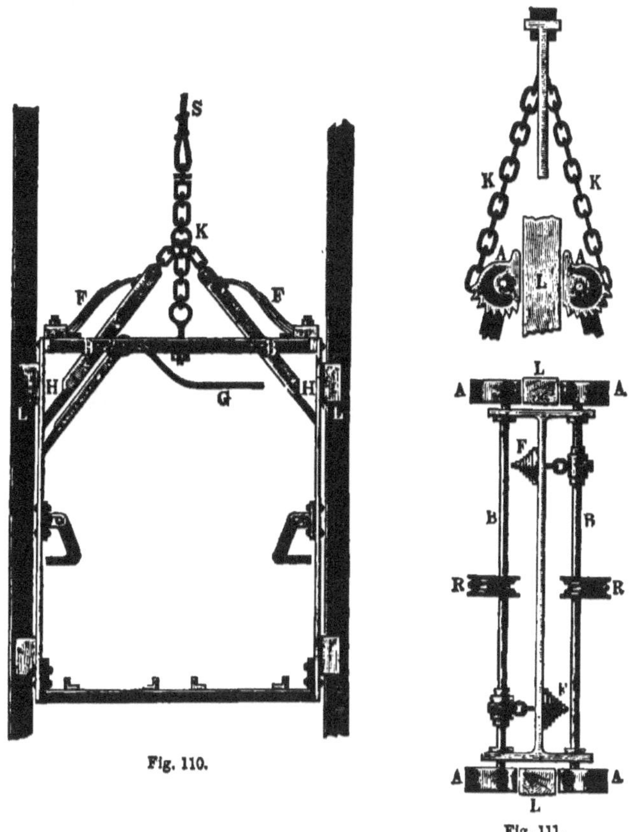

Fig. 110.

Fig. 111.

beams, while the springs F are at the same time kept in tension.
When a rope breaks, on the other hand, the springs cause the
shafts to revolve and thus bring the points of the eccentrics
into action.

The safety-catch just described has proved perfectly reliable
on several occasions, and the principle has been carried out in

various forms.   One of the most interesting designs is un-doubtedly *Hohendahl's* arrangement, Fig. 112, the elastic action of air enclosed in a cylinder C being here used in place of springs for operating the eccentric shafts.   By two brackets or standards E, a cylinder C, open at the bottom, is secured to the cage D, the piston F, through the cross-pieces G and H, and the bars J, being suspended from the ring N, to which the rope

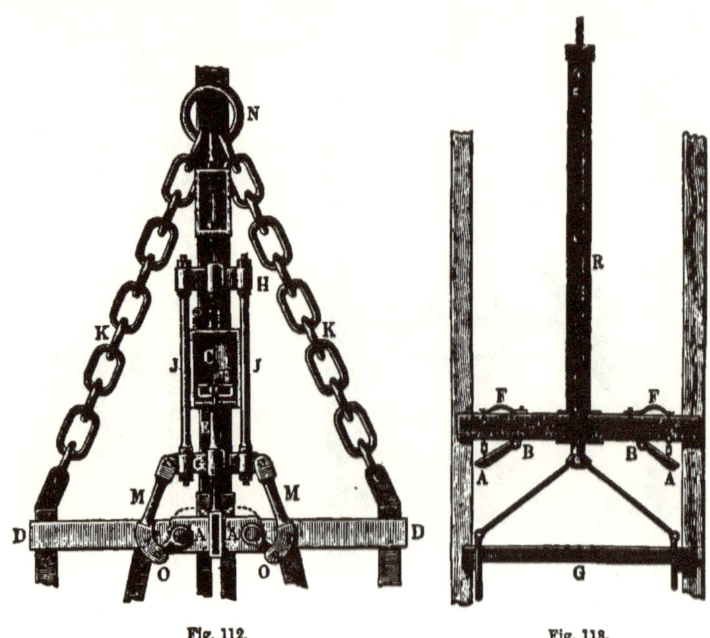

Fig. 112.   Fig. 113.

and chains K are attached.   The lower cross-piece is besides connected to the eccentric shaft by the links M and the levers O.   The cylinder is originally filled with air of atmospheric pressure, but before the cage itself is lifted, the piston F is pulled upward, and by being thus forced into the cylinder, a certain distance raises the pressure of the air to about five atmospheres.   The eccentric shafts A are at the same time turned sufficiently by the action of the links M to allow the guide-beams to pass freely between the eccentrics.   Should the rope happen to break, the enclosed air will operate in the same manner as a spring, and by pushing down the piston bring the eccentrics into action. This apparatus is considered very satisfactory in its operation.

The resistance of the air to the cage in its fall has also been utilised for causing the eccentrics to lock themselves into the guide-beams. In the apparatus constructed by *Krauss* and *Kley*,[1] the eccentric shafts are for this purpose provided with parachutes made of sheet iron, which act retarding on the eccentric shaft, and thus bring about the turning of the eccentric, on account of the greater acceleration of the lower part of the cage.

For throwing the catches into operation a peculiar method has been employed by *Lohmann*. It is based on the principle that a falling body, whose weight is only employed for accelerating its own mass, is unable to exert pressure against surrounding objects. In Fig. 113, A and B are two movable catches connected to the springs F. The force exerted by the springs is only equal to one-half of the weight of the catches A which they carry, so that under ordinary conditions the latter will hang sufficiently low down to clear the guide-beams. At the instant a rope breaks, however, and the cage begins to fall, the weight of the catches A will no longer oppose the tension in the springs, and as a result the latter will cause the catches to enter the guide-beams. The weight of the retarded cage will assist in pressing them still farther into the wood.

The figure also shows an attachment contrived by *v. Sparre* for the purpose of rendering harmless the shock caused by the momentum of the cage when the catches take hold, and which frequently causes the destruction of the whole safety apparatus. To avoid or mitigate the shock, the catches are not attached directly to the cage G, being instead connected to a special girder or framework Q, to the top of which a tube R, 2 or $2\frac{1}{2}$ metres [6·56 or 8·2 feet] in length, is secured. A turned rod S, connected at the lower end with the cage G, and at the top end to the rope, passes through the stuffing box at the bottom of the tube, and has secured to it a piston fitting tight in the latter. The tube is partly filled with air and partly with some soft material, such as sea-grass or horse-hair. When the safety apparatus is thrown into action, the shock produced will be due only to the slight momentum of the frame Q and the tube

---

[1] See *Zeitschr. deutsch. Ing.* 1869, p. 499.

R, the cage being allowed to fall through a distance equal to the length of the latter.   In the meantime the air in the tube is compressed, and together with the elastic material acts like a bunter, which absorbs the shock.

The most satisfactory apparatus, however, for avoiding the hurtful influences of the shock, are the *safety-brakes*, of which, in conclusion of this subject, we will here describe one, as constructed by *Hoppe*,[1] and illustrated in Fig. 114.   Two hardened

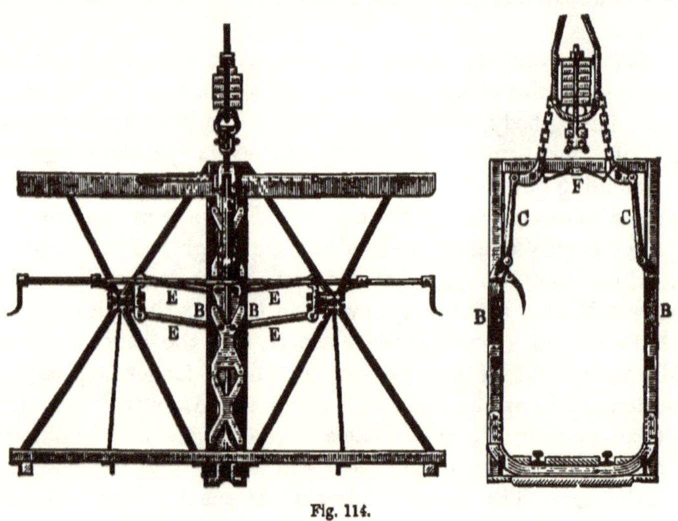

Fig. 114.

brake blocks B on each side are here used for producing the braking action by being pressed against the guides A, made of T-irons.   The pressure is brought about by the torsion spring F, which is thrown into action when the rope breaks, and causes a slight movement of the brake blocks B, the latter being jammed against the guides A by the action of the arms E arranged in the manner of a toggle-joint.   The frictional resistance thus produced gradually brings the falling mass to a stop, and it is only necessary to regulate the retarding action by means of an adjusting wedge, in order to prevent the pressure exerted on the brake blocks from exceeding a given limit, and thus avoid the danger of too severe strains in the framework of the cage.

Zeitschr. deutsch. Ing. 1870, p. 619.

In all safety apparatus it is also necessary to pay due attention to the falling of the portion of the rope which leads from the cage to the point where the break occurs; after the cage has been arrested in its motion, this piece drops on it, and when of great length causes a severe shock, which must be guarded against by providing a strong roof over the cage.

It may here be noted that all safety apparatus, whose action is dependent on a previously occurring break in the rope, furnish no security in such cases as when, for instance, the fall of the cage is caused by a hoisting drum getting loose on its shaft. A brake would then only be active when applied at the rim of a brake wheel attached to the drum itself.

For a study of such other safety apparatus which are employed in hoisting machines to prevent various accidents, as, for instance, the cage being hoisted *over* the rope-pulleys, or striking violently against the bottom of the shaft, and likewise for the construction of controlling apparatus which keeps account of the relative positions of the cages, we refer to special works on mining machinery.

§ 30. **Man-Engines.**—This name is applied to mechanical lifting apparatus for raising or lowering miners in shafts, in order to avoid the waste of time incident upon the use of ladders in the ascent and descent. The first arrangement of this kind was constructed by *Dörrel* in 1833 for the *Spiegelthal* mine in the *Harz*, on the suggestion of *Albert* in *Clausthal*. Since that time a great number of man-engines after *Dorrel's* pattern have been brought into use in Cornwall, Belgium, Westphalia, etc. The apparatus consists principally of one or two vertical rods or timbers which are given a reciprocating motion in the shaft either by steam or water power. We thus distinguish between *single* and *double*-acting man-engines. Rigid pine rods were mostly used in the earlier constructions, while nowadays wire rope is frequently used for the same purpose. The rods are at fixed intervals provided with platforms or steps for the men to stand on, and hand-hooks to take hold of. If a man steps from a stationary staging B to one of the platforms T_1 in its highest position, he will be carried down a distance h equal to the stroke of the rod. If now in this lowest position of the rod a staging B_1 is placed on a level with the platform on which he stands, the workman can step from the movable platform to the fixed one, and has thus, in a single stroke of the rod, been transported downward a distance equal to the lift $h = BB_1$.   He now waits until the up stroke of the rod has been completed, and then steps on to a second platform T_2 located on the rod at a distance h below the first one, and which by this time has arrived on a level with B_1.   In the following down stroke the traveller will be transported another distance h downward, and will at this point be able to step over to the next stationary platform B_2.   By continually repeating the above proceeding, the workman will thus descend from platform to platform, the distance travelled for each double stroke of the rod being equal to the lift h.   It is evident that the same arrangement can also be used for ascending if the stepping is done from the fixed to the movable platform at the moment when the rod is in its *lowest* position. It is readily seen that in a *single*-acting apparatus of this kind, the intervals at which both the fixed and the movable platforms are located must be exactly equal to the lift h, and that $n + 1$ resting places are required, when n is the number of movable platforms communicating between them.   As all these n platforms may be occupied at once, it is also evident that n persons can travel *either up or down* at the same time.   The rod may be utilised for travel both up *and* down at the same time, on different levels, but not to advantage in the same sections, as the time at the traveller's disposal at the turning points hardly admits of a transfer of two meeting parties.

In the *double*-acting man-engines two rods of equal strokes are placed side by side and arranged to move in opposite directions, the reversal of their motion taking place at the same instant for both rods.   If, therefore, the machine is so arranged that two platforms will always be alongside each other on the same level at the turning points, it is easily seen that a person can by successively stepping from one rod to the other, travel up or down according as the stepping is done to the ascending or descending rod.   The relative location of the platforms is given by Fig. 115.   Let A and B represent the two rods, the former being shown in its highest and the latter in its lowest position.   After stepping from the fixed point C to the platform T_1, the person is carried down a distance h to 1.   Passing in this position from A to B, he will in the next down stroke be transported from 1 to P_1, which will then be opposite to the platform T_2 on the rod A, which platform accordingly must be located at a distance $2h$ below T_1. As the same reasoning applies to all remaining steps, it is apparent that all platforms on each rod must be placed at intervals of $2h$, and that the traveller will be transported this distance $2h$ for each double stroke of the engine. Letting l denote the total lift CD between the fixed points C and D, the number of times a passing from one rod to the other must take place is given by $\dfrac{l}{2h} = n$. This quantity also represents the number of people who can ascend or descend at the same time, for the reason that a meeting of two persons· at the same platforms would not be practicable, and consequently it is only possible to occupy all platforms on one rod or the other at the same moment. Arranged in this manner the rods may, of course, be utilised for both ascent and descent at once at different

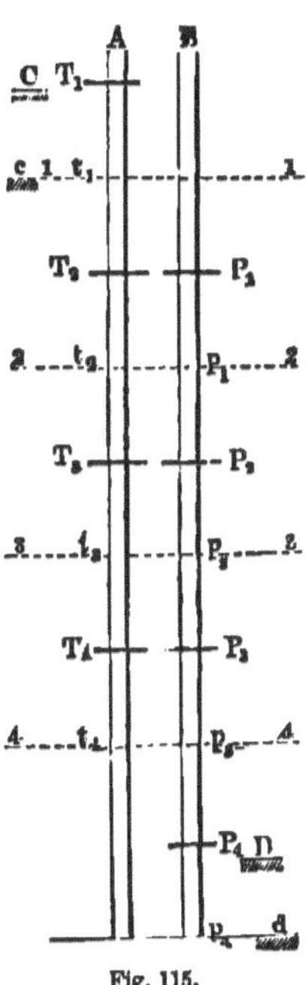

Fig. 115.

levels, but cannot be made to serve both purposes at the same time and place.

If, however, a second set of steps t and p are introduced at intervals of $2h$, and placed half-way between the platforms T and P, then a second double-acting lift will be obtained, whose platforms will correspond to each other only, but not to T and P, and which can therefore be used independent of the latter for either ascent or descent. This set of steps will then maintain the communication between the fixed points c and d, located at a distance h below C and D. The arrangement just described is employed in order to enable the machine to be used for transport both up and down at the same time. When utilised in this manner by a full complement of men travelling in both directions, it is possible to keep both rods equally loaded, which cannot be done when travelling takes place in one direction only.

An interesting construction of the man-engine is presented by that built at the mine Saar Longchamps,[1] which, as seen

by Fig. 116, consists of a combination of one double-acting and two single-acting engines. Here A and B are the two rods carrying at a and b, at intervals of $2h$, the platforms for the double-acting engine, which serves for the ascent, while the platforms c and d placed at intervals equal to h, belong to the single-acting engines which are used for descending into the mine, the fixed resting places being shown at e and f. An ordinary ladder is frequently arranged between the two rods, as is the practice in several mines in the Harz mountains, in order to provide means of leaving or returning to the man-engine at any point.

Fig. 116.

Motion is usually given to the rods by means of two bell-cranks A and B, Fig. 117, coupled together by a rod CD, and receiving their reciprocating motion from a crank shaft, which

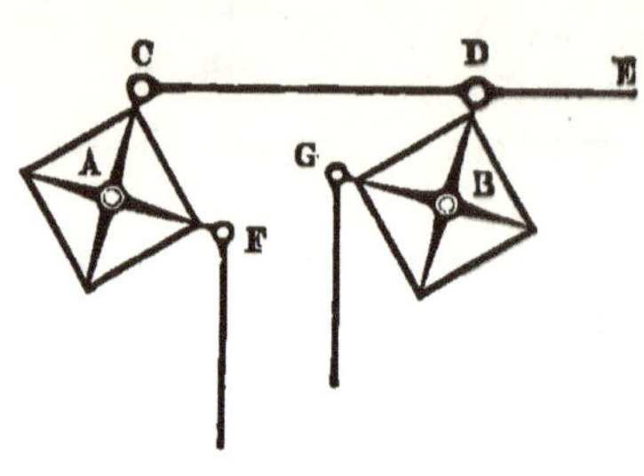

Fig. 117.

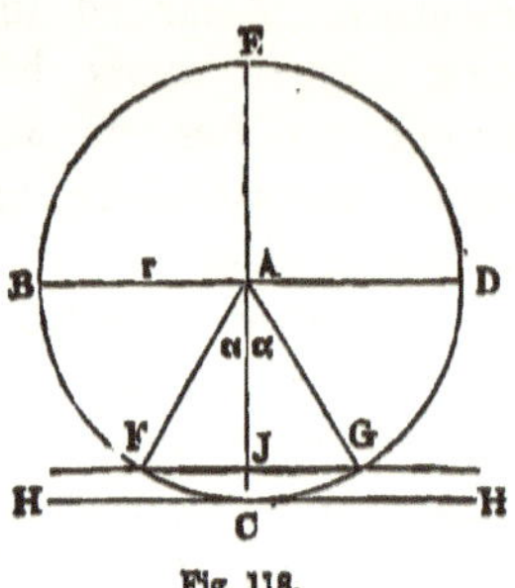

Fig. 118.

operates the connecting rod DE. Although this arrangement, which is the usual method of driving when the prime mover is a water-wheel, does not admit of any period of rest at the turning points, the exchange of platforms is nevertheless connected with no danger, since the velocity of the rods near the dead centres of the crank is very slight.

A clear idea of the reasons for this statement may be had

[1] Serlo, *Bergbaukunde.*

by an inspection of the path of the crank-pin. Let BCDE, Fig. 118, represent the path of the crank-pin, and let the tangent HH at the lower dead centre C be a stationary platform belonging to a single-acting man-engine, then the maximum difference in height between the fixed and movable platforms during the time required for the crank to pass through the angle $FAG = 2a$ will be $CJ = \delta = r(1 - \cos a)$.

As long as this value δ gives rise to no inconvenience in stepping from one platform to the other, this can be done in perfect safety. For, if we assume that the stroke $h = 2r = 3$ m. [9·84 ft.], and that the crank makes four revolutions per minute, then, allowing a time of two seconds for the stepping over, we shall have the corresponding angle of rotation of the crank

$$2a = 2 \times \frac{4 \times 360°}{60} = 48°,$$

and thus the maximum difference in height for this period,

$$\delta = 1·5(1 - \cos 24°) = 0·130 \text{ m. } [5·12 \text{ in.}]$$

By so placing the stationary platforms that their positions correspond to the mean between J and C, the difference in level could be reduced to one-half of the above amount, but in this case the lift for every revolution of the crank would also be reduced, inasmuch as the fixed and movable platforms would then have to be located at intervals of $2\left(r - \dfrac{\delta}{2}\right)$, which would also be the distance travelled for every upward stroke. From Fig. 119 it is evident that for the same stroke and velocity of the crank shaft, in the case of a double-acting man-engine, the maximum difference in height between two opposite platforms, corresponding to an angle of rotation $FAG = 2a$, will be twice as great, that is

$$JJ_1 = \delta = 2r(1 - \cos a),$$

or in the instance above cited 0·26 m. [10·24 in.]

The rods are sometimes driven directly by steam pistons, either by using one double-acting or two single-acting steam-engines. It is then a matter of importance, when the lift is double-acting, to see that the motion of the two rods exactly conform to each other. To attain this object a so-called *hydraulic* regulator is commonly used. This contrivance consists essentially of two cylinders A and B (Fig. 120) open at the top and filled with water, the rods G and H being connected to their pistons K and L. The cylinders communicate at the bottom by the pipe CD, which allows the water from the cylinder B to pass over into A when the plunger L descends by the action of the steam - engine, thus compelling the plunger K to rise, and *vice versa.* The cylinders are also in communication at the top through EF, in order to keep the plungers always covered with water, and thus obtain a tight fit in a simple manner. When the rods are driven directly by a steam-engine, it is evidently possible to interrupt the reciprocating motion at the end of each stroke for a suitable length of time by the use of *cataracts* (see volume on Steam-engines).

In the earlier man-engines employed in the *Harz*, the stroke of the rods was only 1·25 metres [4·1 feet], while in modern constructions it is made 3 m. [9·84 ft.] and more. The number of double strokes can be assumed at 5 per minute, which for a stroke of 3 m. [9·84 ft.] gives an average velocity of 30 m. [98·4 ft.]; when the stroke is longer, a mean velocity of 48 m. [157·48 ft.] is occasionally reached. When the apparatus is driven from a crank shaft, the greatest velocity is naturally obtained in the middle positions, where it may be calculated to be $5 \times 3 \times \pi = 47\cdot1$ m. [154·53 ft.] per minute, for the case that the stroke is 3 m. [9·84 ft.], and that the machine makes 5 double strokes per minute.

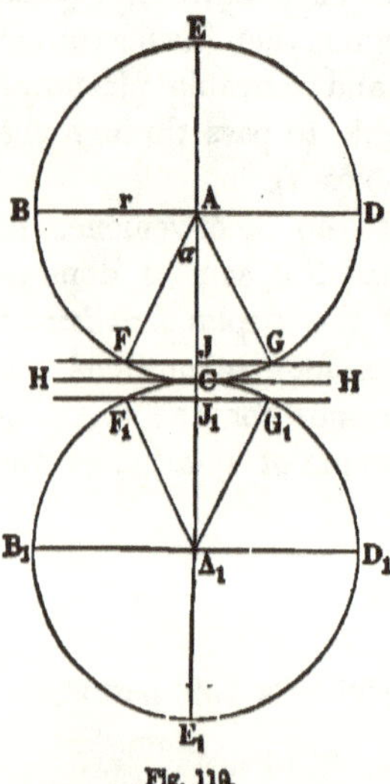

Fig. 119.

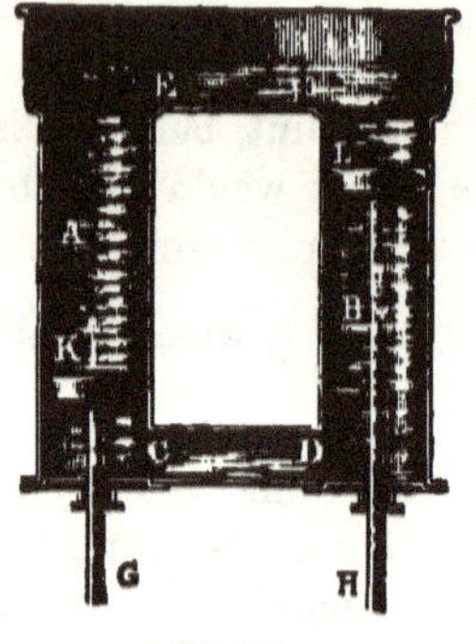

Fig. 120.

From the assumed velocity of the rods and their load in ascending, the driving power required may be obtained in conformity to known rules, if due attention be paid to the frictional resistances between the rods and their guide-rollers. It may here be noted that the rod of a single-acting man-engine is only required to overcome the resisting forces in the ascent, while during the descent, the weight of the occupants tends to accelerate the cranks, which necessitates the application of a powerful brake for this contingency. The weight of the rod is balanced by means of a counter-weight, in the case of single-acting machines, while in the double-acting apparatus the rods balance each other. Perfect balance is also attained in the latter case when both rods are equally occupied, which is the condition at hand when the machine is used for ascent and descent at the same time. On all other occasions only one of the rods is loaded, namely, the ascending one when the travelling is done upwards, and the descending one in downward travelling.

In order to investigate the performance of a man-engine, let h denote the length of stroke, n the number of double strokes or single revolutions of the crank per minute, and l the distance between the top and bottom landings. Then, in the case of a single-acting apparatus, the number of platforms will be

$$z_1 = \frac{l}{h},$$

and for a double-acting machine, on each rod

$$z_2 = \frac{l}{2h}.$$

For each double stroke the traveller is, in the former case, transported a distance h, and in the latter case a distance $2h$, and, consequently, the time required for a complete lifting operation is for the respective machines expressed by

$$t_1 = \frac{z_1}{n} = \frac{l}{nh} \text{ minutes,}$$

and

$$t_2 = \frac{z_2}{n} = \frac{l}{2nh} \text{ minutes.}$$

One person is thus transported twice as quickly on the double-acting as compared with the single-acting machine. If now we denote by N the number of persons to be forwarded, then, since only one person can mount the lift for each revolution of the crank, the platforms will be occupied at intervals of $\frac{1}{n}$ minutes, and thus the last person will mount the platform $\frac{N-1}{n}$ minutes later than the first one. The total duration of a complete hoisting or lowering operation will therefore be given by

$$T_1 = \frac{N-1}{n} + t_1 = \frac{N-1}{n} + \frac{l}{nh}$$

for the single-acting, and

$$T_2 = \frac{N-1}{n} + t_2 = \frac{N-1}{n} + \frac{l}{2nh}$$

for the double-acting man-engine.

From these formulæ it is evident that the greater the number of persons to be forwarded, the more the greater despatch otherwise obtained from the double-acting machine is placed in the background, and for this reason, when a large number of miners are to be transported, it may sometimes be more advantageous to employ two single-acting in place of one double-acting apparatus, as indicated by the example shown in Fig. 116. For this case the total duration of a complete hoisting operation would be only

$$T_1' = \frac{\frac{1}{2}N - 1}{n} + \frac{l}{nh}.$$

Taking, for instance, a working gang of $N = 500$ men and a travelling depth of $l = 300$ metres [984·27 ft.], then, for $n = 5$ revolutions per minute and a stroke of $h = 3$ m. [9·84 ft.], we obtain

$$T_1 = \frac{499}{5} + \frac{300}{5 \times 3} = 99 \cdot 8 + 20 \backsim 120 \text{ min. for a single-acting,}$$

$$T_2 = \frac{499}{5} + \frac{300}{10 \times 3} = 99 \cdot 8 + 10 \backsim 110 \text{ min. for a double-acting, and}$$

[machines.

$$T_1' = \frac{249}{5} + \frac{300}{5 \times 3} = 4 \cdot 8 + 20 \backsim 70 \text{ min. for two single-acting}$$

If we desire to compare these results with the time required when ordinary ladders are made use of, we can assume, on the authority of *Serlo*,[1] a mean velocity for each person of 8 m. [26·25 ft.] per minute in descending, and 4 m. [13·12 ft.] in ascending. Now supposing the ladders to be 6 m. [19·68 ft.] in length, and that three men are always occupying each ladder, that is at distances of 2 m. [6·56 ft.] apart, then each individual will mount the first ladder at intervals of

$$\tfrac{2}{8} = \tfrac{1}{4} \text{ minute, in descending, and}$$
$$\tfrac{2}{4} = \tfrac{1}{2} \text{ minute, in ascending.}$$

The time required for descending will therefore be

$$T_e = 499 \times \frac{1}{4} + \frac{300}{8} = 162\cdot25 \text{ minutes,}$$

and for ascending

$$T_a = 499 \times \frac{1}{2} + \frac{300}{4} = 324\cdot5 \text{ minutes.}$$

Besides taking into consideration the great fatigue incident upon the latter mode of travelling, the great advantage in the use of man-engines is evident, especially when large gangs of men are employed, and the shafts are of great depth.

When a hoisting engine is used for the purpose of raising and lowering the miners, as is the case when no man-engine is at hand, then, taking a hoisting velocity of 4 m. (13·12 ft.) per second, the time required for each lift would be $\frac{300}{4} = 75$ seconds $= 1\cdot25$ minutes. Assuming the stop required for embarking and disembarking to be 1 minute, and that there is room for 5 men in the cage, then, under the above conditions, the time elapsing before the whole working gang has been forwarded will be $T = \frac{500}{5} \times 2\cdot25 = 225$ minutes, which is considerably more than would be the case were a man-engine employed.

Serlo, Bergbaukunde, vol. ii. p. 214.

CHAPTER VI

CRANES AND SHEERS

§ 31. **Cranes.**—A crane is a hoist with facilities for moving
the load horizontally. Cranes are employed chiefly in warehouses, workshops, dockyards, etc., and, for lifting the loads

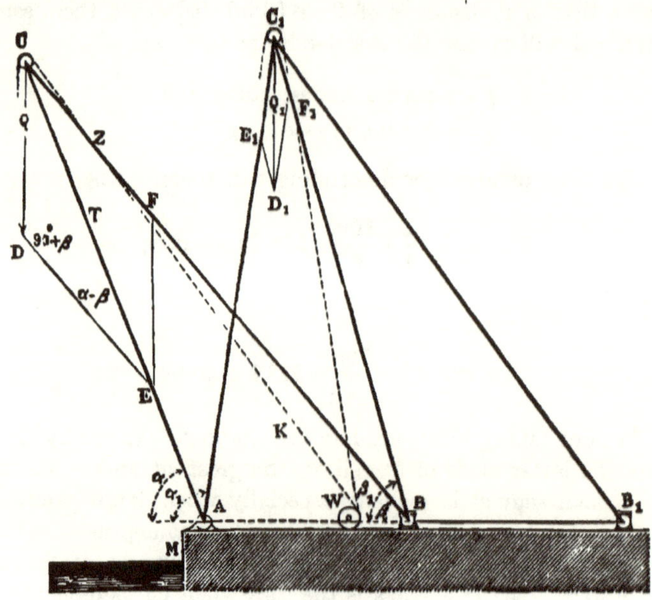

Fig. 121.

are provided with a windlass, the arrangement of which does
not differ from those already described. Various means are
employed for giving a horizontal motion to the load. In the
ordinary *rotary* crane the framework is made in the shape of
a long projecting arm, inclined or horizontal, and may be

rotated about a vertical axis; the long arm or *jib* is provided with a pulley at the extreme end, over which is carried from the windlass the rope or chain which supports the load.

For masting or dismantling ships, lifting the heavier parts of marine engines, etc., large *sheers*, Fig. 121, are made use of, having a triangular jib ABC hinged at the vertices, and carrying a pulley for the chain at the apex C, while the feet A and B rest on the foundation. After the load has been hoisted, it may be conveyed from Q to Q_1, either by moving the foot B to B_1, or, keeping B fixed, by shortening the side BC until it has a length BC_1. It is evident, of course, that the strut AC must be composed of two inclined pieces so as to allow the load to pass between them.

Sometimes the rotary crane is mounted on a four-wheeled carriage travelling on rails, and is then called a *portable* crane; this construction is chiefly employed for railways, shipyards, and wharves. If, instead of being mounted on a carriage, it is supported by a float, it is known as a *floating* crane.

In so-called *foundry cranes* the load is suspended from a small carriage or trolley which travels in and out on a track placed on the horizontal jib; by turning the crane and moving the trolley the load can be conveyed to any point within the circle described by the outer end of the jib, while in an ordinary swing crane, without trolley, the horizontal motion of the load is limited to the circumference described by the extremity of the jib.

Finally, we have the various kinds of *travelling cranes*, in which the windlass is arranged in the form of a truck travelling upon a bridge; the latter is mounted on wheels, so that the whole may traverse a fixed railway placed at right angles to the bridge; with this arrangement, the suspended load may be conveyed to any point within the rectangular area comprised between the rails. Such cranes are especially adapted for use in machine shops, and for conveniently distributing the materials in the erection of large and massive bridges. Where continuous service is not desired, cranes are operated by hand, whereas steam power is made use of in cases where a frequent or uninterrupted service is called for, as in warehouses, for instance, and in the erection of buildings. The crane may then have a steam-engine attached to it, in which case it is known as a *steam crane*, or it may receive power from a stationary engine which is also used for other purposes.

Recently *hydraulic cranes* driven by *accumulators* have been employed to advantage; this arrangement is well suited to cases where the power is used intermittently, and where a simple means of transmitting power is needed. Such is the

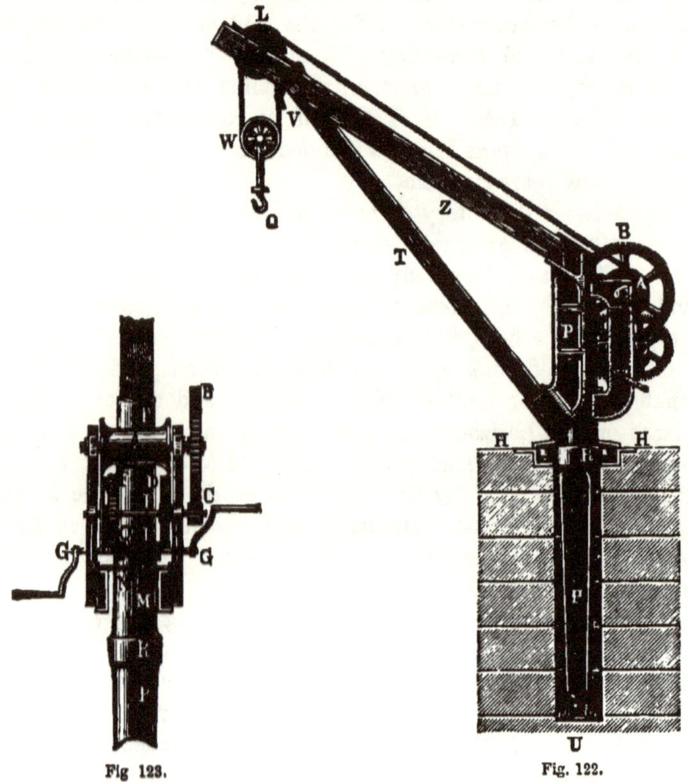

Fig 123.   Fig. 122.

case at docks and other places, where a number of cranes at various points are operated by the same engine. The size of a crane and its capacity for hoisting must, of course, be adapted to the work in hand. The heaviest designs are employed for marine purposes, cranes having been constructed for lifting 100 tons and more.

§ 32. **Rotary Cranes.**—In Figs. 122 and 123 is shown a crane made of wood and iron, and designed by *Cavé* for the harbour at Brest. The cast-iron column P is stepped into a bearing U at the bottom of a well, and is held upright at R by a roller bearing formed within the cast-iron frame H. The jib is composed of a wooden strut T and a tie Z firmly secured to the crane-post P. A pulley L fixed to the outer end receives the rope which hangs down in a bight, and supports the running block W from which the load Q is suspended. The manner in which the rope is wound on the barrel A is obvious from the figure, which further shows how the motion of the winch handle is imparted to the barrel through the medium of two or three pairs of gears.

Referring to Fig. 123, let us imagine the crank shaft shifted to the right; the motion of the pinion K will then be transmitted to the large wheel D on the intermediate shaft, and through the wheels C and B the speed of the winding barrel will be further reduced. By shifting the crank shaft to the left, the pinion O is made to

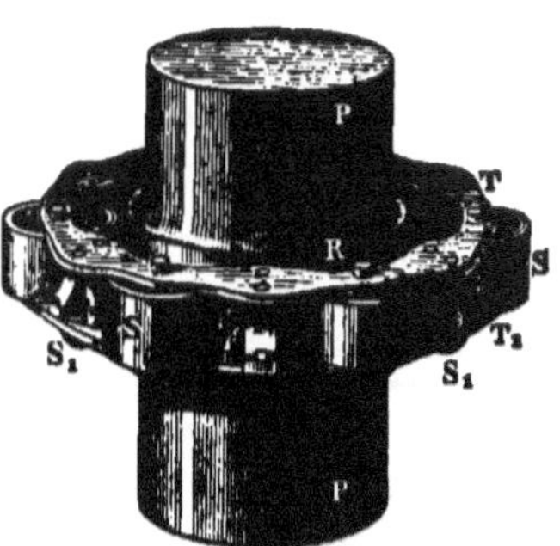

Fig. 124.

engage with the large wheel F on a second intermediate shaft, provided with a pinion E which transmits its motion to D, thus operating the first intermediate shaft. In the figure both wheels O and K are out of gear, which is the case when the load is to be lowered; this requires the application of a brake to a pulley attached to the wheel F, as already explained. In order to arrest the motion of the crank shaft in any desired position, a catch pivoted at N is dropped into one of the three recesses cut in the shaft, and is kept in place by a weight M.

The *bearing for the crane-post* is shown in detail in Fig. 124. Six friction rollers S, working on pins which are fitted into the flat rings T, form a *collar* which embraces the neck R of the crane post P. The action of the load at the end of the jib is to bring a pressure upon one or two of these rollers, which, in turn, is taken up by the inner surface of the annular case H, Fig. 122. Thus, by means of the rollers S, the sliding friction at the bearing R is transformed into the smaller rolling friction. Since a rotation, under these circumstances, is also imparted to the collar, the sliding friction of the lower plate T_1 upon its support is likewise transformed into rolling friction by employing a second set of rollers S_1 working on horizontal pins. The object of this arrangement is to facilitate the swinging of the crane jib, although this operation is rarely effected by the direct application of a lever; a special mechanism is

Fig. 125.

generally required for this purpose, the arrangement of which will be described in the following.

In order to overcome the inconveniences attending the use of a well for the crane-post, and at the same time to gain easy access to the footstep bearing, the construction shown in Fig. 125 of the framework of the crane is frequently used. The hollow cast-iron column S is firmly secured to the plate G, which is bedded in the foundation, and held down by long anchor bolts A. The top of the post is provided with a steel pivot u, from which the movable framework is suspended by means of a cross-piece V, while the collar R on the fixed crane-post serves as a bearing for the friction rollers s, carried by the movable framework. The latter consists of two side frames W which support the windlass; these frames are connected above by the cross-piece V, and below by cross-pieces

Fig. 126.

in which the friction rollers have their bearings. The jib is a wrought-iron tube T pivoted at H, while two wrought-iron ties connect the end L with the cross-piece V. The windlass, consisting of the barrel B, the gear shaft C, and the crank K, is of the ordinary arrangement. The roller J between the tension rods Z support the hoisting chain.

In order to swing the crane conveniently a spur-wheel N is fixed to the pillar S, and gears with a pinion n on a shaft which has its bearings in the movable framework, and is driven by the crank handle *k* through the medium of the bevel wheels *a* and *b*.

In Fig. 126 is represented *Fairbairn's* tubular crane. It is mainly distinguished from the preceding types by the form of the jib T, which is made of wrought-iron plates and angle-irons riveted together and arranged so as to give a body of uniform strength. The jib is supported upon the crane-post

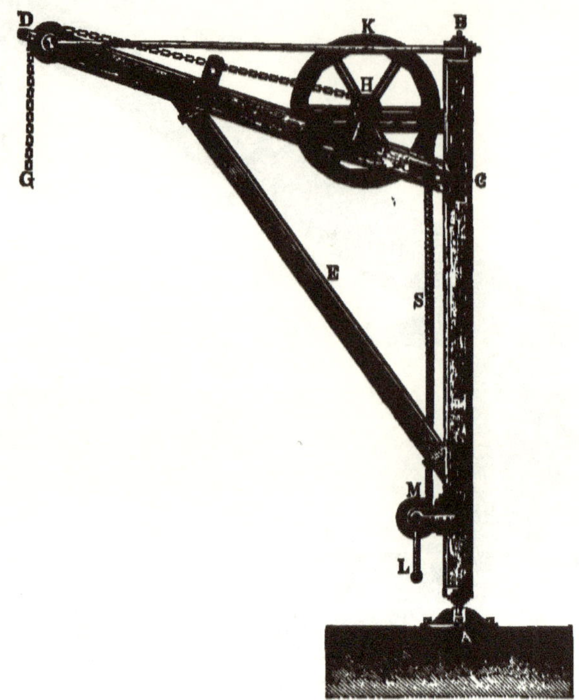

Fig. 127.

by means of the cross-piece V riveted to it, and is guided at the base by friction rollers. For turning the crane the cast-iron foundation plate is fitted with a rim N having internal teeth.

When a crane is stationed within a building, and is not intended for out-of-door use the revolving crane-post may be supported at the top by a joist or beam. In this case the journals may be considerably reduced in size, and the roller bearing dispensed with. A simple arrangement of this kind used on English railroads for loading and unloading goods is shown in Fig. 127. The wooden jib DC is supported by the strut E, and by two iron tie-rods DB. The motion of the crank L is imparted by the rope S to the winding barrel H, upon which the hoisting chain G is coiled. The rope is attached to the barrel M, and is then wound upon the large pulley K, to which it is likewise fastened. The motion of the descending load is controlled by employing coil friction, which is produced by the slipping of a rope about a pulley attached to the barrel M.

Occasionally cranes are secured to the walls of buildings, for instance in warehouses, where the jib-head moves in and out through a door or opening in the wall. The arrangement shown in Fig. 128 may be so applied.

A peculiarity of this apparatus is the manner of operating the hoisting chain which passes over the pulley G and another pulley at the jib-head. The motion is effected by a screw NM, the nut K being connected to the end of the chain by means of two rods HK.

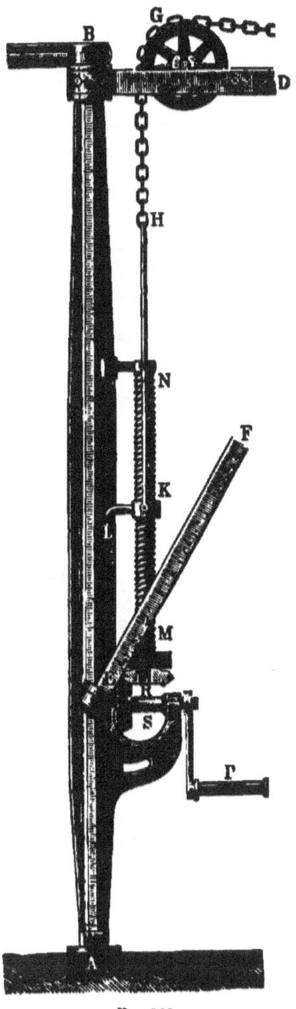

Fig. 128.

The range of lift is limited, and the efficiency, as in all screw-hoists, is small.

Cranes may be arranged so as to weigh the load while it is being lifted. This is the object of the construction in Fig.

129, which depends upon the principle of *Georges'* platform scales.   Here CDE, the jib proper, is connected with the movable crane-post AB by two pairs of iron links PQ and RS, which prevent the crane-jib from overturning; the latter including the suspended load, is supported upon the beam UC at C.   The upper pair of links PQ pull on two knife edges, while the lower pair RS press against two similar edges.   The

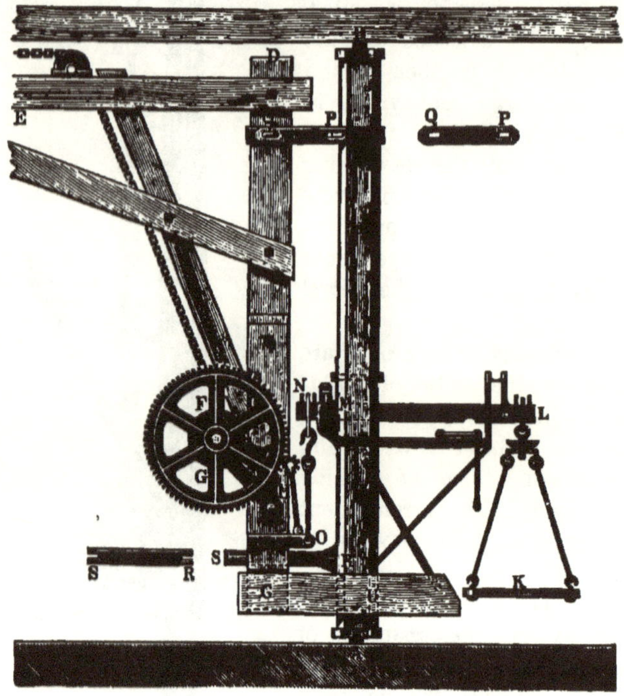

Fig. 129.

mode of weighing the load, including jib, by weights placed on the scales at K, needs no further explanation, and if the ratio ML to MN of the lever arms be 10, a weight of $\frac{1}{10}$ of the load on the scales will suffice for weighing.

Rotary cranes for foundry use, Figs. 130 and 131, are provided with means of altering the distance of load from the crane-post.   In Fig. 130 the load Q is suspended from the lower pulley-block N of a fourfold tackle, the upper block being arranged in the form of a small carriage DF, which travels in and out on the horizontal jib. The latter is made of two parallel beams between which the chain hangs. The manner of operating the latter by means of the windlass PO is evident from the figure. A rope S passing over the fixed pulleys G, H, J, and coiled round a barrel L, is employed for moving the carriage or trolley DF conveniently from below, rotation being imparted to the barrel by means of a crank M and a pair of gears. Since both ends of the rope are attached to the carriage, it follows that by turning the barrel L to the right or left the load will travel toward or away from the crane-post.

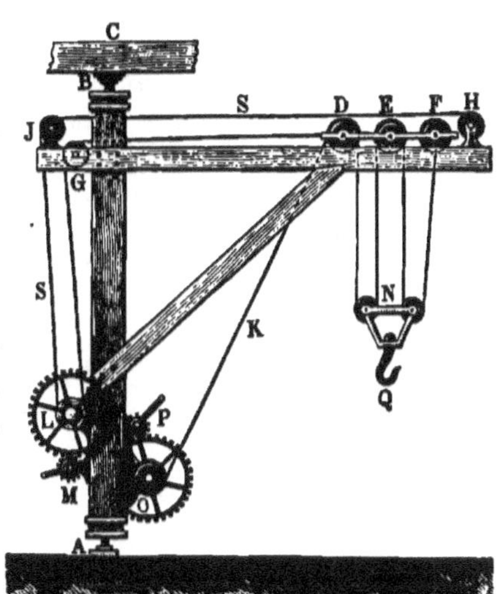

Fig. 130.

In the crane, Fig. 131, the travelling motion along the jib is obtained by means of a pinion on the shaft G, which engages a rack EF, to which the pulley ED is attached. The manner of transmitting

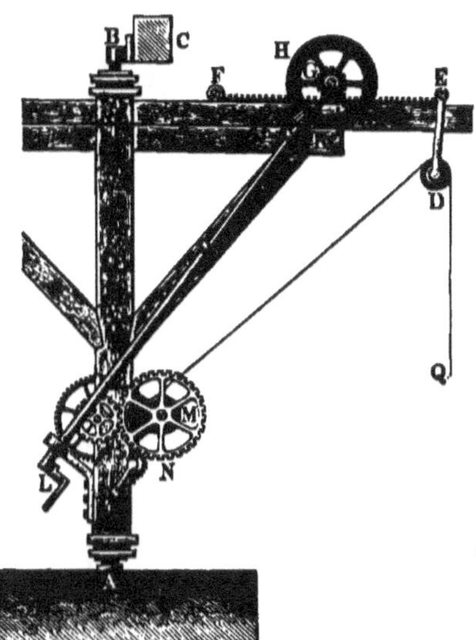

Fig. 131.

the rotation of the crank L to this pinion by interposing the bevel wheels K and H is evident from the figure.

A defect of the two last-mentioned cranes is the alteration
in the height of the load as the trolley moves along the jib,
inasmuch as the arrangement of the hoisting chain causes the
load Q to descend as it approaches the crane-post, and to rise
as it recedes from it.   This variation of height is considerable
in the crane shown in Fig 131, but is less in that given in
Fig. 130 on account of the great purchase of the tackle.   Fig.
132 shows an arrangement for obviating this defect by attach-

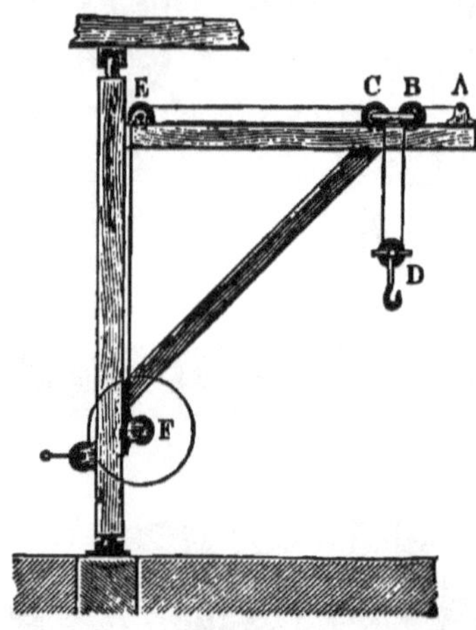

ing the chain at A, and
allowing it to pass over the
pulleys B, D, C and E, thus
making the length of the
chain constant between F
and A as the load travels
along the jib.

§ 33. **Condition of Equilibrium of Rotary Cranes.**—
For the purposes of this
investigation let ABC, Fig.
133, represent the triangle
formed by the axes of the
crane-post, strut, and tie-rods,
and let a, β, and β_1 be the
respective inclinations to the

Fig. 132.

horizon of the strut AC, the tie-rods BC, and the hoisting chain K.   For the present the weight of the movable crane is neglected.   Further, let P denote the effort
applied to the chain to lift the load Q.   This force acting in
the chain is given by the arrangement, as when the load is
directly suspended from the chain, we may place $P = Q$, and
when a movable pulley is employed we may put $P = \dfrac{Q}{2}$, if
the wasteful resistances of the pulleys are neglected.   To
ascertain the thrust T in the strut and the tension Z in the
tie-rod, we may use the following equations expressing the
equilibrium of forces acting at C.

$$Q + P \sin \beta_1 + Z \sin \beta = T \sin a,$$

and

$$P \cos \beta_1 + Z \cos \beta = T \cos a,$$

from which we obtain

$$Z = \frac{Q \cos a - P \sin (a - \beta_1)}{\sin (a - \beta)},$$

and

$$T = \frac{Q \cos \beta - P \sin (\beta - \beta_1)}{\sin (a - \beta)},$$

or, assuming the ordinary case, for which $\beta = \beta_1$ we have

$$P + Z = Q \frac{\cos a}{\sin (a - \beta)},$$

and

$$T = Q \frac{\cos \beta}{\sin (a - \beta)}.$$

For this last case, that is, for $\beta = \beta_1$, let us make $C_1G = Q$ and $C_1F = P$, and let us resolve the resultant C_1D of Q and P

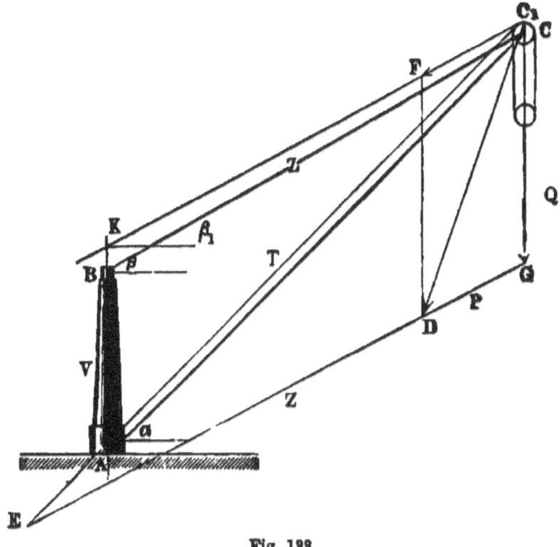

Fig. 133.

into the components acting along C_1E and EG, then the length C_1E will represent the thrust T, and the length EG the sum of the forces $Z + P$ acting along the tie-rods and chains; and since the vertical pressure V upon the crane-post BA is equal to the load, we have $Q = C_1G$. Hence we conclude that *the stresses along the individual pieces of the triangular frame of a crane are as the lengths* AC, BC, *and* AB *of these pieces.*

The action of the force Z is to tear the tie BC, while that of the force exerted along the axis of the strut is to crush it; consequently, in calculating the dimensions of the latter, it must be treated as a long column rounded at both ends, and corresponding to case II., vol. i. § 273, Fig. 437, Weisb. *Mech.* Additional stresses are produced in the rods and strut by the bending action of their own weight, and these must be taken into consideration in accordance with the rules laid down in

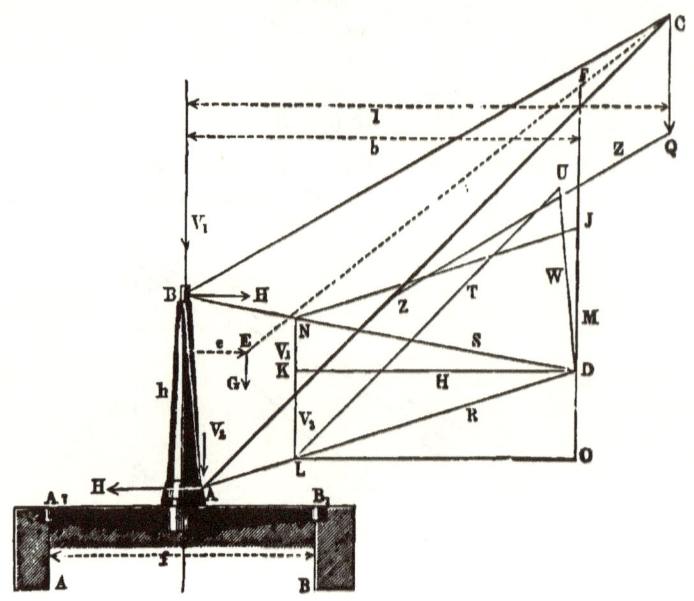

Fig. 134.

vol. i. § 278, Weisb. *Mech.*   The crane-post AB is subjected to a thrust Q, and also to the bending moment produced by the load Q and the weight of the structure combined.   In order to ascertain the stress to which the post is subjected, it is necessary to consider the weight G of the movable crane-jib, and the mode of arranging the bearings.   Let G be the weight of the rotating framework, acting at its centre of gravity E, Fig. 134, at a distance e from the pillar, let M denote the resultant of the weight G and the load Q, so that $M = Q + G$, and let FJ be the line of action of this resultant.   The jib bears against the crane-post at A, and AD normal to the acting surface at this point, may be taken to represent the direction [1] of the reaction of the pillar. As this reaction R, together with the reaction S of the pivot B, must keep the load G + Q at FJ in equilibrium, it follows that the resistance offered by the pivot B must be exerted in a direction traversing the point of intersection D of the forces M and R—that is, in the direction DB. Let us therefore resolve the resultant M = Q + G = JD into components acting along DA and DB, then DL will represent the thrust R of the strut at A, and ND the pull S on the pivot B. Furthermore, let these two forces be resolved into their horizontal and vertical components acting at A and B, then OL = KD will be the two horizontal forces H forming a couple acting at A and B which tends to break the post. Let h be the vertical distance between the points of application A and B, and let l, b, and e be the horizontal distances of the forces Q, M, and G from the axis of the crane-post, then the bending moment acting on the post is

$$Hh = Mb = Ql + Ge.$$

The vertical pressure on the post AB is expressed by

$$V = NK + KL = Q + G,$$

of which $NK = V_1$ is supported directly by the pivot B, and $KL = V_2$ by the conical bearing A. When the latter bearing has a cylindrical form, which makes AD horizontal, the whole pressure V = Q + G falls upon the pivot B; on the other hand, when the inclination of the conical bearing A is such as to bring the intersection D (of the reaction R, and the load Q + G) on a horizontal line passing through B, no part of the vertical load will fall on the pivot, nor on the crane-post between A and B. A further elevation of D would even give rise to an upward pull on B.

To ascertain the amount of pull W acting in the movable framework between A and B, we must resolve the reaction R = LD along the directions of the strut AC and connecting line AB, which gives UD = W as the required force in the frame, and LU as the thrust T in the strut. Also resolving

[1] Strictly speaking, the direction of the reaction makes an angle with the normal at A equal to the angle of friction.

the load $Q = CQ$ along the directions CA and CB, will give $ZQ = Z$ as the stress in the tension rods and chain.

To prevent the crane from overturning, the anchor bolts A_1 A_1 must be sufficiently strong to resist the force X deduced from the equation $(Q + G)b = Xf$, where f is the horizontal distance between the anchor bolts A_1 and B_1. The weight of the masonry in which the foundation bolts are secured must be proportional to this value of X, in order to give the crane the necessary stability.

As a rule, graphical methods give the simplest solution of crane problems, owing to the fact that the analytical formulas differ for each individual case, and generally are very inconvenient for use.

The steps for determining the effort P required at the crank for lifting the load Q are exactly the same as those followed in §§ 8 and 11.

Let η_1 denote the efficiency of the pulley in the jib head (the efficiency of the hoisting tackle, if such is used, being included in this symbol), let η_2 denote the efficiency of the drum, and η_3, η_4 . . . represent the efficiencies of the successive pairs of gears; then the efficiency of the entire machine is

$$\eta = \eta_1 \, \eta_2 \, \eta_3 \, \eta_4 \, \cdots$$

The resistances due to friction of the rollers supporting the chain may, in most cases, be neglected as inappreciable. If we wish to include them in the calculation, it is only necessary to increase the tension S in the portion of chain which passes from the drum to the pulley in the jib head by an amount $\phi \dfrac{r}{r} G_1$, where G_1 is the weight of this chain, and $\dfrac{r}{r}$ the ratio of the radii of the pin and rollers, so that $\dfrac{S}{S + \phi \dfrac{r}{r} G_1}$ expresses the efficiency of the supporting rollers, and may be entered in the formulas as such; this term will generally differ but little from unity. Let n represent the ratio between the velocity of the crank handle and load Q, then the effort will again be expressed by

$$P = \frac{1}{\eta} \frac{Q}{n}.$$

In the ordinary wharf crane with capacity to lift 100 to 200 cwt., and provided with movable pulley and a double-geared windlass, the mean value of the efficiency ranges from 0·75 to 0·80. In every case the efficiency may easily be found by the aid of the preceding formulas and tables, or by graphical methods.

Detailed information has already been given in treating of windlasses, § 11, concerning the arrangement of brakes and the force required for their operation, a repetition being therefore needless. (See also vol. iii. 1, chap. 9, Weisb. *Mech.*)

A special investigation is yet to be made, however, for determining the force required to swing the crane. In the turning operation the frictional resistances at the two bearings of the crane-post must be overcome; the resistances offered by the inertia of the masses may generally be neglected in consequence of the small velocities of the moving parts.

As before, let the crane-post support the load $V = Q + G$, and assume a cylindrical bearing at A, which will bring the whole load to bear upon the pivot B, whose radius we will represent by r. The lever-arm of the pivot friction ϕV is $\frac{2}{3}r$. Besides, the horizontal forces H constituting the couple will generate friction at the sides of the pivot, acting with a moment ϕHr, and at the lower end of the column at A, where owing to the rolling bearing the moment may be expressed by $\phi Hr\frac{r_1}{r_1}$, when r is the radius of the column at A, r_1 that of the roller, and r_1 that of the pin on which it turns. Hence the moment of all frictional resistances which oppose the swinging of the crane is

$$M = \phi V \tfrac{2}{3}r + \phi H\left(r + r\frac{r_1}{r_1}\right) = \phi(Q + G)\tfrac{2}{3}r$$

$$+ \phi\frac{Q+G}{h}b\left(r + r\frac{r_1}{r_1}\right) = \phi(Q+G)\left[\tfrac{2}{3}r + \frac{b}{h}\left(r + r\frac{r_1}{r_1}\right)\right].$$

Now let (Fig. 135) $AE = a_1$ be the radius of the toothed wheel AE, which is secured to the crane-post A, and gears with a pinion B of radius b, keyed to the shaft FG, which has its bearings attached to the movable frame LL. At the end of the latter shaft there is a bevel wheel C of radius c, gearing with the pinion D of radius d on the crank-shaft K. The effort applied to the crank-arm K of length l is found as follows. Let P_1 and P_2 be the pressures between the teeth at E and H respectively ; then in the absence of wasteful resistances, $P_1 b_1 = P_2 c$, and hence the transverse pressure upon the shaft FG is expressed by $P_3 = P_1 + P_2 = P_2 \dfrac{b_1 + c}{b_1}$. If now this bearing pressure P_3, acting at a distance $a_1 + b_1$ from the axis

Fig. 135.

of the crane-post is to be sufficient to swing the crane, we must have $M = P_3(a_1 + b_1) = P_2 \dfrac{b_1 + c}{b_1}(a_1 + b_1)$. Hence in the absence of friction of the toothed gearing, the effort applied to the crank K must be

$$P = \frac{d}{l} P_2 = \frac{d}{l} \frac{M}{a_1 + b_1} \frac{b_1}{b_1 + c},$$

where M denotes the above moment of friction,

$$M = \phi(Q + G)\left[\tfrac{2}{3}r + \frac{b}{h}\left(r + r\frac{r_1}{r_1}\right)\right].$$

Taking into consideration the wasteful resistances of the teeth and journals in accordance with § 3, would necessitate an increase in the driving force at the crank of about 10 or 15 per cent.

EXAMPLE.—It is required to determine the dimensions of a crane for lifting a maximum load of 6000 kg. [13,230 lbs.] at a radius of 6 metres [19·7 ft.], the inclination of the strut to the horizon being 45°, and the height of the crane-post $h = 2$ m. [6·56 ft.]

The respective stresses Z and T along the ties and strut are proportional to the corresponding sides of the triangular frame of the crane ; the lengths of the ties are found to be $\sqrt{4^2 + 6^2} = 7\cdot21$ metres [23·66 ft.], and the length of the strut is $\sqrt{2 \times 6^2} = 8\cdot48$ m. [27·8 ft.]   Hence we have

$$Z = Q\frac{7\cdot21}{h} = 6000\frac{7\cdot21}{2} = 21{,}630 \text{ kg. [47,700 lbs.]},$$

and

$$T = Q\frac{8\cdot48}{h} = 6000\frac{8\cdot48}{2} = 25{,}440 \text{ kg. (56,100 lbs.]}$$

If we assume that the whole strain Z is to come upon the tension rods, and none upon the chain, then for a tensile stress of $k = 6$ kg. [8500 lbs.] the sectional area of each tie is $\dfrac{21{,}630}{2 \times 6} = 1803$ sq. mm. [2·8 sq. in.], which corresponds to a diameter of 48 mm. [1·89 in.]

The strut is to be regarded as a long column rounded at both ends ; its area of section F is computed according to vol. i. § 274, Weisb. *Mech.*, which gives

$$P = F\frac{K_{II}}{\dfrac{l^2}{\pi^2 uv} + 1}.$$

Here l designates the length, K_{II} the ultimate resistance to crushing, which for wrought iron is $K_{II} = 22$ kg. [31,300 lbs. per sq. in.] ; further v expresses the ratio $\dfrac{E}{K_{II}}$, where $E = 20{,}000$ [27,400,000] is the modulus of elasticity, and

$$u = \frac{W}{F} = \frac{\text{Moment of Inertia of Cross Section}}{\text{Sectional Area}}.$$

In the present case $v = \dfrac{20{,}000}{22} = 910$; for an annular cross-section with external diameter d and internal diameter $d_1 = 0.96d$, we have

$$F = \frac{\pi}{4}(d^2 - d_1^2) = \frac{\pi}{4}(1 - 0.96^2)d^2 = 0.0615d^2,$$

and

$$u = \frac{W}{F} = \frac{\frac{\pi}{64}(d^4 - d_1^4)}{\frac{\pi}{4}(d^2 - d_1^2)} = \frac{1}{16}(1 + 0.96^2)d^2 = 0.120d^2.$$

Introducing this value in the above formula, and assuming a factor of safety of 6, we have

$$6 \times 25{,}440 = 0.0615d^2 \frac{22}{\dfrac{8480^2}{9.87 \times 0.120d^2 \times 910} + 1}.$$

Following the method of solution proposed for the example in § 274, vol. i. Weisb. *Mech.*, we find

$$d = 336 \text{ mm. } [13.23 \text{ in.}], \text{ hence } d_1 = 0.96 \times 336 \doteq 322 \text{ mm. } [12.68 \text{ in.}],$$

so that the thickness of the material of the strut will be

$$\delta = \frac{d - d_1}{2} = 7 \text{ mm. } [0.28 \text{ in.}]$$

In order to determine the dimensions of the crane-post we must take into consideration the weight G of the movable jib. Let us, as a rough estimate, call this weight 1500 kg. [3300 lbs.], and suppose its centre of gravity to be at a distance of 1.5 m. [4.9 ft.] from the axis of the crane-post; then for the vertical pressure upon the latter we have

$$V = Q + G = 6000 + 1500 = 7500 \text{ kg. } [16{,}550 \text{ lbs.}]$$

and for the horizontal force H of the couple, which tends to break it off, we have

$$H = \frac{6000 \times 6 + 1500 \times 1.5}{h} = 19{,}125 \text{ kg. } [42{,}170 \text{ lbs.}]$$

Suppose D and $D_1 = 0.7D$ to represent the external and internal diameters of this hollow crane-post at the point of attachment to the foundation plate; then for a permissible stress $h = 3$ kg. [4270 lbs. per sq. in.], the diameter D is given by

$$H \times h = 19{,}125 \times 2000 = \frac{\pi}{32} \frac{D^4 - D_1^4}{D} k = \frac{\pi}{32}(1 - 0.7^4)3D^3,$$

from which we get D = about 560 mm. [about 22 in.] This gives for the internal diameter $0.7 \times 560 = 392$ mm. [15.4 in.], and thus the thickness of material at this place is $\dfrac{560 - 392}{2} = 84$ mm. [3.3

in.] The tendency of the force H is also to break off the pivot at the top of crane-post. Assuming the ratio

$$\lambda = \frac{l}{d} = 1,$$

for a cast-steel pivot, we find the diameter, according to vol. iii. 1, § 3, Weisb. *Mech.*, to be

$$d = 2 \cdot 26 \sqrt{P \frac{\lambda}{k}},$$

or for $k = 10 [14{,}220]$, and $P = H = 19{,}125$ kg. [42,200 lbs.],

$$d = 0 \cdot 72 \sqrt{19{,}125} = 100 \text{ mm. } [3 \cdot 94 \text{ in.}]$$

Since the sectional area of this pivot is $\frac{\pi}{4} 100^2 = 7854$ sq. mm. $[12 \cdot 2$ sq. in.], it follows that the vertical pressure $V = 7500$ kg. [16,550 lbs.] increases the stress in the outermost fibres on one side, and diminishes it on the other side by $\frac{7500}{7854} = 0 \cdot 95$ kg. per sq. mm. [1350 lbs. per sq. in.], so that the maximum stress in the outermost fibres is $10 \cdot 95$ kg. [15,570 lbs. per sq. in.]

In order to fix the dimensions of the winding gear, let us suppose that the load is suspended from a movable pulley, and that the efficiency of the tackle consisting of this movable and the fixed pulley in the jib head is $0 \cdot 95$, then the chain leading to the winding barrel is subjected to a tensile stress of $\frac{1}{0 \cdot 95} \frac{6000}{2} = 3160$ kg. [6970 lbs.] To resist this stress we require a diameter of iron for the chain of $\delta = 0 \cdot 326 \sqrt{3160} = 18 \cdot 3$ mm. $[\delta = 0 \cdot 00864 \sqrt{P} = 0 \cdot 72$ in.] (see vol. iii. 1, § 119, Weisb. *Mech.*)

Assuming, say 10δ or $0 \cdot 18$ m. $[7 \cdot 21$ in.] as the radius of the drum, and $0 \cdot 90$ as the efficiency of each of the two pairs of gears by which the windlass is worked, and that four men are to exert 15 kg. [33 lbs.], each on cranks $0 \cdot 40$ m. $[15 \cdot 75$ in.] long, then the velocity ratio n of the gearing may be computed from the equation :

$$4 \times 15 \times 0 \cdot 4 \times 0 \cdot 90 \times 0 \cdot 90 = \frac{3160 \times 0 \cdot 18}{n} ;$$

which gives $n = 29 \cdot 3$. As a suitable estimate of the velocity ratio for the first pair of wheels, let us take $1 : 5$, and for the second $1 : 5 \cdot 9$, which gives the desired purchase $n = 29 \cdot 5$.

In order to determine the proportions of the turning gear, we will first calculate the moment of friction; assuming a coefficient of friction $\phi = 0 \cdot 1$, this moment becomes

$$M = 0 \cdot 1 \times V \times \tfrac{1}{3} \times 0 \cdot 100 + 0 \cdot 1 \times H \times \tfrac{1}{2} \times 0 \cdot 100 + 0 \cdot 1 \times H \frac{r}{r} \times 0 \cdot 28,$$

when $\frac{r}{r}$ expresses the ratio, say $\frac{1}{4}$, of the radii of the friction rollers and their pins.   Introducing the numerical values :

$$M = 0{\cdot}1 \times \frac{7500}{3} \times 0{\cdot}100 + 0{\cdot}1 \times \frac{19{,}125}{2} \times 0{\cdot}100 + 0{\cdot}1 \times \frac{19{,}125}{4} 0{\cdot}28 = 254{\cdot}5 \text{ mkg.}$$

$$= [1840{\cdot}8 \text{ ft. lbs.}]$$

Now let the spur wheel attached to the crane-post have a radius $a = 0{\cdot}35$ m. [13·78 in.], and the radius of the pinion gearing with it be $b = 0{\cdot}08$ m. [3·15 in.] ; further, assume the radii of the bevel wheels to be $c = 0{\cdot}25$ m. [9·84 in.] and $d = 0{\cdot}05$ m. [1·97 in.], and the length of the crank arm 0·35 m. [13·75 in.]   If we take the efficiency of the turning gear at 0·85, the effort required at the crank for swinging the jib will be

$$P = \frac{1}{\eta}\frac{d}{l}\frac{M}{a+b}\frac{b}{b+c} = \frac{1}{0{\cdot}85}\frac{0{\cdot}05}{0{\cdot}35}\frac{254{\cdot}5}{0{\cdot}43}\frac{0{\cdot}08}{0{\cdot}33} = 24{\cdot}1 \text{ kg. [53 lbs.]}$$

This force can be easily applied by two workmen.

§ 34. **Sheers.**—This type of hoisting apparatus is used for

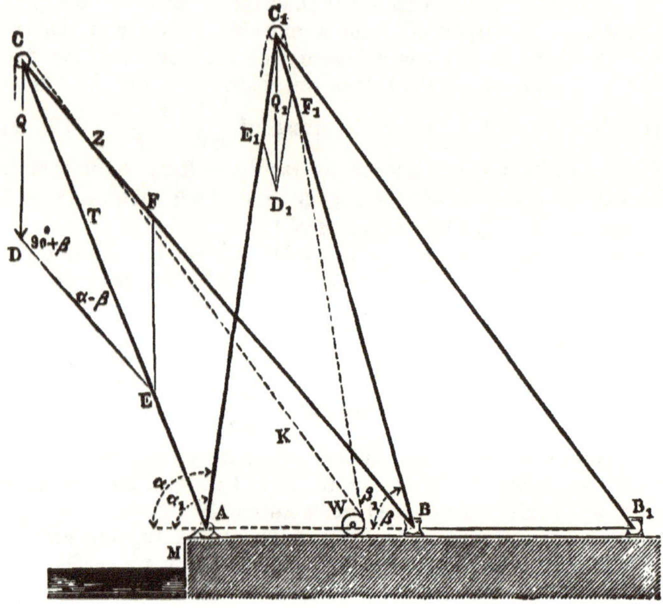

Fig. 136.

masting ships, for lifting boilers and engines on board steam-ships, etc.   It consists of two long spars AC reaching out from the wharf wall M, Fig. 136; the spars lean towards each other and are joined at the top C by a cross-piece; from this piece a chain or stay (or a pair of stays) CB is carried to an anchor at B.

A hoisting chain K, worked by a windlass W, is carried to the fixed pulley or hoisting tackle suspended at C. This arrangement allows the load to be moved in the vertical direction only, as the spars AC are exposed to a thrust, and the stay BC to a pull. More recently, however, this framework has been so constructed as to admit of the load being moved horizontally. This is accomplished either by making the foot B of the third leg BC movable in the direction BB_1, by which means the apex C can reach C_1, and the load Q may be brought into the position Q_1, or by keeping the foot B stationary, and shortening the third leg to a length BC_1. In both cases this leg must be a rigid member, since at the moment it passes the vertical position it will be subjected to a thrust.

Let a denote the inclination to the horizon of the plane of the legs AC in the extreme position, and let β represent the inclination of the tie-rod BC, then, if the pull along the chain K is neglected, the stresses produced by the load Q, when resolved along CA and CB, are expressed by

$$CE = T = Q\frac{\cos \beta}{\sin (a - \beta)} \text{ along the strut AC,}$$

and

$$CF = Z = Q\frac{\cos a}{\sin (a - \beta)} \text{ along the spar BC.}$$

The same formulas hold for the position AC_1B_1, and the corresponding forces T_1 and Z_1 are obtained by merely introducing a_1 and β_1 to represent the inclinations of AC_1 and B_1C_1. We see that Z changes its sign when a exceeds $90°$, that is Z changes from a pull to a thrust. Hence the spar BC is to be treated as a strut, and its dimensions must be proportioned to sustain the thrust

$$Z_1 = Q\frac{\cos a_1}{\sin (a_1 - \beta_1)},$$

for the stress produced by a thrust in such a long member is more unfavourable than the tensile stress due to the force

$$Z = Q\frac{\cos a}{\sin (a - \beta)}$$

acting when the sheers are in the extreme position, although Z is greater than Z_1. The thrust T, on the other hand, has its greatest value for the extreme position, and if 2γ denotes the angle between the legs AC at C, the sectional area of each sheer must be able to bear the thrust

$$\frac{T}{2 \cos \gamma} = \frac{Q}{2} \frac{\cos \beta}{\cos \gamma \sin (\alpha - \beta)}.$$

In this calculation we must make use of the formulas for compound stress (vol. i. § 278 and the following), as, on account of the great length of the spars, the bending strain due to their weight must not be neglected.

The manner of transporting the load horizontally is seen from the two following figures.

Fig. 137 shows the earlier form of a sheers at *Pola*, in which the forked end B of the spar CB is jointed to the nut of a powerful screw S by means of two gudgeons. The nut M slides in a rectangular guide G, which prevents the nut from turning, while it moves from B to B_1, thus carrying the load from Q to Q_1. The screw receives its motion from a steam-engine N by means of the bevel gearing H; the engine also drives two winches W, whose chains K are passed over the fixed guide-pulleys R to the hoisting tackle suspended from C. Owing to the great length of chain, each winch is provided with two drums similar to those in Fig. 53, and the wells J receive the free portion of the hoisting chain.

In order that the screw S may not be exposed to a lateral stress, the guide G must be suitably constructed and secured to the foundation, so as to resist the vertical upward pull $V = Z \sin \beta$ which acts in the extreme position of the sheers. In this position a direct pull $H = Z \cos \beta$ is exerted along the screw, and in proportioning the screw, this pulling action must be considered. From what precedes, it follows that this force gradually diminishes to zero in the vertical position of the sheers AC; any further motion in the same direction changes Z to a thrust, so that the nut is now pressed downward with a force $V = Z \sin \beta$, while the screw is likewise subjected to a thrust $H = Z \cos \beta$. The screw is therefore under the action of a variable force whose initial intensity is $Z \cos \beta$, then becomes zero, and finally increases to the value $Z_1 \cos \beta_1$.

Let a denote the horizontal distance of the nut M in the extreme position of the sheers from the line joining the two bearings A, and let t and z denote the lengths of the legs AC and BC; then the distance f through which the nut must

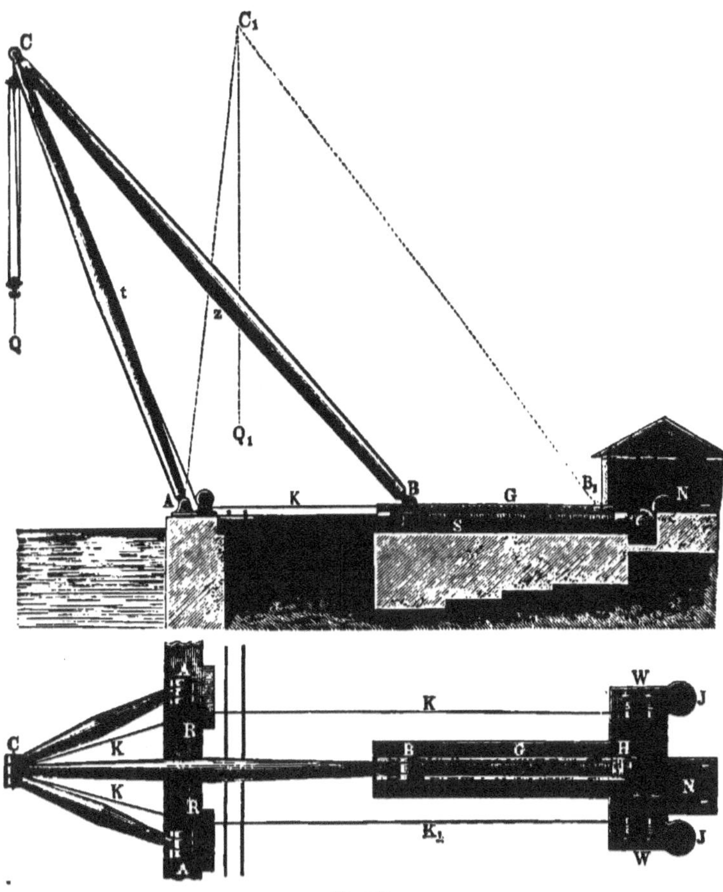

Fig. 137.

advance in order to change the inclinations a and β of the legs to values a_1 and β_1 is found from

$$a = z \cos \beta - t \cos a,$$

and

$$a + f = z \cos \beta_1 - t \cos a_1,$$

by subtraction, which gives

$$f = z \,(\cos \beta_1 - \cos \beta) - l(\cos a_1 - \cos a).$$

The great sweep required for this kind of hoisting apparatus generally makes f very large, and as it is impossible to support the screw except at its ends, it follows that the screw must be made very heavy, which greatly increases the frictional resistances.   This evil is less noticeable in the construction,[1] Fig.

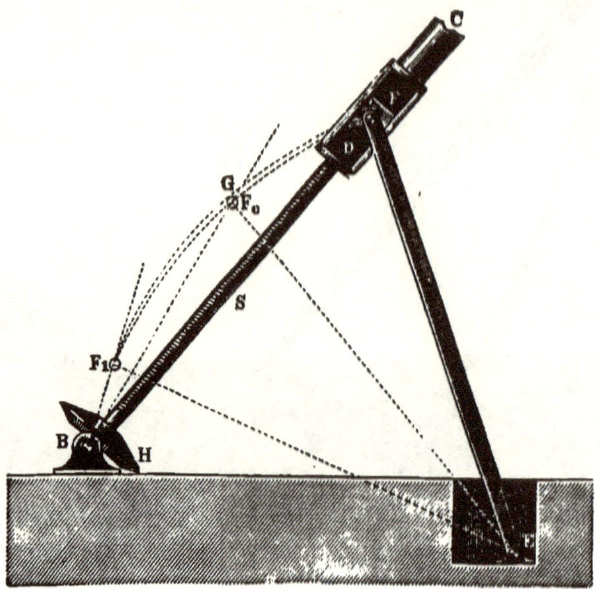

Fig. 138.

138, which was used for a large masting sheers designed at the machine works at *Waltjen*, in Bremen, for *Wilhelmshaven*. Here the screw S for moving the load toward the wharf is pivoted at the fixed bearing B, and works in a nut attached to the lower end of the leg CF.   If now the motion of the engine is communicated to the screw S by means of a bevel wheel H, the nut advances toward B, which is equivalent to a shortening of the spar BC.   It is evident that the length of screw required is considerably less than in the form given in Fig. 137, other things being equal; for drawing in Fig. 136

[1] See Rühlmann, *Allgemeine Maschinenlehre*, vol. iv.

the horizontal line BB_1 through the centre at B, and making $C_1B_1 = CB = z$, then the nut M in the former case advances through a distance $f = BB_1$, while in the present form of arrangement the advance only amounts to $f' = CB - C_1B = C_1B_1 - C_1B$, which value, representing the difference between two sides of a triangle, is always less than the third side.

The resistance to be overcome by the screw in the present

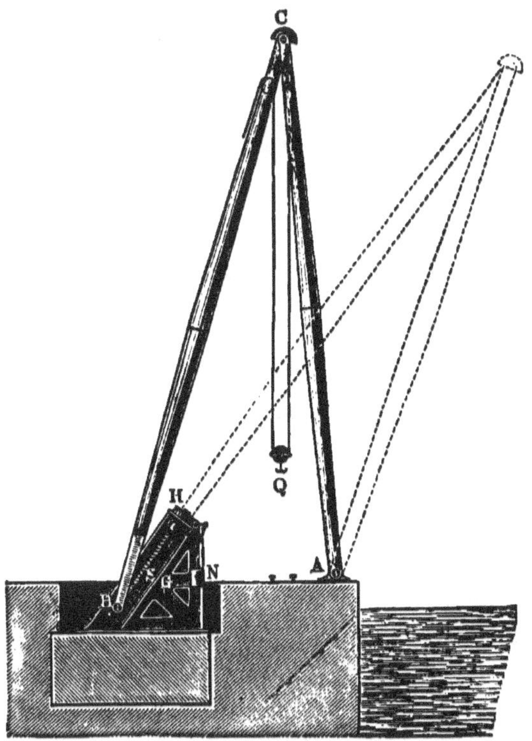

Fig. 139.

case is represented by the force Z exerted along the spar, which force as before acts as a pull until the sheers reach the vertical position, when it changes to a thrust.

It is with reference to this last stress that the end F of the spar is guided by a link FE pivoted at E, the link being so arranged as to cause the arc FF_0F_1, described by the pin F, to coincide as nearly as possible with the path FGF_1 in which the end of the spar CD moves.   Since the path of the latter point does not, however, coincide exactly with an arc of a circle, it is

Fig. 140.

necessary to provide the link EF at F with a sliding piece, which is capable of a slight motion in a slotted head attached to the spar CD.

A third construction by Clark,[1] Fig. 139, needs no explanation. The screw S is supported within the fixed frame G, and its nut, which is provided with two gudgeons, is grasped by the forked end B of the spar. The screw S is turned by a worm wheel H, the pertaining worm being directly set in motion by a steam-engine.

§ 35. **Hydraulic Cranes** are largely employed at the present day. Fig. 140 represents such a crane, as constructed by *Armstrong*. The crane-jib UV turns around the hollow cast-iron pillar KL, the former carrying two chain sheaves H_1 and H_2 for supporting the chain which passes through the centre of the pillar. The hoisting chain, provided with a counter-weight W to facilitate the overhauling (paying-out) of the chain, is passed over the pulleys H, F, and G, and attached to one end of the movable carriage DD. If water is allowed to enter the cylinder A, the carriage with its pulley F will move up the inclined railway, and, in consequence of the arrangement of the chain and pulley tackle, the load Q will be lifted to a height equal to three times the space passed over by the piston. Owing to the inclination of the rails E the weight of the carriage causes it to return to its original position, while the chain is kept taut by the counter-weight W. The lifting cylinder A is to be regarded as a single-acting

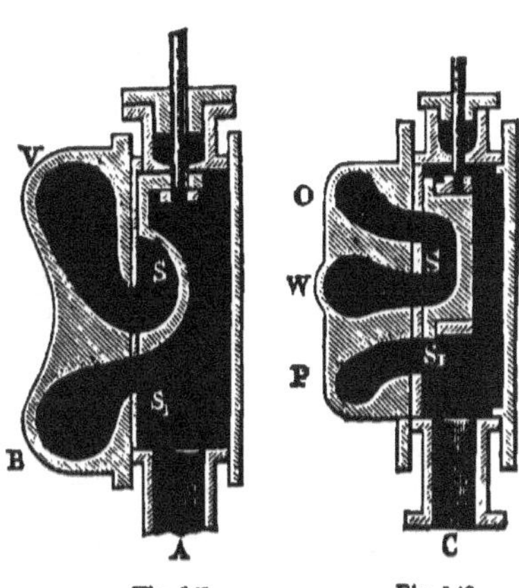

Fig 141.   Fig. 142.

water-pressure engine. Another water-pressure engine, with a double-acting cylinder MN, is used for turning the crane jib. According as the water from the accumulator passes to one or the other side of the piston, the piston-rod RR moves in or out of the cylinder, and thus causes the rack S to turn the toothed wheel T on the framework in one or the other direction.

The regular operations of the crane are effected by means

<hr>

[1] See *Excursionsbericht* von Riedler, Skizze 71.

of a valve gear worked by hand. Fig. 141 illustrates the valve chest of the lifting cylinder, and Fig. 142 that of the cylinder for swinging the crane. When the slide valve S is moved into the position shown in Fig. 141, the pipe A opens communication with the supply pipe from the accumulator,

Fig 143

and allows the water to enter the lifting cylinder through the pipe B. If, however, the valve is brought into the position S_1, the water in the cylinder can escape from B through the cavity of the valve into the discharge pipe V. When lowering the load the descent can be controlled by throttling the passage S_1, so that in this case, as in all hydraulic hoisting apparatus, a brake is unnecessary. It is also unnecessary to use a ratchet wheel and paul for sustaining the load at any elevation, since for this purpose the valve S can be brought into its middle position, where the lower edge of the valve just covers the post S_1, thus cutting off communication with the cylinder by either A or V.

The arrangement of the distributing valves of the water-pressure engine for swinging the jib is understood from Fig. 142 without further explanation, if it be noticed that the driving water enters through C, and is discharged through W, whereas the two pipes P and O are in communication with the ends of the double-acting cylinder. It is evident that these valves are similar to those of the single and double-acting steam-engines.

The force required to move the valve is considerable, since the frictional resistance between the face and seat arising from the pressure of the driving water upon the back of the valve must be overcome. For this reason it is impossible to move the valves simply by means of levers, as is the case in the hydraulic hoisting apparatus, Figs. 73 and 77, but some convenient mechanism must be introduced with a view to increasing the force exerted by hand. Fig. 143 shows how the valve is moved by the rods E and E_1, which terminate in screws fitting in the hollow heads of the valve rods T and T_1. In order to ascertain the positions of the valve from the outside, the hands Z and Z_1 are employed, which turn above two horizontal dials, and the apparatus is so arranged that each of these hands will make less than one revolution while the corresponding valve traverses the distance between its extreme positions. For this purpose the cranks K are fastened to the rods E and E_1, whereas the hands Z are attached to sleeves which are placed loosely on the rods. By means of the toothed wheels a, b, c, d, of which a is attached to the rod E, d to the sleeve of the pointer, and b c on a separate stud, the rotation of the cranks is reduced to that of the pointer, as in the so-called back-gearing of lathes.

In order to avoid the hurtful effects of the shocks (§ 18) which occur when the water is suddenly shut off, and which are especially great in the turning apparatus, on account of the comparatively great horizontal velocity of the jib and its load, special relief valves are introduced in the pipe connections O and P between the turning cylinder and its valve chest. A pair of these valves, X and X_1, Fig. 144, are placed in each of the valve chambers represented by X and Y, Fig. 143. Of these two clacks X acts as a safety-valve, being held down by the weight of the water in the supply pipe A_1, while X_1 acts as a suction-valve, inasmuch as it communicates with the discharge pipe W (Fig. 143) by means of W_1. It is evident

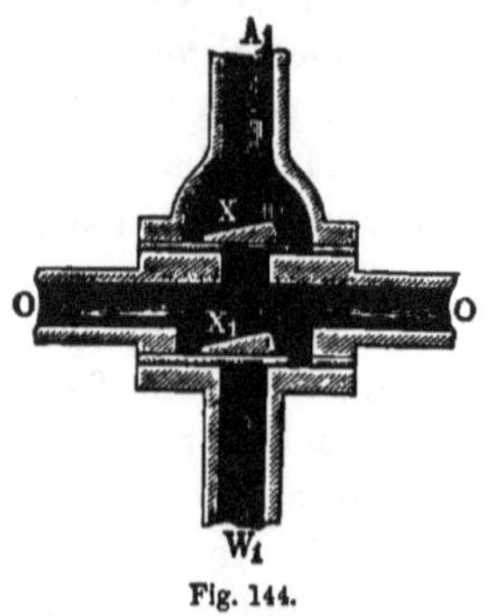
Fig. 144.

from what was said in § 18, that by suddenly closing the slide - valve of the turning apparatus the water which is being discharged from the cylinder, through the pipe O for instance, opens X by virtue of the living force of the jib. The result is that a small quantity of water is forced back from the cylinder to the accumulator, while in consequence of the vacuum occurring in the pipe P, the clack X_1 is lifted, and water is sucked up through W from the waste-water cistern. In addition to the cranks for working the slide-valves, a third crank is frequently employed to move a throttle valve in the supply pipe.

Plungers which are only packed at the stuffing-boxes, as in hydraulic presses, are generally used in hydraulic cranes instead of pistons, Fig. 140, which are packed at both cylinder and stuffing-box, as in steam-engines. Plungers are also employed for the turning-gear, and as they are single-acting, it is necessary to combine two such cylinders in order to turn the crane jib in both directions.

Such an arrangement is shown in the ten cranes constructed for the harbours of *Gustemünde*,[1] Fig. 145. The crane jib B is constructed according to *Fairbairn's* system, and turns with the cross-piece C on the hollow pivot of the fixed crane-post A, through the centre of which the chain K ascends. The latter is carried over the guide-pulleys L_1 and L_2 to the end of the jib, and is provided with a hook for the load Q and a counter-weight G, which serves to keep the chain tight when the hook descends empty. The horizontal

---

[1] *Zeitschr. d. Hannov. Arch.- u. Ing.-Vereins*, 1866.

cylinders D lying side by side are employed for raising weights, and are so arranged that the middle one is applied for loads not exceeding 20 cwt., and the two outer ones for loads not exceeding 30 cwt., or all three cylinders for loads not exceed-

Fig. 145.

ing 50 cwt. This plan is preferable on account of the great economy in the use of the driving water. The three plungers are provided with a cross-head E, which, with the pulleys attached to it, constitutes the movable block of a chain-and-pulley tackle. The fixed block of this tackle is likewise provided with three pulleys, and is represented by H. Thus the motion of the piston is multiplied sixfold. For the purpose of bringing the plungers to their original position when the empty hook descends, the cross-bar E is connected with the piston of a counter-cylinder D_1, which is always in communication with the supply pipe. By this means the pressure of the water in this cylinder causes the return of the plunger whenever the water, after use in the working cylinder, is allowed to escape, without the need of a special valve-gear for the auxiliary cylinder. Of course, the effective pressure in the lifting cylinder is only equal to the difference between the pressure in this cylinder and that in the counter-cylinder. For swinging the crane the jib is provided with a chain pulley M, whose acting surface is adapted to the shape of the chain K_1. The ends of this chain are attached to the cylinders J and J_1 of the turning apparatus, after being carried over the movable pulleys N and N_1, which are fixed to the cross-heads of the plungers O and O_1. Let us suppose a slide-valve similar to the one in Fig. 142 to be so connected with these two cylinders that one of them is always in communication with the discharge pipe while the driving water is being admitted to the other. Then, as one of the plungers O is driven out of its cylinder J, the other plunger O_1 is pushed into its cylinder J_1, thus turning the crane jib. The combined action of the *two single-acting cylinders* J and J_1 is therefore equivalent to a *double-acting* water-pressure engine, and allows the jib to swing in either direction. The effect of interposing movable pulleys N and N_1 is that the velocity of the chain K_1 is twice that of each plunger, and thus the crane jib revolves through an angle

$$\omega = \frac{2s}{2\pi r}$$

corresponding to the length of stroke s, where r represents the radius of the chain-pulley M.

The stroke of the turning piston is calculated to give the jib a circular motion of 0·6 of a turn to each side of the middle position in which the jib-head is farthest from the wharf, so that the total range of motion is 1·2 of a complete revolution.

Accordingly, for a radius $r = 0·38$ metre [14·96 in.] of the chain-pulley, the length of stroke of each working plunger is estimated at

$$s = \tfrac{1}{2} \times 1\cdot 2 \times 2\pi r = 3\cdot 77 \times 0\cdot 380 = 1\cdot 43 \text{ m. } [4\cdot 69 \text{ ft.}]$$

In order to keep the chain K_1 always taut, the cross-heads of the plungers O and O_1 are connected by an additional chain

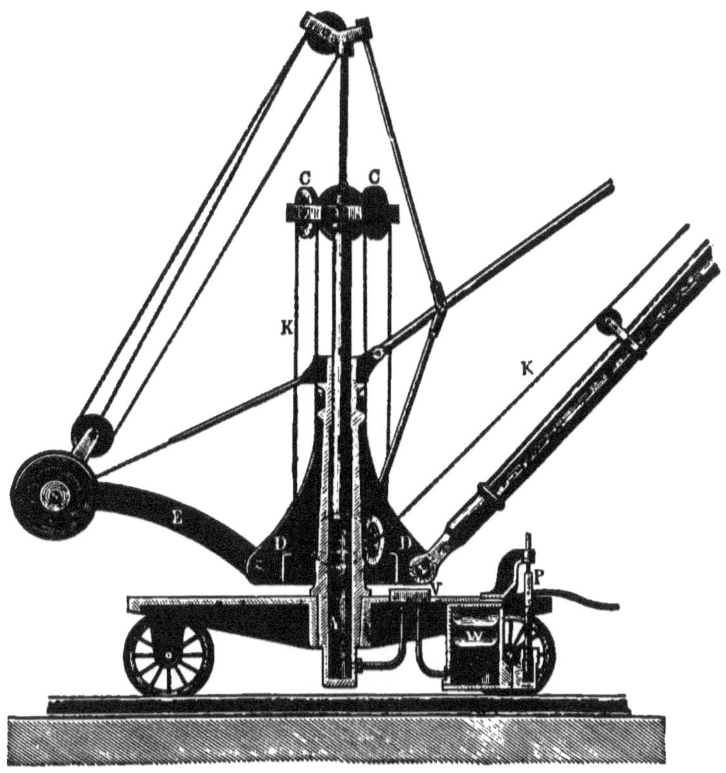

Fig. 146.

K_2, which passes over the fixed pulley N_2. This chain comes into action when the swinging of the jib is stopped by suddenly shutting off the driving water. In this case the inertia of the jib carries it a little further, the result being that, although one of the plungers is forced into its cylinder by the chain K_1, the other plunger would not move unless the chain K_2 exerted a pull upon it. The valve gear for the turning cylinder is a slide-valve of the form shown in Fig. 142, while lift valves are employed to work the hoisting apparatus.

A peculiar arrangement of a hydraulic crane is that designed by *Ritter* of *Altona*.   Here the crane-post A is also the hoist cylinder, and the cross-head B is provided with three pulleys C which make angles of 120° with each other.   The driving-chain K is carried over these pulleys and the three fixed pulleys D, so that the motion of the plunger is multiplied sixfold.   The water is drawn from a reservoir W by means of a hand-pump P, and forced into the hoist cylinder.   The reservoir is closed at the top, and answers the purpose of an air-chamber.   During the descent of the load Q the water passes from the hoist cylinder back into the air-chamber, thus compressing the air, and at the same time lifting a counter-weight G which swings with the lever E.   Through the action of the counter-weight and the compressed air, the hook for attaching the load and the working plunger ascend after the load is removed.   This arrangement offers special advantages for unloading and lowering loads.   V is a valve for drawing the water from the cylinder A and forcing it into the reservoir W or the reverse.   This crane is of the portable type, which class of machines will be more fully described in the following paragraph.

EXAMPLE.—A hydraulic crane, Fig. 145, is to lift a load of 2000 kilograms [4410 lbs.], the pressure of water in the accumulator corresponding to a head of 500 metres [1640 ft.].   It is required to determine its dimensions.

Let F denote the area of the working plunger, and f that of the counter-plunger, and let us suppose that the motion of the plunger is multiplied sixfold by interposing a chain-and-pulley tackle between the piston and the hoisting chain.   Then assuming an efficiency of $\eta = 0{\cdot}75$ for the hoisting apparatus, we have

$$(F - f) \times 500 \times 1000 \text{ kilograms} = \frac{1}{0 \cdot 75} \times 6 \times 2000,$$

$$\left[(F - f) \times 1640 \times 62 \cdot 5 = \frac{1}{0 \cdot 75} \times 6 \times 4410 \right],$$

from which we find the difference between the area of the working plunger above that of the counter-plunger to be

$$F - f = 0 \cdot 032 \text{ sq. m. } [0 \cdot 344 \text{ sq. ft.}]$$

The area f is determined by the consideration that the pressure $f \times 500 \times 1000 [f \times 1640 \times 62\cdot5]$ of the driving water upon the counter-plunger must be sufficient to force the water after use in the hoist cylinder into the waste-water cistern, besides overcoming the friction of the stuffing-box.

Let 10 metres [33 ft.] represent the head of water which is equivalent to the hurtful resistances in the waste pipe. This, added to the head (10 metres) of the waste-water cistern, gives the following equation for determining f,

$$f \times 500 \times 1000 = F(10 + 10)1000,$$

$$[f \times 1640 \times 62\cdot5 = F(33 + 33) \times 62\cdot5];$$

hence

$$f = 0\cdot04F,$$

so that

$$F - f = 0\cdot96F = 0\cdot032 \text{ sq. metres},$$

or

$$F = 0\cdot0333 \text{ sq. metres } [0\cdot358 \text{ sq. ft.}]$$

and

$$f = 0\cdot04F = 0\cdot00133 \text{ sq. m. } [\cdot0143 \text{ sq. ft.}]$$

This gives $0\cdot206$ metre [$0\cdot676$ ft.] for the diameter of the working plunger, and $0\cdot041$ metre [$0\cdot135$ ft.] for that of the counter-plunger. If the load is to be lifted to a height of 12 m. [39 ft.], the length of stroke of the plunger will be 2 metres [$6\cdot56$ ft.].

In order to fix the dimensions of the turning apparatus, let us suppose the moment M of friction to be computed according to the method of § 33. Let us assume that we have found $M = 300$ metre-kilograms [2171 ft. lbs.] Let the radius of the chain-pulley M (Fig. 145) be $0\cdot3$ m. [$0\cdot98$ ft.], and let the cross-head of each plunger of the turning apparatus be provided with a movable pulley for the purpose of doubling the motion. Then, according to the above data, the difference between the pressure of the driving water upon one plunger and the pressure of the waste water upon the other will be equivalent to a head of $500 - 20 = 480$ metres [1575 ft.] Furthermore, assuming the efficiency of the turning apparatus to be $0\cdot80$, the area F of each plunger will be found from

$$0\cdot80 \times F \times 480 \times 1000 \times 0\cdot3 = 2 \times 300,$$

which gives

$$F = 0\cdot0052 \text{ sq. metres } [\cdot056 \text{ sq. ft.}];$$

this corresponds to a diameter of $0\cdot081$ m. [$0\cdot266$ ft.] The length of stroke of the plunger depends upon the angle through which the crane-jib is to turn. It is the usual rule to allow the jib to swing through an angle of somewhat more than half a turn, say about $200°$ to each side of its middle position; accordingly the total distance moved over by a point on the circumference of the chain-pulley is

$$\frac{2 \times 200}{360} \times \pi \times 0\cdot6 = 2\cdot094 \text{ m. } [6\cdot87 \text{ ft.}],$$

and each plunger, owing to the interposition of the movable pulley, must have a length of stroke of 1·047 metres [3·44 ft.]

§ 36. **Portable Cranes.**—For building purposes, railway yards, and docks, portable cranes are often employed. In this class of cranes the post and jib, instead of being permanently secured in one place, are mounted upon a low carriage which travels upon rails. It is, of course, understood that the load may. also be transported along the railway, in which case the crane takes the place of a car. But this use of cranes is the exception, as, for instance, when the crane is used to transport the heavier parts of machinery in large erecting shops.

Portable cranes are constructed to work both by hand and steam power. The hydraulic crane of *Ritter*, Fig. 146, which comes under the head of hand cranes, is an exceptional case. The *steam cranes* proper, which have their own engine and boiler mounted upon the carriage, must be distinguished from *power cranes*, which are driven by a stationary engine, which may also be used for other purposes. At the present day the latter kind of portable cranes are operated to advantage by means of *cotton ropes.*

To the two mechanisms which affect the raising or lowering of the load, and the turning of the crane jib, must now be added a third for transporting the crane along the railway. In the lighter hand cranes this travelling motion is obtained by direct hauling on the part of the workmen, or by the employment of horse-power. In the heavier designs a common arrangement is to provide one or both carriage axles with a toothed wheel, which receives its motion through the medium of gearing worked by a crank shaft. In steam cranes motion is occasionally imparted to the crank shaft by hand, as connecting the latter with the engine which revolves with the crane gives rise to many constructional inconveniences. At times the windlass of the crane is utilised for propelling the car; this is done by paying out the hoisting chain and attaching its hook to some fixed point of the railway, so that the winding of the inclined chain upon the drum will cause the desired motion. This means, however, should be employed with care, as there is danger of upsetting or breaking the jib, which is usually not designed to resist such a stress. Portable cranes for railroad use may be best conveyed from place to place by attaching a locomotive.

In all portable cranes particular attention must be paid to

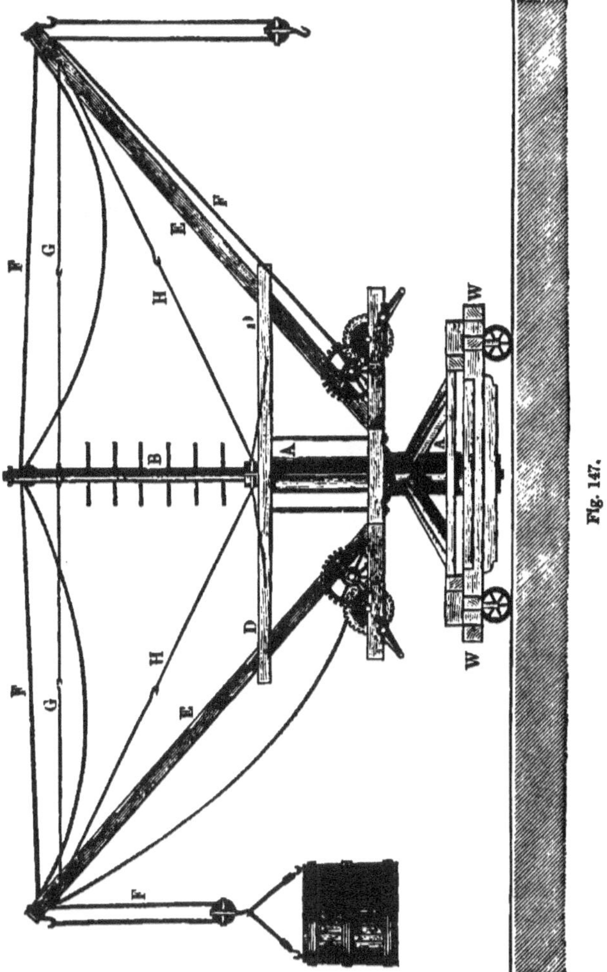

Fig. 147.

the question of stability; this may be obtained by a proper distribution of the weights. When the fulfilment of this condition gives rise to great difficulties the crane may be secured to fixed points of the railway, as, for instance, by holding-down clips which are attached to the top of the rails. This plan is only to be used in case of necessity, and is of course impracticable when the crane is to be transported with the jib and suspended load standing at right angles to the railway. In order to prevent the overturning of the crane under the action of the suspended load, counter-weights are generally used. For this purpose either actual weights, or, in the case of steam cranes, the weight of the boiler and engine,

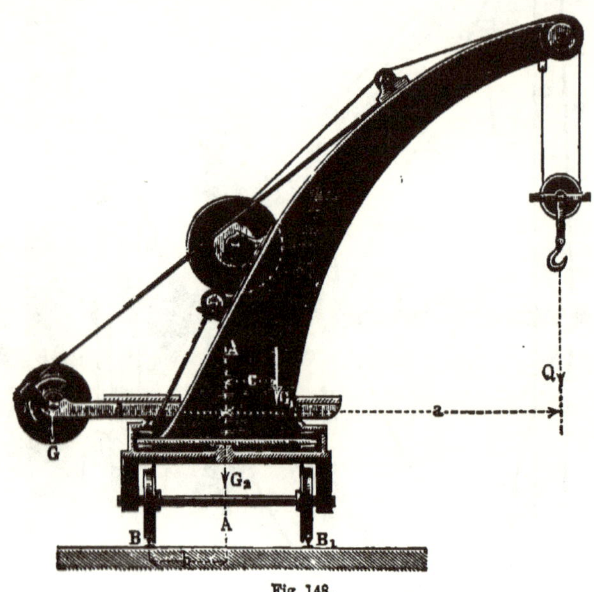

Fig 148.

may be advantageously employed. Since these balance-weights and the load Q must always be placed on opposite sides of the crane-post, it follows that the former must be connected with the movable jib, and not with the carriage.

For certain kinds of work the crane is made double, having two jibs and cranks, in which case sufficient stability is secured.

Such a crane is shown in Fig. 147. Here AA is a hollow crane-post fixed to the four-wheeled carriage W, while B is a vertical spindle which is free to rotate within the post. The two braces E, united by the wooden ties D, are connected with the spindle B by tension rods G and H, and rest upon the platform CC, which in turn is supported by a collar on the fixed post A.   The other parts of the construction need no explanation.

In order to test the stability of a portable crane let Q be the greatest weight to be lifted, and a its horizontal distance from the axis of the crane-post AA, Fig. 148, and let G be the counter-weight acting at a distance d from the centre of crane-post.   Further, let G_1 be the weight of the movable crane jib, including all parts (excepting the counter-weight) connected with it, and c the distance of their centre of gravity from the post.   Finally, let G_2 denote the weight of the carriage, including all pieces connected with it, as, for instance, the crane-post, which in the present case is reduced to a mere pivot, and let $2b$ be the width of the track BB_1.   To insure the stability of the crane, whether loaded or unloaded, it is necessary that the resultant of all the weights shall fall well between the rails.

Let us assume that the distance from the crane-post is not to exceed the quantity νb, where ν is less than one, a suitable value of same lying, say, between 0·8 and 0·9.

Under this supposition the condition of equilibrium for the loaded crane is :

$$Q(a - \nu b) = G_1(\nu b - c) + G_2 \nu b + G(d + \nu b) \quad . \quad . \quad (1)$$

and for the unloaded crane

$$G(d - \nu b) = G_1(c + \nu b) + G_2 \nu b \quad . \quad . \quad . \quad (2)$$

By subtraction we obtain

$$Q(a - \nu b) - G(d - \nu b) = -2G_1 c + G(d + \nu b),$$

or

$$G d = Q\frac{a - \nu b}{2} + G_1 c . \quad\quad . \quad . \quad (3)$$

Substituting this value for Gd in (2) we have

$$Q\frac{a - \nu b}{2} + G_1 c - G\nu b = G_1(c + \nu b) + G_2 \nu b,$$

from which we find

$$G = Q\frac{a - \nu b}{2\nu b} - G_1 - G_2 \quad . \quad . \quad . \quad (4)$$

and from (3)

$$d = \frac{Q}{G}\frac{a - vb}{2} + \frac{G_1}{G}c \quad . \qquad . \qquad . \qquad . \quad (5)$$

EXAMPLE.—Suppose 2500 kilograms [5500 lbs.] to represent the maximum load to be lifted by a portable crane ; let the radius of the jib be $a = 3\cdot5$ metres [$11\cdot5$ ft.] ; the weight of the movable jib $G_1 = 1500$ kg. [3300 lbs.] ; the distance of its centre of gravity from the crane-post $c = 0\cdot3$ m. [$0\cdot98$ ft.], and the weight of the carriage $G_2 = 1200$ kg. [2651 lbs.], how large must the counter-weight G be, and what must be its distance d from the crane-post, assuming the distance between the rails to be $2b = 1\cdot44$ m. [$4\cdot72$ ft.], and the distance of the centre of gravity of the crane from the crane-post not more than $0\cdot85b$?

The weight G is

$$G = 2500\,\frac{3\cdot5 - 0\cdot85 \times 0\cdot72}{2 \times 0\cdot85 \times 0\cdot72} - 1500 - 1200 \backsimeq 3200 \text{ kg. [7100 lbs.]},$$

and the distance of its centre of gravity from the axis of the crane-post is

$$d = \frac{2500}{3200}\,\frac{3\cdot5 - 0\cdot85 \times 0\cdot72}{2} + \frac{1500}{3200} \times 0\cdot3 = 1\cdot27 \text{ metre} = [4\cdot16 \text{ ft.]}$$

Instead of connecting the counter-weight rigidly with the jib, it is sometimes constructed in the form of a small carriage,

Fig. 149.

which travels upon a horizontal railway fixed to the jib, so that the distance of the counter-weight from the crane-post may be altered to correspond to any variation in the weight of the load.

The railway and balance-carriage have also been so arranged

that the action of the load itself causes the necessary movement of the carriage. The manner in which this is accomplished is shown in the sketch, Fig. 149.

The hoisting chain is carried from the drum W over the fixed pulley R_1, and hangs down in a bight which supports a movable pulley with load Q. It is then passed over the fixed pulleys R_2 and R_3, and finally attached to the counter-weight G, which is mounted on wheels, so as to travel along the curved railway L. Let Q represent any load to be lifted, and G the counter-weight; further, let P be the tension in the chain K_2, which tension, in the present case, must be taken equal to $\frac{1}{2}Q$ in the absence of friction at the pulleys. Then the action of the force P is to draw the weight G up the curved track until it reaches the point at which

$$P = G\frac{\sin \alpha}{\cos \beta},$$

where α denotes the inclination to the horizon of the path described by the centre of gravity of the counter-weight at the instant mentioned, and β the angle included between the direction of the pull and this path.

The form of the railway is therefore fixed by the condition that the moment of the counter-weight G for every position must be in equilibrium with the moments of the load and jib. This mode of arrangement, however, has never been extensively used, owing to the great length of chain required and other inconveniences.

A portable steam crane is shown in Fig. 150. As heretofore, the movable crane jib is supported by a cross-piece B upon the pivot of the crane-post A, which is fixed to the carriage W. The two side frames D extend back of the jib, and carry the vertical boiler K and the tank V, which at the same time serve as counter-weight. The small (4 to 6 horse-power) steam-engine E is provided with a link-reversing gear, and transmits its motion by means of the pinion F to the toothed wheel G attached to the winding-drum W. Motion is imparted to a vertical shaft through an intermediate pair of bevels not represented in the figure, which turn the crane as described in Fig. 135. The travelling motion is obtained from the engine by means of an endless chain N, which connects the chain-

pulley on the shaft of the engine with another on the rear axle
L of the carriage.   It is obvious that this chain gearing cannot
be employed unless the two shafts connected are parallel, that is
to say, unless the crane jib stands in the vertical plane through
the centre line of the track.   In order to transport the crane,
whatever the position of the jib, a shaft may be arranged to
coincide with the axis of the crane-post.   This shaft may receive

Fig. 150.

motion from the engine through a pair of bevels at its upper
end, and impart it to one of the car axles by means of a pair
of bevels at its lower end.   Z are holding down clips for clamp-
ing the crane to the rails, should the weight of the boiler and
engine be insufficient to balance the heavier loads.

The direct action of steam has also been advantageously
applied to steam cranes, as, for instance, in the excellent steam
crane constructed by *Brown* of London, which is distinguished

for its simplicity of arrangement and driving mechanism, and
is to-day largely used in practice. (In Hamburg alone there
are more than forty of these cranes in use.) The raising of
the load is effected, as in the hydraulic hoisting apparatus, by
interposing an inverted chain-and-pulley tackle, which multi-
plies the motion of the piston. Fig. 151 represents such a
crane with derrick motion. The crane jib T is guided by
means of friction rollers a upon the conical roller path B of the

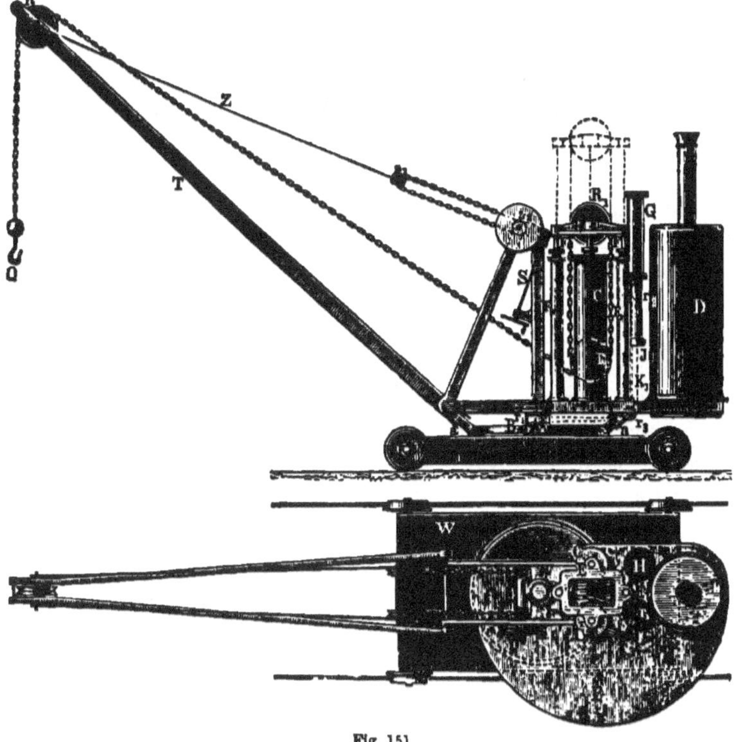

Fig. 151.

carriage W, while the weight of the jib is largely supported by
the lower pivot A_1, as in the case of turn-tables. The boiler
D also serves as a counter-weight. For raising the load there
are two cylinders C; the piston-rods are connected above with
a cross-head E, the central portion of which is removed to
receive the three pulleys R_1 placed side by side. The cross-

head represents the movable block of a six-pulley tackle, the fixed pulleys being placed at R_2; thus the height of lift of the load is six times the stroke of the piston. To prevent the load from running down (in consequence of the gradual condensation of the steam below the piston), the cross-head E is provided with two vertical plungers F, which during the upstroke of the main piston are lifted out of their cylinders F_1, and draw in water from the reservoir H through a suction-valve. As this valve closes when the plungers are pressed down by the action of the load Q, the water checks the descent until the valve is again opened; thus, by adjusting the area through which the water in the cylinder F_1 is forced into the reservoir H, we may create any required resistance and allow the load to descend uniformly.

For turning the crane jib the direct motion of a piston working in a double-acting cylinder G is employed; for this purpose the head J of the piston-rod has attached to it both ends of a chain K_1, which, after passing over the fixed pulleys r_1, r_2, and r_3, is carried around the pulley r_4 attached to the fixed pivot A_1. From this it is evident how the motion of the head J causes the chain to be unwound from the pulley r_4, and as the latter cannot turn, and the slipping of the chain is made impossible, it follows that the crane jib A with platform P is turned round the pivot A_1; the motion is to the left or right, according as the piston in the cylinder G moves up or down. In this apparatus there is no special arrangement for obtaining the travelling motion; as a rule, a method already mentioned is used, which consists in securing the hoisting chain to some fixed point on the railway, and then giving an upward motion to the lifting piston.

The relations between the forces and motions of this crane are to be determined as in hydraulic rotary cranes. Inasmuch as the tackle used has a sixfold purchase, the length of stroke of the lifting pistons must be made equal to one-sixth of the maximum height of lift of the load, and the areas of the pistons must be so proportioned that the total pressure of the steam upon both surfaces shall exceed $\dfrac{1}{(\eta)}6Q$ after deducting the friction of the piston and stuffing-box, where (η) represents the efficiency corresponding to the reverse motion of the tackle

(See table, § 8). The stroke of the piston for the turning apparatus is to be estimated, as in the hydraulic crane (§ 35, Fig. 145), from the radius of the chain-pulley r_4, and the angle through which the crane-jib is to swing. The present crane is made to swing one and a half times round, namely, three-quarters of a circle to each side of the middle position. In the cranes used at Hamburg with capacity to lift 40 cwt. each hoist-cylinder has a diameter of 0·40 metres [15·75 ins.]; and a stroke of 1·8 metres [5·9 ft.] The pressure of the steam in the boiler varies from 6 to 8 atmospheres, according to the load. To adapt the crane to the work in hand, the radius is made to vary by the following arrangement. The chain for the derrick motion, after passing over the movable pulley z_1 at the end of the tension-rod Z, has one of its ends made fast to the windlass frame, while the other is attached to the derrick barrel z_2, so that by turning the latter by the aid of a worm-shaft S, the radius is altered; the maximum radius of the jib is 11 metres [36 ft.]

The derrick motion requires that the strut T be pivoted at the base of the movable frame. When in consequence of high water in the harbour the total lift is small, it is evident that for the lowest position of the load the lifting piston would be a considerable distance from the cylinder bottom, and that the clearance spaces below the piston, which must be filled with steam at each lift, would be very great. To prevent the consequent waste of steam it is only necessary to shorten the hoisting chain by taking in the end attached to the frame, so as to bring the pistons nearly to the bottom of the cylinder when the hook is in its lowest position; that is, the stroke should always begin at the lower end of the cylinder, the length of stroke, of course, corresponding to the smaller lift.

§ 37. **Travelling Cranes.**—This term applies to the class of hoisting machines which, like portable cranes, travel upon a pair of rails, and in which the load, besides being lifted, has a horizontal motion at right angles to the motion of the entire apparatus. As regards construction, this kind of hoisting machine is chiefly distinguished from the portable crane by the absence of the rotating jib; for this reason the term *crane*, strictly speaking, is less applicable to it, but taking into consideration its general usage, we shall retain the word. Also as

T

regards its application, the travelling crane differs from the portable crane, inasmuch as the object of the former always is to convey the load lifted in horizontal directions at right angles to each other, while in the latter class the travelling motion is used merely for shifting the position of the crane, and rarely to transport the load. Accordingly, the travelling motion of the entire machine is always obtained by means of a suitable mechanism, whose employment for portable cranes, as before mentioned, is the exception, and not the rule. Travelling cranes are principally employed in foundries, machine and erecting shops, and in the construction of large engineering works, as, for instance, for distributing the tools and materials in building piers and massive bridges. It is evident that with the travelling cranes the load may be conveyed to any point within the rectangle formed by the total travel in each direction. In the smaller cranes, and for transferring light loads, the motion of the bridge and winding gear, as also the raising of the load, is effected by hand-power, while for heavy work engine-power is applied either directly or by means of rope transmissions, electric motors, etc.

A requisite of every travelling crane is a strong *bridge*, which is fitted with rails for a truck or trolley, and which has a motion of its own along a track perpendicular to its length. According to the location of the track along which the bridge travels, we may distinguish two forms of travelling cranes. When circumstances permit the track to be placed at the same level to which the load is to be lifted, as, for instance, in workshops and buildings, and in high scaffoldings, a bridge constructed of two parallel frames of timber bolted together, and mounted at each end upon two wheels, is all that is required. In cases where a firm scaffolding is impossible to obtain, as in many building operations and in railroad yards, the tracks are laid upon the ground, and the wheels are fixed to the base of two high frames or trestles, which support the bridge. The travelling crane is then known as a *gantry*.

Fig. 152 shows such a gantry as used at freight depots, and for distributing the materials in building piers. The two frames or trestles ABC, which travel upon the rails DE, support the two timbers SS_1, fitted in turn with a pair of rails, upon which the trolley with winding-gear WW_1 moves. This trolley contains the two pulleys R and R_1 for the two ropes or chains, which, after being carried over the fixed pulleys CC_1, are coiled upon the barrels NN_1 of two winches worked by a single pair of toothed wheels. A simultaneous rotation of the barrels, which wind on or unwind the rope at the same rate of speed v, will cause the load to rise or lower with this velocity v.

Fig. 152.

Neglecting wasteful resistances, the load $Q = 2S$, where S denotes the tension in each rope. To cause the load to travel horizontally from one end of the bridge to the other, the same winches NN_1 are used; this is accomplished by winding the rope on to one of the winches, and unwinding it from the other at the same time and rate. If one of the winches only is turned, the load will describe an inclined path. It is obvious that for heavier weights the load Q may be raised by the application of two movable pulleys, one for each rope, the ends of the latter being secured to the trolley W ; or the single pulleys RR_1 may be replaced by two or three pulleys mounted side by side on the same shaft.

From the figure we see how the crane can be made to travel

Fig. 153.

along the rails H by turning the winch-handles FF_1, which drive a pair of toothed wheels GG_1.

In order to prevent the bridge from binding during the longitudinal motion of the crane, motion must be given at the same time to one of the wheels at each end of the bridge.

Fig. 153 represents a travelling crane employed in shops. The bridge, which is made of two wrought-iron girders A connected at each end, travels along the rails B, the latter either resting upon an offset in the masonry or supported from below by iron columns. The crab W carries the chain barrel T, which receives its motion in the usual manner from the winch-handle K and the double windlass CC_1 and DD_1. The friction wheel F and ratchet wheel E require no further explanation. The longitudinal and transverse travelling motions are obtained from a second winch-shaft k, which may be shifted in its bearings. Thus the pinion a can be made to engage with the spur wheel b on the shaft c, which in turn drives the wheels of the truck W through the intermediate pair of gears d and e; or by throwing into action the bevel gears f and g, motion may be given to the shaft h connected with the bridge. This shaft is provided at each end with a pinion i, which gives motion to the bridge-axles n by means of intermediates o and l. In the former case the crab moves, in the latter the bridge. As the shaft h is too long to be without support between the extreme bearings, supporting levers are introduced, which are pivoted at m, and always tend to assume a vertical position by the action of the weight n. In travelling back and forth the bevel g pushes the lever far enough aside to allow it to pass. This gear is provided with a feather, which slides in a key-way running the whole length of the shaft h, and consequently can be made to rotate this shaft in any position of the truck. H represents a platform for the workmen, supported by the brackets G. When, as in foundries, the motions of the trolley and bridge must be obtained from below, the cranks K and k are generally replaced by chain-pulleys and endless chains, the loops of which can be conveniently grasped by the workmen.

Little is to be added concerning the formulas applicable to these cranes. The proportions of the hoisting apparatus are to be determined from the rules laid down for windlasses. The force required to transport the load horizontally is obtained as follows: Let G represent the total weight of the crab, including the suspended load, or the weight of the bridge, including the workmen, crab, and load; further, let R denote the radius of the wheels carrying the crab or bridge, and r the radius of the journals. Since we may suppose the entire weight to be concentrated upon one axle, the resistance to be overcome at the circumference of the carrying wheel will be expressed by

$$W = \phi\, G\, \frac{r}{R} + \frac{f}{R}\, G = \frac{G}{R}(\phi r + f),$$

where ϕ is the co-efficient of journal friction ($0\cdot08$), and f the co-efficient of rolling friction, which, according to I, § 197, Weisb. *Mech.*, may be taken at $f = 0\cdot5$, when R is expressed in millimetres [$f = \cdot01968$ when R is expressed in inches]. In order to overcome this resistance W acting with a lever-arm R, the toothed gearing required for transporting the apparatus must be arranged as in a windlass, which has a load W hanging from the winding barrel of radius R.

The girders of the bridge are to be considered as in the condition of a beam supported at both ends, which, in addition to its own weight, is to support a movable load, consisting of the useful load and the weight of the crab. The application of graphical statics to this problem is explained in vol. i. Appendix, § 45, Weisb. *Mech.*

§ 38. **Power Cranes.**—The necessity of giving a more rapid motion to the load than can be obtained by muscular effort has caused the introduction of other motive power. Although the power derived from a stationary motor may be easily transmitted by means of rope or toothed gearing to a fixed winch or tackle, special arrangements must be employed to adapt the driving gear to cranes. In rotary cranes the power must be communicated through the axis of the movable jib, while in travelling cranes the mechanism must be so arranged that the crane shall not be thrown out of connection with the prime mover by the shifting of the truck and bridge.

Fig. 154 illustrates a crane in use at the railway station at Liverpool, and driven by a continually running shaft. The winding barrel E is supported by fixed standards, and the rope, after passing over the pulley F, located opposite the axis of the crane jib FL, is carried down through the centre of the latter, and over the two pulleys G and H. The shaft B, which receives its motion by means of frictional gearing from the continuously running shaft A, transmits it through the medium of the toothed gearing DC to the winding barrel. For this purpose the bearing R of the shaft A is arranged upon the lever U; by pulling upon the cord X, attached to the lever, a small friction-wheel at the end of A is caused to press against the inner surface of the pulley T, by which the shaft B is driven. By releasing the cord, contact between the friction-pulleys ceases, as the lever U is withdrawn by the action of a counter - weight. A brake - strap embraces the wheel T, and sustains the load while the jib is being swung around. The brake is applied by pulling on the cord S, attached to the lever N, and carried down over the pulleys W and P to the handle MZ, to which it is fastened. When the

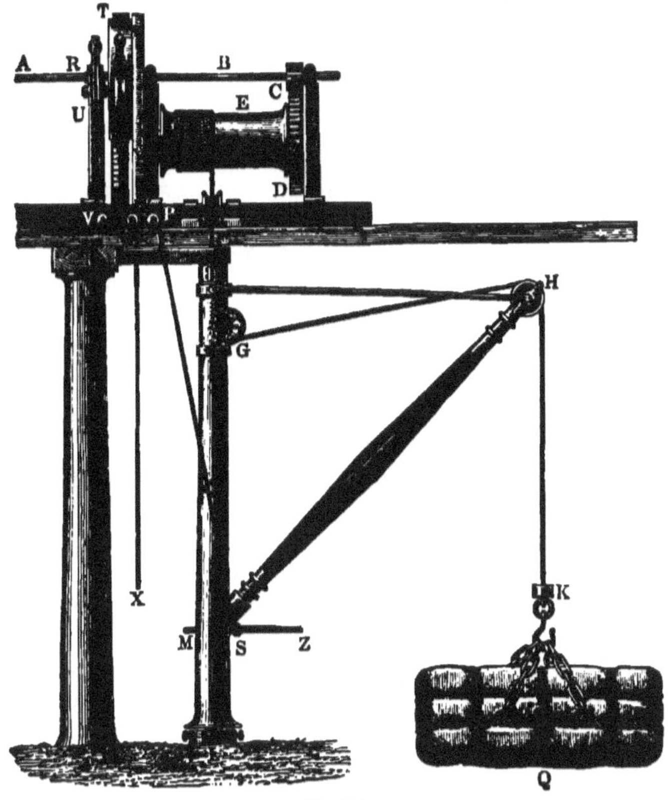

Fig. 154.

tension is removed from the cord S, the brake is released by the counter-weight Y, and the load descends.

For transmitting power to a travelling crane from a fixed motor, the original plan adopted was to employ two long shafts, one of which, A, was secured to the bridge, and parallel to it, while the other, B, was run at right angles to the former, and was supported either from the wall or by the columns carrying the longitudinal track. Each shaft was provided with a key-way receiving a feather set in the hub of a pinion, by which the latter was made to take part in the rotation of the shafts. In this case, if the wheel on the shaft B is a bevel b, which gears with another bevel d on the shaft A, then in any position of the bridge the shaft A will receive motion from B. By employing a reversing gear, it is a simple matter to obtain the longitudinal travelling motion.

Similarly, the right or left-handed rotation of the crane barrel, and of the shaft for shifting the position of the truck, is obtained from A by the shifting spur-wheels a_1 and a_2, and the two reversing gears c_1 and c_2, which are connected with the truck. This makes quite a complicated and awkward arrangement, however, for the reason that the two shafts A and B are too long to be without support between the end bearings. It is therefore necessary to employ a series of intermediate bearings, which, however, cannot be stationary, as the passage of the traversing wheels a and b must not be interfered with. The plan adopted was the lever arrangement described in Fig. 154.

These inconveniences are avoided by the use of suitable rope-driving, concerning the general arrangement of which a detailed account may be found in iii. 1, § 58, Weisb. *Mech.* In the following we will only mention an application of rope-driving to travelling and portable cranes as made by *Ramsbottom*,[1] and used to advantage in the locomotive works at Crewe.

Fig. 155 shows the plan and side view of the travelling crane. Here AA are the girders of the bridge, constructed of wood and iron, and mounted on wheels a_1 for travelling along the rails a_2, while the truck B, mounted on wheels b, rolls on the bridge. The latter carries the four vertical pulleys c_1, c_2, c_3, c_4, about which the endless rope s_1 s_4 runs in the direction indicated by the arrows. This rope is carried over two pulleys of 1·2 metre [3·94 ft.] diameter, at each end of the building, one of which is driven by a steam-engine, while the other, which is weighted and movable horizontally, acts as a tightener, so as to keep the endless rope at the necessary degree of

---

[1] For further details on this point see the article by G. Lentz, *Zeitschr. deutsch. Ing.* 1868, page 289.

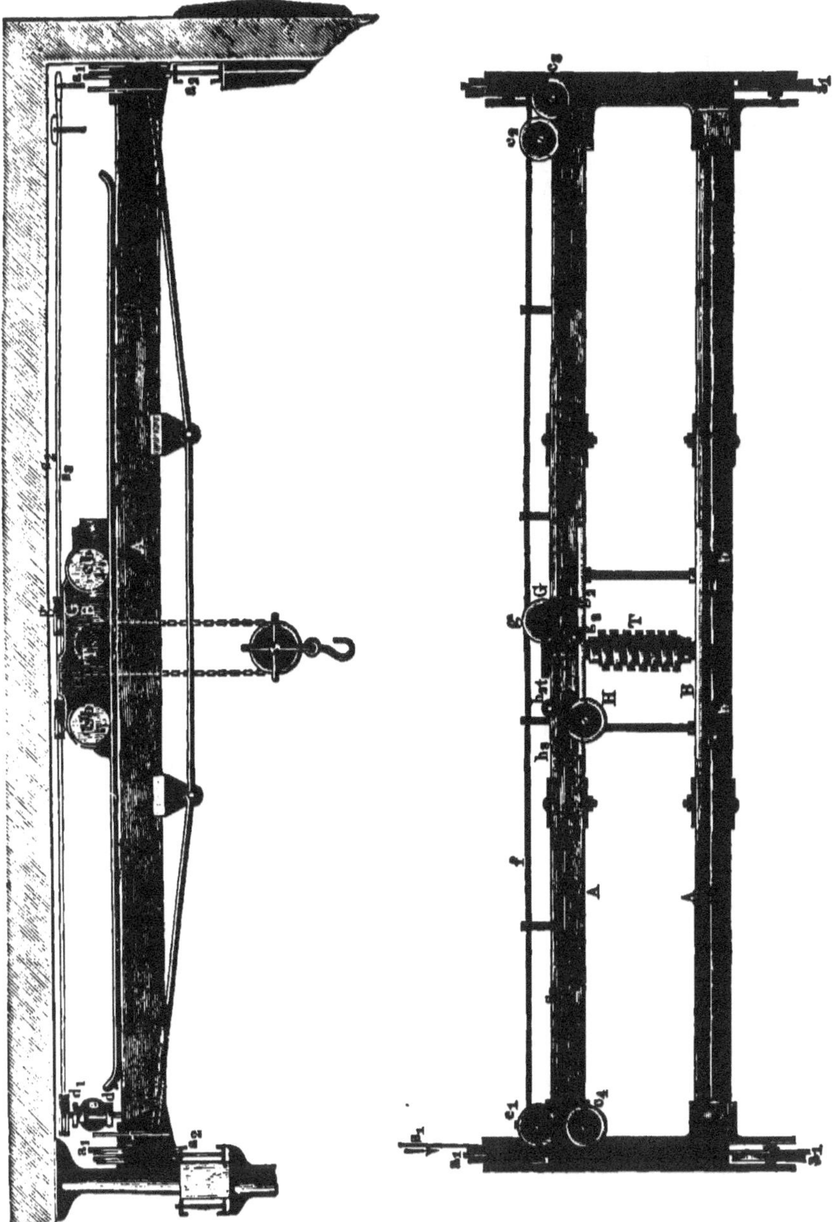

tension. The rope-pulleys c therefore turn continually, and the pulley c_1, owing to the connection established between the friction-wheels $d_1\ d_2$ and c_1, serves to shift the crane when, by means of a hand-lever, either the upper wheel d_1 or the lower d_2 is brought in contact with a friction-wheel e, carried on a horizontal shaft. Thus the latter shaft can be turned to the right or left, and being connected by gears with the shaft f on the bridge and the wheels a_1, it may be made to communicate the corresponding motion to the bridge.

The crane-barrel T is driven by a pulley g on a short upright shaft G, with a worm engaging a worm-wheel t on the barrel-shaft. Ordinarily the pulley g does not touch the driving-rope s; it is only when, by means of the roller g_2, the rope s_2 or, by means of the roller g_3, the rope s_3 are forced into grooves in the pulley g that motion is communicated to the shaft G; it is obvious that these motions will be in opposite directions, and they are therefore adapted for raising or lowering the load. The grooves in the pulley g are of different radii, the smaller being utilised for lowering the load rapidly, while the larger is used for hoisting.

Motion can be imparted to the truck B through the pulley h on the vertical spindle H; this is effected by pressing either of the ropes s_2 or s_3 into the single groove of the pulley h by means of the respective rollers h_2 or h_3. The spindle H, by means of a worm and worm-wheel, drives one of the axles of the truck B. The driving-rope is made of cotton, and is 16 millimetres ($\frac{5}{8}$ in.) in diameter, and by the action of the tightener-pulley it is subjected to a tension of 50 kilograms [110 lbs.]. It travels at the high velocity of 25 metres [82 ft.] per second. In order to diminish as much as possible the wear of the rope and the resistances due to bending the latter around the pulleys, the diameter [0·455 metres] of the pulleys is about thirty times that of the rope, which gives to them a velocity of 1000 revolutions per minute. This high speed requires careful balancing and good lubrication for the journals. The velocity ratio of the hoisting-gear is likewise very large, and has a value of 1:3000 for a maximum load of 25 tons, so that the rate at which the load is lifted is 0·495 m. [1·62 ft.] per minute; while for smaller loads and a velocity ratio of about 1:800 the ratio is, say, four times as great [1·96 m.] The longitudinal and transverse travelling motions have a velocity of about 9·14 metres [30 ft.] per minute. Experiments gave 17 lbs. as the force required at the circumference of the driving - pulley corresponding to a load of 9 tons and a velocity ratio of 3000. As the theoretical force required to lift the load is

$$P_0 = \frac{18,000}{3,000} = 6 \text{ pounds,}$$

the efficiency is

$$\eta = \frac{6}{17} = 0·353 ;$$

this small value is probably due to the use of the worm-gearing. The span of the travelling crane is 12·37 metres [41·5 ft.], and the longitudinal travel 82 metres [269 ft.] Within this length two travelling cranes are employed, so that a heavy load, as, for instance, a locomotive, may be lifted by the combined action of the two cranes. It is also evident that the load may ·be lifted and transported at the same time.

Fig. 156 illustrates the portable crane for the car-wheel works, also driven by rope transmission. The cast-iron crane-post A is set upon a box-like bed B made of boiler iron, which is mounted on two carrying-wheels travelling on a rail D running the whole length of the shop. An iron tube E fits over the crane-post A, and is guided above by a roller F working between two ⊢-shaped rails F_1, which are secured to the stringers overhead ; the crane jib G, which is carried by E, is guided by a conical roller path on the post A by means of a roller H. The hoisting-chain K is coiled upon the crane-barrel T, which receives motion from the worm c through the medium of the toothed wheels t_1 and t_2 and the worm-wheel b. The worm is driven by a vertical shaft, which passes through the centre of the hollow crane-post, and is provided above with a grooved pulley a_1. The driving-rope s_1 s_2 is wound half round this pulley, as shown by Fig. 156, III. ; this is accomplished by the two guide - pulleys a_2 and a_3, arranged at the upper end of the crane-post. In consequence of this arrangement, the driving-rope communicates its motion to the vertical shaft a, whatever the position of the crane, and this motion may be at will imparted to the crane-barrel T or the carrying-wheels C.   For driving the barrel the worm-shaft is provided with a conical friction-wheel e, while the shaft a carries a pair of similar bevels e_1 e_2, which may be shifted on a feather in the shaft.   A hand-lever k serves to

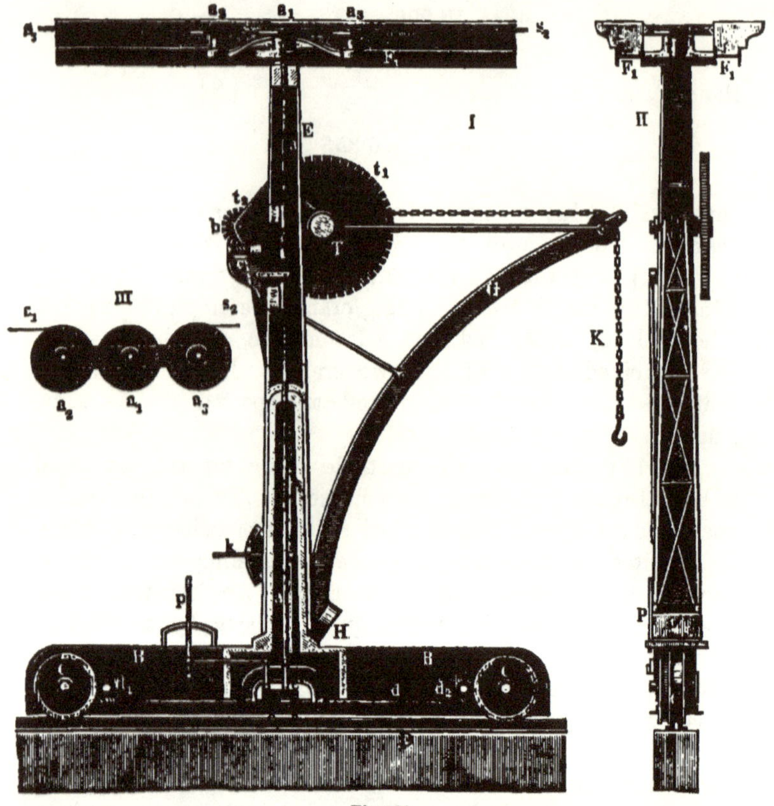

Fig. 156.

raise or lower this pair of bevels, and in this manner motion in either direction may be given to the worm c.

In a similar manner motion is transmitted to the car-wheels·C through the worm-wheels d_1 and d_2, and the intermediate shaft d; for this purpose a reversing gear is employed, consisting of the three friction cones e, e_1, and e_2, which are operated by means of a lever p.   The swinging of the jib is

accomplished by hand-power. This crane has a motion longitudinally of 36 metres [118 ft.], and is capable of lifting
80 cwt. at a radius of 2·59 metres [8·5 ft.] The speed at
which the load is lifted is 1·76 metres [5·77 ft.] per minute,
corresponding to a velocity ratio of about 1:1000. The diameter
of the driving-rope is also 16 mm. [$\frac{5}{8}$ ins.]

Such cranes are suitable only where a large amount of
work is to be done, as otherwise there would be an unnecessary
waste of power for driving the rope alone.

CHAPTER VII

§ 39. **Excavators.**—These hoisting arrangements, which are similar to cranes in their construction and mode of action, are largely used at the present day, especially in America, both as *dredging machines* in excavating canals and building-grounds, and also for similar purposes in railroad building. In their main features they agree with the *handle* or *scoop dredges*, which have been in use for a long time. Like the latter apparatus, they carry as their essential constituent a *scoop* or *dipper* provided with a handle, the combination being operated by the driving motor in such a manner that for every revolution the scoop cuts out a fixed quantity of material, which, after being lifted, is delivered to a vessel for further transportation. The work to be done by these machines, therefore, consists not only in a hoisting operation, but also in a *digging* or *cutting* process, which requires that a suitable shape and movement be given to the bucket.

When actual *dredging*, that is, increasing the depth of a waterway, is to be done, the excavator is placed on a barge or vessel, whereas for excavating purposes *on land*, it is placed on a car travelling along a special temporary track in the manner of portable cranes. In the latter manner these machines were used on a large scale in the construction of the *Pacific Railroads*.

In Fig. 157 the arrangement of a handle dredge, as built by *Otis* in New York, is shown in its essential features. The barge A, which is made of wood or iron, and is of rectangular plan and cross-section, carries the steam-engine with its tubular boiler, and has one end arranged to receive a movable crane jib BCD, the vertical post BD of the latter being provided with gudgeons, which turn in bearings secured to the barge. At the end of the jib are placed the pulleys c_1 c_2, around which the chain k is passed in such a manner as to carry in its lower bight the movable pulley e, from which the scoop G is suspended. The latter consists of an iron cylinder of oval section, open at the top, and provided at the upper edge g_1 with a steel

Fig. 157.

cutter, while the bottom g_2 is hinged at g_3, and may be thrown open by drawing out a bolt or latch at g_4 operated by a rope. The scoop G is fast to a long handle F, which passes through the double jib, and has at its lower side a rack f, which serves as a means of moving the handle longitudinally when the shaft f_1, with its pinion, is for this purpose rotated by means of the chain-wheel f_2 and the chain k_1. By this arrangement the handle F, besides having a sliding motion, may obtain a simultaneous rotary movement about the shaft f_1. The jib is revolved horizontally by means of a chain-pulley W, at the circumference of which the two ends of a chain k_2 are fastened, the latter being operated by a windlass mounted on the deck of the barge, and arranged to be turned in either direction.

As will be seen from the following considerations, the scoop G can, by the mechanism just described, be given the same kind of motion which is required for digging by hand. Let us assume that, by the action of the hand-lever H, the connection between the chain-wheel f_2 and the pinion on the shaft f_1 is interrupted, so that the former will run loose on the shaft, then the scoop with its handle will assume a position such as to bring its centre of gravity in the vertical tangent of the pulley c_1. By paying out the chain k this centre of gravity will sink in the same vertical VV, the handle F at the same time moving downwards a corresponding distance. Hauling in the chain k will naturally cause a vertical rise of the centre of gravity, and a corresponding upward movement of the handle F. By this means the scoop may thus be brought to any height that may be required in the dredging operation. If now the chain-wheel f_2 by the lever H is rigidly connected to the pinion-shaft f_1, then paying out the chain k will cause the pulley c_1 and the chain-wheel f_2 to revolve to the left, and thus bring about an *upward* movement of the handle F. The extent of this movement will be expressed by νh, if h denotes the increased length of the portion c_1 G of the chain and ν the velocity ratio of the pulley c_1 and the pinion f_1. Likewise, by hauling in the chain, a *downward* motion of the handle F will take place. In consequence of this combination, the scoop will describe a certain curved path, depending on the velocity ratio ν, and indicated in the figure at G_1 G G_3 for the point G, under the assumption of a ratio $\nu = \frac{1}{4}$. From the figure is evident how excavating as well above as below the water-level can be done by a motion of this kind at any height of the dipper. It may here be noted that, since the handle is lowered by the action of gravity, the extreme position of the dipper will be at the point G_2, which is the lowest position of its centre of gravity.

If it is required to pull the dipper further backwards, as, for instance, to G_1, this may be done by a special chain k_3 attached to its rear end, and operated by a winch; such was also the original arrangement in this kind of machines.[1] The same result may also be accomplished by deriving the motion of the shaft f_1 directly from the motor, and not from the pulley c_1. This is the method employed on the dredges built by the *Philadelphia Dredging Company.* It is possible in this case to continue the backward motion of the dipper close to the barge at G_1 by the simultaneous movement of the chain k and the pinion at f_1.

After the dipper has been lowered and pulled back to its extreme position, it is filled by cutting out a quantity of material in the above described manner, and is then, by the continued motion, raised to a certain height. The jib is subsequently swung around by throwing into action the winch for the chain k_2, and a pull on the latch-rope allows the bottom door to fly open, thus causing the contents of the dipper to drop into a special barge or car. The operation is repeated after the jib has been pulled back and the handle again lowered.

These machines work very rapidly, the whole operation in many cases requiring but one minute, provided that the person in charge is in possession of the necessary experience. In an apparatus of this kind used for improvements of the *Drau,*[2] the cubic contents of the dipper was about 0·6 cub. m. [21 cub. ft.], the swing of the jib was 7·32 m. [24 ft.], the length of the barge was 18·59 m. [61 ft.], and its width 7·32 m. [24 ft.] A steam-engine of 14 h. p. was used, and the earth excavated was 310 cub. m. [405·5 cub. yds.] in ten hours for a greatest dredging depth of 4·88 m. [16 ft.] below the water-level, and a maximum discharging height of 4·27 m. [14 ft.] above it.

Another kind of excavator is that in which the receptacle for the masses to be removed is simply lowered to the bottom by means of a winch and chains, and then, after being filled, hoisted again above the surface to the desired height for discharge of the material. To this end the receptacle consists of two scoop-shaped parts (*clam shells*), which are so connected by hinges that they can come together or move apart in the

---

[1] See *Zeitschr. deutsch. Ing.* 1872, page 269.
[2] See *Zeitschr. d. österr. Ing.- u. Arch.-Ver.* 1871, page 181.

same manner as the jaws of a pair of tongs.   This principle is illustrated in Fig. 158.   The two shells A_1A_2 are at a_1a_2 hinged to an iron frame B, in which the shaft C has its bearings.   On this shaft C is secured a large chain-wheel D, from the circumference of which a chain K is carried upwards to a

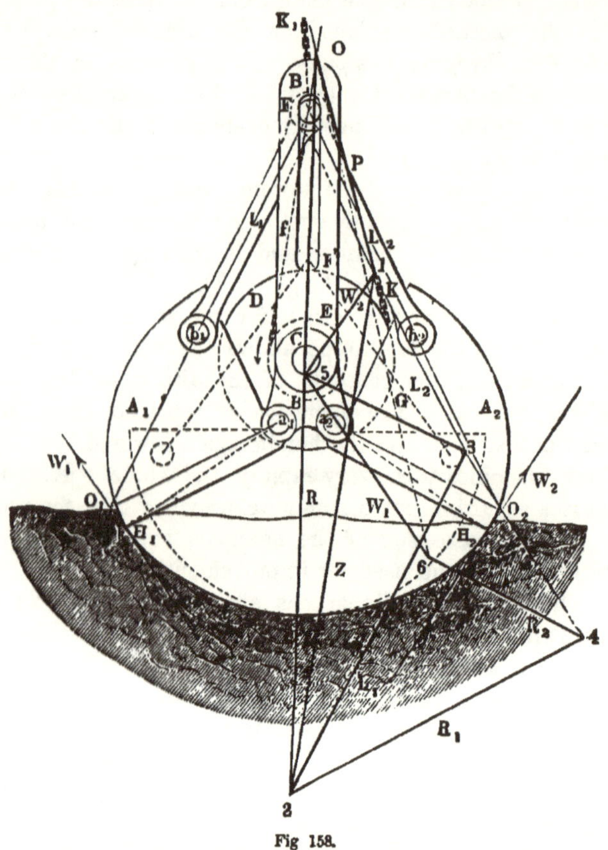

Fig 158.

guide-pulley, and thence to a winding-drum, which may be revolved at will.   Besides this larger chain-pulley D two smaller pulleys E are also placed on the shaft C, a chain f leading from each to a cross-shaft F higher up.   When the shaft C, by hauling in the chain K, is revolved in the direction of the arrow, the cross-shaft F, being guided at each end by slots in the framework B, will therefore be approached to the shaft C through the action of the chains f.   When the shaft F has arrived to the point F_1, the shells A_1 and A_2 will be closed by the agency of the links L_1 and L_2, as indicated by the dotted lines in the figure.   The whole contrivance is suspended from a second chain K_1 attached to the cross-shaft F, and carried to a second wind-drum.

Let us assume that the whole apparatus is lowered by pay-ing out the last-mentioned chain K_1, then the shells will be separated under the influence of the comparatively great weight of the frame B and the shaft C. The two cutting edges H_1 and H_2 will then penetrate into the bottom, a certain depth depending on the weight of the apparatus and the nature of the ground.   If, now, the previously slack chain K is hauled in, the shaft C will be revolved, and, operating in the manner already explained, will cause the cutting edges H to enter still further into the bottom, until the jaws are completely shut, and contain the slice of ground in the space enclosed.   Further hauling in the chain K will no longer cause the shaft C to revolve, but instead cause the whole apparatus, together with the piece excavated, to rise as soon as the tension in the chain has reached the value $G + Q$, when G is the weight of the apparatus, and Q that of the material contained in it. During the ascent the shells will remain closed by virtue of the pull in the chain K, thus preventing the contents from dropping out.   When the bucket has been raised to the desired height it is swung aside, so as to hang over the transporting barge, and the material is discharged simply by hauling in the other chain K_1, which operation slackens the chain K, and causes the shells to open.

The action above described presupposes that the weight G of the excavator is sufficient to prevent that a mere hoisting takes place without a previous entering of the shells into the bottom.   The weight G required depends, in the first place, on the resistance W offered by the bottom, and, secondly, on the construction of the apparatus.   It is evident that the greater the resistance, the greater must be the weight of the excavator.

In order to form an idea of the weight required, as well as of the force P to be applied in the chain K, we will make use of the following graphical method.   Let any driving force P act in the chain K, then this force will produce in the chain f

and in its direction a tension Z, which must intersect the former in O. These two forces P and Z will be in equilibrium with the reaction R produced by the frame G on the journals of the shaft C, and for this reason this reaction must have a direction OC. Making, therefore, to any scale O 1 equal to P, we obtain the tension Z in 1 2, and the reaction R at the journals in O 2, if 1 2 is drawn parallel to the chain f. Now resolving the tension $Z = 1\ 2$ along the directions of the push-rods Fb_1 and Fb_2, we obtain the pressures acting in these rods, namely, L_2 in 1 3, and L_1 in 3 2. Now, to determine the resistance of the bottom, let us recollect that the cutting edges H_1 and H_2 have a tendency to revolve about a_1 and a_2, and for this reason the resistance of the bottom must be assumed perpendicular to the respective radii $H_1 a_1$ and $H_2 a_2$, that is, in the directions $H_1 W_1$ and $H_2 W_2$. As these resistances W_1 and W_2 intersect the forces L_1 and L_2 in O_1 and O_2, we further obtain in $O_1 a_1$ and $O_2 a_2$ the reactions R_1 and R_2 of the frame on the journals a_1 and a_2. Resolving, therefore, the force $L_1 = 3\ 2$ acting in the push-rod along 3 4 parallel to $W_1 H_1$ and 4 2 parallel to $a_1 O_1$, we shall have in 4 3 the resistance offered by the bottom at H_1, and in 2 4 the reaction R_1 of the frame on the journal a_1. In the same manner we resolve the force $L_2 = 1\ 3$ along directions parallel to $W_2 H_2$ and $a_2 O_2$, and thus obtain in 5 1 the resistance of the bottom at H_2, and in 3 5 the reaction of the frame on the journal a_2. We have thus constructed the polygon of forces 1 0 2 4 3 5 1, and if for easier survey we remove $W_1 = 4\ 3$ to 6 5, and $R_2 = 3\ 5$ to 4 6, that is, if we draw the parallelogram 4 3 5 6, the polygon 6 5 1 0 2 4 6 will give a clear view of all the forces acting on the apparatus. The external forces W_1, W_2, and P are here represented by 6 5 1 0, and the reactions R, R_1, and R_2 of the frame at the journals C, a_1 and a_2 by O 2 4 6. It will therefore be seen that for a resistance at the bottom expressed by $W_1 = 6\ 5$, the driving force P in the chain K must have the value 1 O, and that the resultant reaction which the frame must exert on the journals is given by the line O 6. This line thus represents the weight G required.

In the figure the two forces W_1 and W_2 are made of different size, and it is therefore evident that in the position assumed only the edge H_1, at which the greater force acts, will

enter the bottom.   A change in the position of the apparatus is
hereby brought about, together with a change in the ratio of the
resistances W_1 and W_2, in consequence whereof the second edge
H_2 is also made to enter.   It is also apparent that the machine
will always so adjust its position that the resultant reaction
O 6 of the frame at the journals will always be vertical, since
this reaction is due to the weight of the frame, and thus is

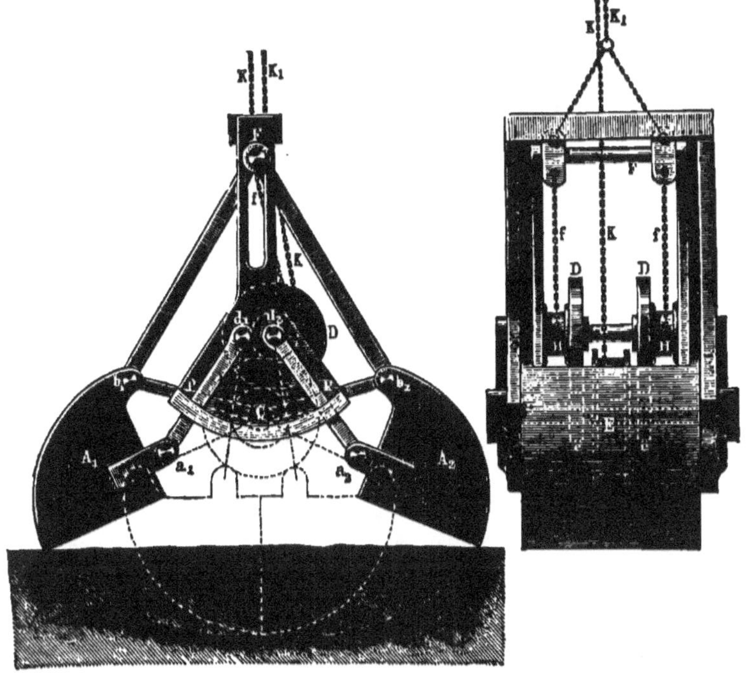

Fig. 159.

able to act in a vertical direction only.   In the above presenta-
tion the hurtful resistances, such as journal friction, etc., are
neglected.   Were these to be taken into consideration, the
only change necessary in the above construction would be to
draw the reactions R, R_1, and R_2 tangential to the corresponding
friction-circles, in place of drawing them through the centres,
as has been previously explained.   (See vol. iii. 1, Appendix,
Weisb. *Mech.*)

While for dredging in soft bottoms, which only slightly

oppose the pressure of the cutting edges, the weight of the machine itself is sufficient to bring about the desired result, it is necessary, when excavating hard clay bottoms, to add considerable extra weight, amounting to several tons for the larger apparatus. Owing to this circumstance, the efficiency of the machine in such cases is very slight. For, letting Q denote the weight of the detached mass, and G. the weight of the machine with its ballast, then the final pulling force in the chain must be equal to $G + Q$, and thus we obtain for the actual hoisting operation, neglecting all wasteful resistances, an efficiency of $\eta = \dfrac{Q}{Q + G}$ only.

When the total weight G of the apparatus is insufficient, an imperfect action of the excavator results, inasmuch as the jaws then continue to enter the bottom only until the tension in the chain becomes equal to the weight of the machine plus the resistance which the bottom offers to tearing off the under-cut mass. In such cases the bucket is never completely filled, the contents being two lumps of earth only. In order to overcome this difficulty the machine has been constructed in various ways, one of these, as originated by *Both*,[1] and illustrated in Fig. 159, is here given. Here a nearly constant cutting resistance is obtained by the use of the two guide-links C*b* and *da* on each side, in place of the fixed gudgeons for swinging the shells A_1 and A_2, the closing of the jaws being as before accomplished by the push-rods F*b*. The drums H for the chains *f* are not placed directly on the shaft operated by the chain K, being instead attached to another shaft d_2, and revolved from C by the gears *ee* and D*D* on each side. On account of this arrangement the motion of each shell may in every moment be considered as an infinitely small rotation around the corresponding pole or instantaneous centre P, which will be located at the point of intersection of the links C*b* and *da*.

When the bottom is rocky, and for raising the stone fragments after blasting under water, the jaws of the excavator are given the form shown in Fig. 160, which represents the apparatus designed by *Holroyd*, and used by the American Dredging Company. When the machine is lowered (I) it is suspended from

[1] *Zeitschr. deutsch. Ing.* 1874, page 35.

the double chain K_1, which is attached to the shaft C of the shear-shaped blades ACB, while the jaws are closed and the enclosed material lifted by a pull in the chain K connected to the cross-shaft F (II).   The links H merely serve to give greater stiffness to the construction.

§ 40. **Dredging Machines.**—The most common mode of dredging consists in using machines provided with endless chains in the manner of *Paternoster lifts* or *elevators*, the receptacles or buckets being attached to the link-chains, and

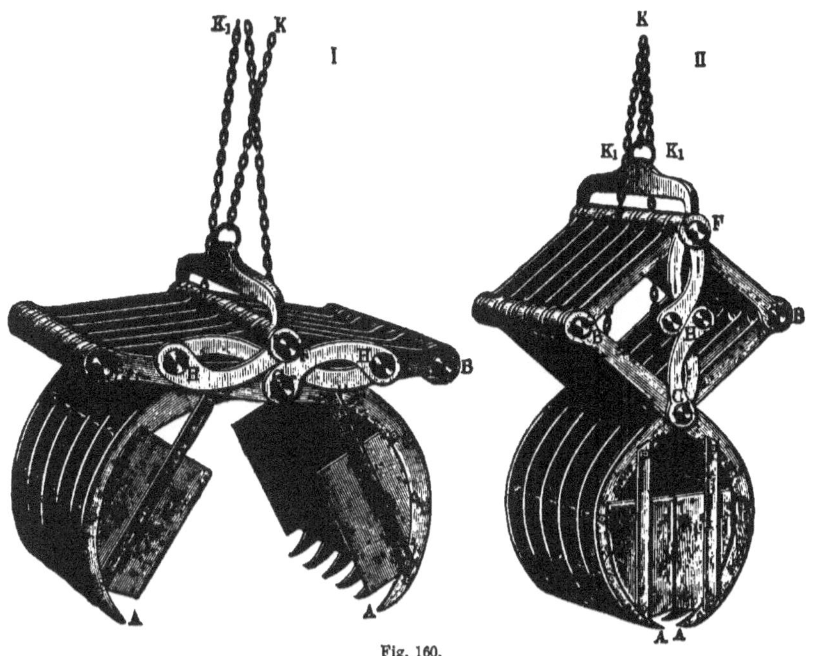

Fig. 160.

provided with knife edges for detaching the masses.   These chains are carried over two horizontal drums, which are shaped like regular four or six-sided prisms.   The shafts of these drums have their bearings at each end of a long wooden or iron frame (*ladder*), which hangs in a vertical or inclined direction from the dredge platform or vessel.   By revolving the upper drum the bucket-chain is given a continuous motion, the arrangement being so contrived that for every revolution of the drum a number of chain-links will pass over it corresponding to the number of its sides. The width of the latter must evidently be exactly equal to the length of the links. These . dredging machines are of two kinds : those having *vertical* and those having *inclined* bucket-chains. The former are chiefly employed for dredging at building grounds, the upper drum in this case being movable on the fixed pile planking in the manner of a travelling crane. For dredging in canals, rivers, and harbours, on the other hand, the inclined chains are mostly used. In this case the whole apparatus is placed in a hull, which may easily be moved to any point where dredging is to be done. Only in small dredging machines used for slight depths is the upper chain-drum revolved by hand-power, whereas for heavy service and great depths steam is always the motive power. The lower drum is never operated directly, but derives its rotation from the motion of the chain. The *horse dredge*, formerly used in Holland, is rarely employed at the present day, and the same thing can be said of the machines used at an earlier period for dredging in large rivers and operated by *ship-mill wheels*, the principal objection to the latter being that the dredging, as a rule, was desired in places where the current was of slight force.

· Along with the movement of the bucket-chain a simultaneous advancing of the whole apparatus is evidently necessary, which movement in steam-dredging machines is also accomplished by steam-power. As it is not always possible, where excavating is to be done to any greater depth, to accomplish the desired result by passing over the ground only once, it is necessary to make the apparatus adjustable vertically within certain limits. In vertical machines this is accomplished by increasing the length of the bucket-chain and frame, while in inclined machines the pitching of the chain to the horizon may be varied, whereby the same result is produced.

The bucket is filled simply by being forced against the bottom by the advancing motion of the whole machine, and upon reaching the drum at the summit, discharges its contents by overturning, the latter action being occasionally assisted, when the material is excavated from tough clay bottoms, by light blows on the bucket. The material is discharged into an inclined chute leading to the transporting-scow or barge. The slope of the chute must be greater than what would correspond to the natural point of sliding for the excavated mass. For sand an inclination of not less than 30°, and for clay as much as 45° to the horizon is chosen. It is thus evident that the matter must be lifted considerably above the actual hoisting height, and the more so the longer the chute, that is, the farther the receiving vessel is removed from the upper drum. In some cases, when the inclination of the chute is slight, it is necessary to facilitate the removal of the masses by adding water, which is carried along in the bucket, while ordinarily the useless lifting of the water is avoided by allowing the latter to run out through holes in the bottom of the buckets. In the building of the Suez Canal[1] water was employed on a large scale for washing away the excavated material, which was discharged into troughs measuring about 70 m. [230 ft.] in length, and having only a slight inclination. The latter was provided at each end with a drum, over which moved slowly an endless chain, whose links carried transporting plates, dished out for the reception of the material, which was further assisted in its motion by a simultaneous supply of water. With this arrangement an inclination of the troughs of 4 to 5 per cent proved sufficient when sand was removed, and 6 to 8 per cent when clay was carried. The quantity of water required equalled about one-half of the excavated volume for sand, and less for clay.

In order to prevent the mass which is being discharged from the bucket from falling back to the building-ground when vertical dredging-chains are used, it is necessary to slide the chute close up to the chain, so that it will come in the way of the bucket for a short period of time. After the discharge the trough must be pulled aside, so as to allow the bucket to pass, and then be pushed back again to the former position. For small machines this operation is performed by a workman, while in larger machines an automatic contrivance is introduced for the purpose. Each bucket is then provided with a projecting stud, which at the proper moment engages a lever, which communicates the desired motion to the chute. When the chain is inclined, no mechanism of this kind is required, since a fixed chute in this case offers no hindrance to the returning bucket.

See Oppermann, *Portfeuille économique*, 1869, Pls. 15, 16.

As the principle embodied in these machines can be utilised for excavating on land as well as for dredging below the water-level, it has more recently been carried out to advantage in so-called *land dredges*, which were in operation at the construction of the Suez Canal, and also at the works carried on for improving the course of the Danube near Vienna. A land machine of this kind, as first employed by *Couvreux* at the Suez Canal,

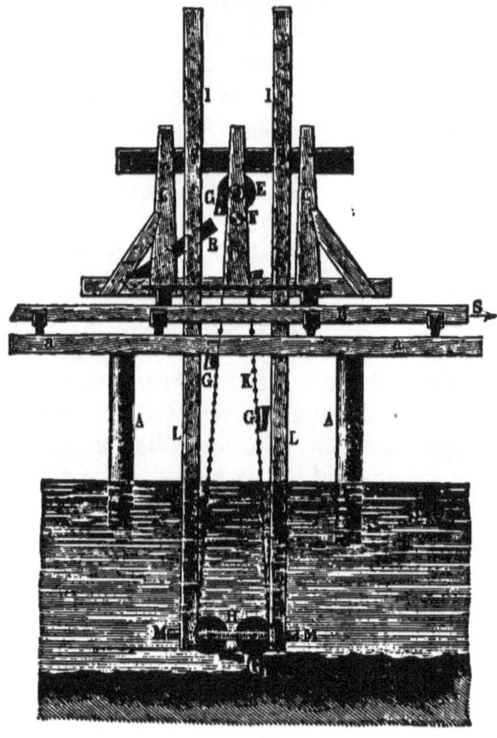

Fig. 161.

may in the main be designated as a *portable steam crane*, in which the hoisting drum is replaced by the upper drum of an inclined bucket-chain, the framework of the latter being at its lower end attached to a tackle suspended from the crane jib.

A vertical hand dredge, as used in excavating for bridge piers, is shown in Fig. 161.[1] The platform B, resting on the surrounding piles A, is movable on the stringers *a* in the

[1] See Hagen, *Wasserbaukunst*, vol. iii. part iv.

direction of the excavation, and carries six uprights C with two cross-girts D. The upper chain-drum E has its bearings in the middle posts C, and receives its motion from a crank-shaft F through a pair of gears. Two endless link-chains K pass over the four-sided drum E, and are at the bottom carried around two cylindrical drums H. The sheet-iron buckets G, secured to some of the connecting pins between the link-chains, enter the ground at G_1, and discharge their contents at G_2 into the chute R, the latter, as mentioned above, being for each arriving bucket pushed sufficiently aside to allow of a clear passage. A frame, consisting of the four vertical timbers L, connected by cross-girts M, and bolted to the cross-pieces D, serves to keep the chain tight and in its proper position, side motion of the frame being prevented by guides at the platform. From the figure it is evident how the posts L may be placed lower or higher, when a change in the dredging depth is desired, by moving the bolts to different holes l in the former, a corresponding number of links being at the same time either inserted in or removed from the bucket-chain. The change in the dredging depth is therefore not arbitrary, depending, as it does, on the length of the links. If during the dredging operation the platform, together with the frame, is slowly advanced by means of a rope S operated by a windlass, a groove of a width equal to the width b of the bucket, and of a depth t corresponding to the cut, will be excavated; when the groove is of the desired length the apparatus is returned to its original position, and after moving it a distance b to one side, a new groove may be excavated, etc. The buckets perform no work during the return motion, the machine being always so directed as to do its work while advancing *against* the ground to be cut out. If the first cut fails to give the desired depth, it is necessary to lower the frame and lengthen the chain, whereupon the dredging operation is repeated over the same ground.

An inclined bucket-chain is usually employed in the larger steam dredging machines, the framework for the former being, as a rule, arranged in a lengthwise slot in the carrying hull, as shown in Fig. 162, where AB represents the frame and CD the vessel. The arrangement may either be that shown in (I), where the upper chain-drum is located at the end A of the hull, the matter being discharged on the scow M in the direction of the arrow, or the lower drum B may, as in (II), be placed at the end of the vessel, the material raised being in this case precipitated at A into the two chutes R and R_1 alternately. Hence it falls either into the scow M or M_1, the direction being governed by the movable trap-door K. With the latter arrangement no loss of time occurs, as in arrangement (I) when an empty scow M is substituted for a loaded one, since in (II) it is always possible to keep an empty scow in waiting on one side of the vessel, while another is being filled on the other side. This circumstance is of great advantage, especially in heavy surf or rough sea, as an exchange of scows is then very difficult, and connected with great loss of time. A further advantage of style (II) is that it can be utilised for procuring the necessary depth of water required for its operation. This is owing to the fact that the lower end B of the bucket-chain protrudes beyond the end of the hull, a circumstance which enables the machine to be employed for excavating low shores or projecting necks of land, that is, to

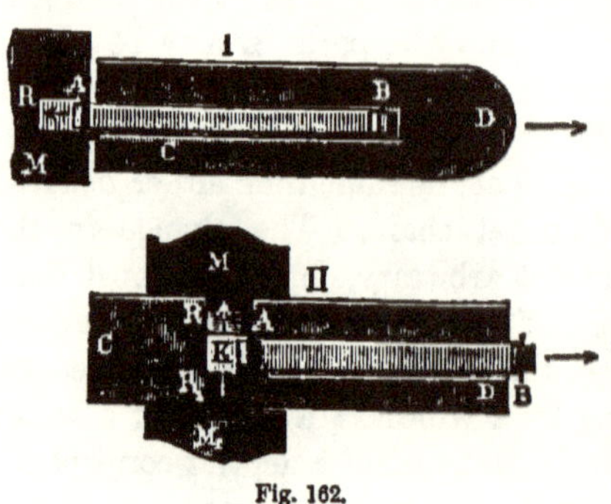

Fig. 162.

act in a measure as a *land dredge*, which is not possible with construction (I). On the other hand, style No. II suffers from the disadvantage that the excavated matter here must be raised to a greater height than is the case in No. I, on account of having to cross half the width of the vessel upon leaving the buckets, which requires a corresponding inclined piece to be added to the chutes R and R_1. For this reason more power is absorbed in lifting the material. As, however, by far the greater amount of power is in this class of machines required for the process of excavating and not for lifting, the disadvantage just mentioned is of small consequence (see below).

The bucket-chain is occasionally placed crosswise at one end of the dredge-boat, as was the arrangement in the earlier horse dredges at the ports of the Baltic.[1] This construction

<hr>

[1] Hagen, *Handbuch der Wasserbaukunst*, vol. iii. part iv.

may under some condi-
tions be recommended for
use in dredging for bridge
piers.

Finally, steam dredg-
ing machines with two
bucket-chains [1] have been
constructed, having the
latter placed lengthwise
at each side of the vessel,
and intended to fill two
scows at the same time.
This arrangement has not
proved very satisfactory,
however, the results
obtained from a double
dredge never equalling
those derived from two
single dredges of the same
proportions. The reason
may be traced to the
frequent interruptions in
the working of the machine
due to the fact that the
two receiving scows are
never full at exactly the
same time; it is there-
fore necessary to stop the
whole machine at frequent
intervals for the sake of
replacing one of the scows.
Double dredges are on this
account passing out of
use.

In Fig. 163 a sketch

[1] For drawings of a double
dredge as used on the Clyde see
*Proceedings of the British Institu-
tion of Civil Engineers*, 1864, and
Rühlmann's *Allgem. Maschinenlehre*, vol. iv.

Fig. 163.

taken from *Hagen's* work is shown, representing the steam dredging machine "Greif" used on the River Oder. The chain frame L, consisting of two strong iron girders connected by braces, is hinged on the upper drum-shaft at A. By the tackle F, which is attached by means of a chain to the lower end of the framework, the latter may be given a greater or less inclination according to the depth required. The lower ends of the girders carry the bearings for the lower drum B, the former being so contrived that the chain may be given any degree of tension by means of an adjusting screw. The upper half of the chain which carries the loaded buckets is supported by a number of rollers w introduced between the girders, while the lower slack half of the chain hangs freely in an arc. The tightening of the chain is done in order to prevent the slack portion from dragging on the bottom b_1 before the working bucket reaches the point of excavation at b_2. Care must be taken, however, not to make the *verse sine* of the lower chain so small that undue strain is brought on the links by the influence of the weight of the buckets and chain.

The motion of the chain-drum A is derived from the gears C attached to its shaft, and driven by additional gearing from the vertical steam-engine D provided with two cylinders. The connection between the steam-engine and the upper drum is made in such a manner as to yield when the power transmitted exceeds a certain limit. For this purpose either a friction-coupling is introduced, or the gears C are secured to the shaft a by means of wooden keys, the dimensions being so chosen that for a given maximum strain the keys are shorn off, whereby the motion of the drum is discontinued, although the gears continue to revolve. This feature is necessary in order to prevent break-downs, which would inevitably result at times were the connection made rigid. If, for instance, a bucket should strike a large boulder or a piece of driftwood too long to pass crosswise into the opening of the hull, etc., the exceptionally great resistance offered to the momentum of the rotating masses would of necessity cause a breakage at some point.

From the figure it is evident how, by turning the trap-door from the position K_1 to K_2 and *vice versa*, it is possible to conduct the discharging material to one side or the other. The raising and lowering of the chain-frame is also accomplished by the steam-engine, the chain f leading from the tackle being carried to a windlass, which may be turned in either direction.

For moving the vessel in the required manner several other hoisting appliances are arranged on deck, the mode of operation being as follows. In the first place, at one end of the vessel is placed a winch W, from which a chain k several hundred yards in length is carried up the river, and there secured to a main anchor. This chain is kept taut by the resistance of the bottom during the dredging, and by being gradually hauled in, causes the buckets to enter the ground. From a second winch or capstan W_1 in the stern of the hull a chain k_1 is carried to another anchor located down the river, and serving chiefly to prevent the machine from drifting in case of a change in the

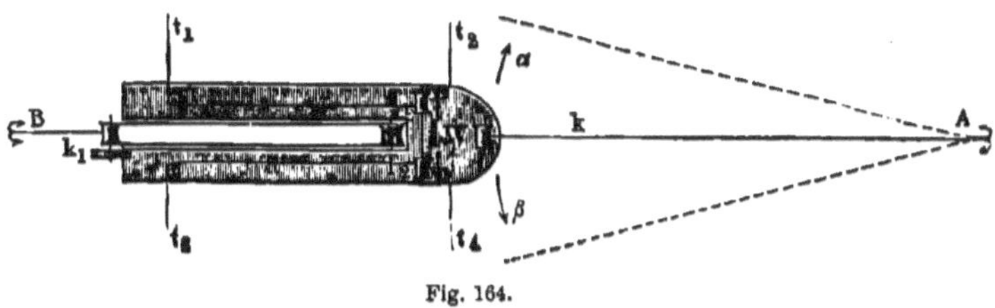

Fig. 164.

current. The mode of operation is different from that employed with the vertical dredge, Fig. 161, the direction of the groove excavated not coinciding with the plane in which the bucket-chain moves, but being at right angles to it. The proceeding is evident from Fig. 164. Here A represents the main anchor with the chain k, of about 300 or 400 metres (yards) in length, leading to the winch W. Besides the winch W_1, already referred to and connected to the second anchor B, two winding drums T_1 T_2 are also located on the deck, from which the cables t are carried to fixed points on each shore, or to auxiliary anchors. Revolving the winches T by the steam-engine in such a manner that the cables t_1 and t_2 will unwind from T_1 at the same rate as t_3 and t_4 are wound on the drum T_2 will cause the whole vessel to move sidewise in the direction of the arrow a, and thus describe an arc of a circle with A for centre. After arriving at the end of the furrow, the machine is advanced slightly by hauling in the chain k and paying out k_1, whereupon the winches T are reversed and made to move the vessel back in the direction of the arrow β, a new furrow concentric to the preceding one being now excavated. When the chain k has been reduced in length to about 150 or 200 metres (yards), the main anchor A is removed to a point farther up the river.

This modern method of *transverse* or *radial* dredging offers many advantages over *longitudinal* dredging, one of these being that the loss of time incident with the latter method, when the vessel is returned to start a new furrow, does not occur in the employment of the former method. Besides, the one-sided resistance during the longitudinal dredging was very apt to cause the bucket to return into an already excavated furrow, and if an attempt was made to overcome this difficulty by applying a sidewise pressure to the vessel, the furrows were liable to become separated by remaining ridges of ground.

The inclination of the bucket-ladder to the horizon ranges from 45° for great depths to 15° when the buckets are not intended to operate, as during transportation of the machine. The length of the ladder therefore depends entirely on the depth at which dredging is to be done, and amounts in the machine shown in Fig. 164 to 18·4 m. [60·36 ft.] In this case the upper drum, which is generally made four-sided, makes from 5 to 8 revolutions per minute, according to the quality of the bottom, thus allowing in this time 20 to 30 chain links to pass, or half this number of buckets, if the latter are attached to every other pair of links. The velocity of the bucket-chain may, on an average, be taken at 0·30 m. [1 ft.] per second, and the side motion of the vessel for each advancing bucket may be assumed to be 0·10 to 0·12 m. [4 to 5 ins.] The latter velocity largely depends on the nature of the bottom, however, as well as on the depth of cut to be taken. For hard clay bottoms the depth of cut is about 0·5 m. [1·64 ft.], while dredging in loose sand is often done with a cut of 2 m. [6½ ft.] in depth, and more.

The size of bucket required naturally depends on the work to be done by the machine in a given time and for a certain velocity of the chain. It may be assumed that the buckets are never filled to more than ½ or ⅔ of their cubic contents. As regards the power required for a dredging machine, a calculation for determining it, based on the work to be performed during the lifting process, would not give a sufficiently large result, even though all wasteful resistances in chains, rollers, gears, etc., which here are very considerable, were to be taken into account. The greater portion of the power is instead absorbed in the work of detaching the material, and it is evident that this amount may easily be estimated from empirical results.

According to *Hagen*, the best steam dredging machines rarely give a performance exceeding 4·45 cub. m. [5·82 cub. yds.] per horse-power per hour. Assuming a dredging depth of 6 m. [19·68 ft.] below, and an additional lift of 5 m. [16·4 ft.] above the water-level, the specific gravity of the raised material being 2, we can calculate the work per second corresponding to the hoisting operation, taking also into consideration the reduction of weight below the surface, to be

$$L = 4\cdot45\frac{2 \times 1000 \times 5 + (2-1)1000 \times 6}{60 \times 60} = 19\cdot8 \text{ kg. m. } [143\cdot2 \text{ ft. lbs.}]$$
$$= 0\cdot264 \text{ horse-power.}$$

Hence we conclude that more than 73 per cent of the total power absorbed are required for detaching the material and overcoming wasteful resistances. It may be noted, however, that according to other authorities the performance of steam dredging machines falls considerably short of that cited above. For further information on this subject we refer to Rühlmann's *Allgemeine Maschinenlehre*, vol. iv.

According to *Hagen*, $\frac{4}{7}$ of the total driving-power are required for the actual lifting, $\frac{1}{3}$ for overcoming wasteful resistances, and the remainder for excavating the material, when the latter consists of sand. At any rate, it is evident from these figures that an increase in the lift above the surface, with a view to attaining a quicker and surer precipitation of the excavated matter into the scow, will but slightly influence the efficiency of the apparatus. It is therefore always advisable to provide ample lift, more especially as the expense connected with the operation of the dredging machine is generally greatly exceeded by that incurred for removal of the excavated matter.

In conclusion, a *land dredge*, as employed by *Couvreux*[1] in the building of the Suez Canal, is shown in Fig. 165. Here

---

[1] See Oppermann, *Portef. économique d. Mach.* 1865, and Rühlmann, *Allgem. Maschinenlehre*, vol. iv.

a frame attached to a car W carries the upper chain-drum with its shaft A, around which the ladder AB is movable, the lower end of the latter being suspended from the crane jib CD by means of the tackle F.   As in a portable crane, a vertical tubular boiler K and a steam-engine E are placed on the car W for the purpose of operating the chain prism by the aid of

Fig. 165.

suitable gearing.   The buckets cut the material from the slope G, and discharge it through the chute R into the transporting car T.   From the figure it is evident how the whole apparatus may be moved along the track S, and also how the pitch of the bucket-ladder may be varied by means of the tackle.   The daily performance of a machine of this kind driven by a 20 horse-power engine is claimed to be 1000 cub. m. [1308 cub. yds.] in 10 hours.

CHAPTER VIII

§ 41. **Pile-Drivers** also belong to the class of machines which
are employed to raise a weight, here called the *ram, hammer,*
or *monkey,* to a certain height, for the purpose of letting it fall
again and drive a pile into the earth by the momentum acquired.
The piles are either used to bear part of the weight of a structure, or to enclose a space, in which latter case they are driven
in close contact, whereas the bearing piles A, B, . . ., Fig.
166, are driven from 20 to 40 inches apart, their heads
being secured to longitudinal timbers or string-pieces C, D, E.

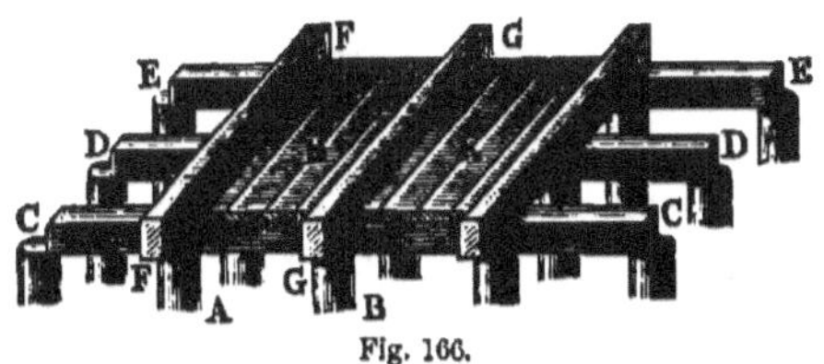

Fig. 166.

Upon these, shorter cross - pieces or sleepers F, G . . . are
placed, together with planking H, K . . ., which fill the
space between the adjoining cross-pieces, the whole furnish-
ing a foundation for the masonry.   The close piles are provided
with grooves, in which tongues are fitted, which assist in giving
a perfect or almost water-tight joint.   The close piles range
from 13 to 33 ft. in length, and from 9 to 18 inches in diameter
at the top, and their feet are pointed or provided with a shoe,
in order that they may more easily penetrate the ground.

The ram with which the piles are driven into the ground
is either made of firm oak or of cast-iron, and weighs from 5
to 15 cwt.   The wooden ram requires iron hoops to prevent it

from splitting. The ram is either raised directly by hand, or by means of a rope passing over a pulley. In the former case it is known as the simple *rammer*, in the latter as the *ringing pile-engine*. In the ringing engine the hauling part of the rope branches out into a number of ropes held by men, who pull together whenever the ram is to be lifted. In the so-called *monkey-engine* the ram is lifted by suitable mechanisms, as, for instance, toothed gearing, etc.

The rammer is but an imperfect contrivance for driving piles. It consists of a block AB, Fig. 167, of oak, provided

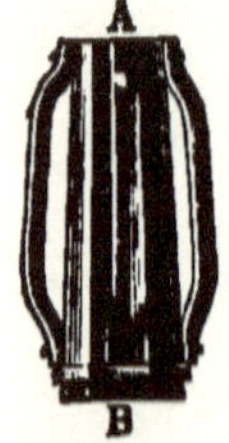

with four long handles, by which four workmen grasp and lift it. Such a ram should not weigh more than 60 kg. [132 lbs.], since 15 kg. [33 lbs.] is all that should be allotted to each man. They are therefore only adapted for driving small piles.

In the ringing engine, Fig. 168, the ram R is provided with arms, which embrace the upright timbers, and serve to guide it when rising and falling. The framework ABC rests upon a portable base ADE, which may be provided with planking for the workmen to stand upon. A pulley H at the top of the upright serves to carry the rope RHK from the ram to the platform. The winch LM is used to place the pile in position.

Fig. 167.

Fig. 169 shows a very simple and efficient pile-engine used in Holland. The frame is composed of three timbers AD, BD, and CD, secured at the top by a bolt; its feet are fitted with iron pins, resting on two planks, as shown in the figure. The ram Q is provided with eight lugs, which embrace the light uprights EF, whose iron feet are either set into the ground or into planks. The frame is made steadier by a guy DG, which passes from the apex of the frame to a post G driven into the ground. The employment of muscular force to work a ringing pile-engine is a very unsatisfactory mode of driving, as after only a short period of such severe exertion the workmen are obliged to take an equal interval of rest.

The efficiency of the pile-driver increases with the weight of the ram, and the height to which it is raised, and therefore, since a large number of men cannot work to advantage at the same time, and the greatest height to which a ram can be raised is only 1½ metres [5 ft.], it follows that the ringing engine is a very imperfect means of performing the work of pile-driving.   These imperfections are, in a great measure,

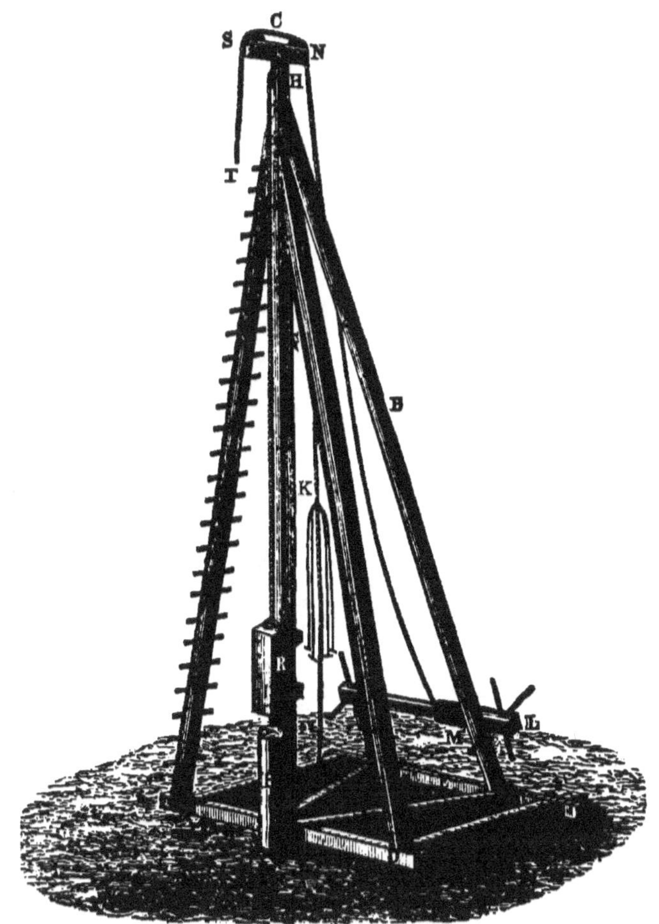

Fig. 168.

avoided in the monkey-engine, as here the workmen can be employed to better advantage in the work of turning cranks, and it is also possible to increase the weight and fall of the ram at will by interposing suitable gearing.   Consequently, this type of pile-driver is decidedly to be preferred to the ringing pile-engine.

The arrangement of a simple monkey-engine is seen from Fig. 170. By means of two winch handles, motion is communicated to the shaft B, and in turn imparted through E and F to the barrel G. Let us suppose the ram Q to have been lifted to a certain height; the crank B is then shifted endwise by means of the lever CDE, the toothed wheel E being thereby thrown out of gear with the wheel F, so that the ram Q can freely fall upon the pile P. A disadvantage of this contrivance is that the rope

Fig. 169.

which sustains the ram rapidly unwinds from the barrel during the fall, and consequently the arrangement is not only liable to get out of order, but the moving parts are besides subjected to great wear. The effect of the blow is also considerably diminished by the frictional resistances of the rapidly-revolving barrel. For this reason it is preferable to attach the rope to the ram by means of a hook, which is automatically detached, and allows the ram alone to fall after it has reached a certain height. The use of a pair of nippers, as shown in Fig. 171, is very expedient. The ram Q is held by a staple in a pair of tongs HOK, which are

secured to the block F at the end of the hoisting-rope. Two
steel springs I, I, pressing against the handles H, H of the

Fig. 170.

nippers, keep the latter closed ; as the block reaches the
summit of the uprights, the handles of the nippers are pressed
together by the slot in the cross-piece above, thus releasing
the jaws KK, and allowing the weight to drop upon the

head of the pile. At this moment the lever CDE throws the toothed wheels out of gear, freeing the block F, which falls, and as it strikes the ram, causes the jaws to open and engage the staple.

With the monkey-engine here described, three to six men are capable of lifting rams weighing from 300 to 800 kg. [660 to 1760 lbs.] to a height ranging from 5 to 10 metres [16 to 33 ft.]

Monkey - engines were formerly driven by tread-wheels, whins, or water - wheels, but at the present day are more generally worked by steam-power. In the classification of steam-pile drivers we must distinguish those that are *direct-acting*, the ram being lifted by the piston-rod as in the steam hammer, from those which are ordinary *monkey pile-drivers* operated by *steam-engines*.

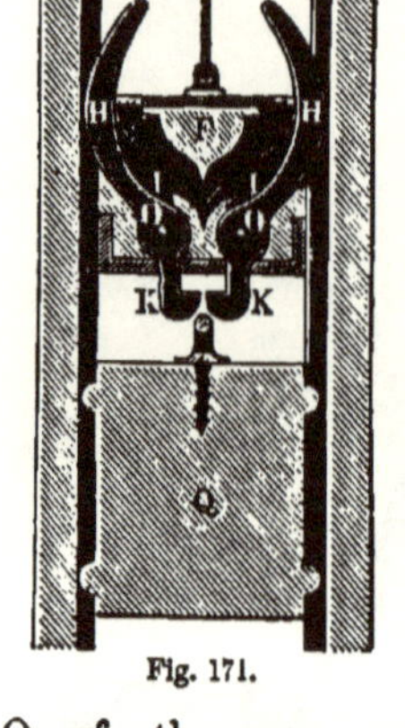

Fig. 171.

§ 42. **Direct - acting Steam Pile-Drivers.**—This type of engine, first constructed by *Nasmyth*, has demonstrated its usefulness and efficiency in heavy service. It is distinguished from the monkey-engine chiefly by the feature that a ram of great weight is lifted to a small height, and that the blows are allowed to follow each other in quick succession. Since the energy of the blow depends upon the product $Q\,h$ of the weight Q of the ram and the height h from which it falls, it follows that no loss of energy is incurred by diminishing h if Q is increased in the same ratio. The arrangement, on the contrary, offers the decided advantage that it admits of making the engine direct-acting, that is, so as to have the ram lifted directly by the piston-rod, an impracticable construction in monkey-engines, owing to the great height to which the rams must be hoisted. The chief benefit derived from the direct-acting pile-driver is due to its rapid action, experience having proved that the piles enter more readily when the blows follow each other in quick succession. The hammer sometimes weighs as much as 50 cwt., and makes from 70 to 80 blows per minute, with a fall of one metre [3¼ ft.]   In the monkey-engine, on the other hand, where only a small num-ber of men can work at the same time, the operation is very slow, such machines, in fact, giving no more than 10 to 40 blows per hour.

The platform A of *Nasmyth's* direct - acting engine, Fig. 172, is mounted on four wheels, and arranged to travel along a railway.   The vertical guide-post C is bolted to the platform A, and is further stiffened by a stay E and a guy D.

The driving apparatus T, consisting of the hammer and steam cylinder, is suspended from a chain K, which passes over the pulley R to the barrel of a windlass W.   The lower part of T is provided with a conical enlargement rest-ing on the head of the pile P; by turning the winch W, the driving apparatus is enabled to follow the pile in its downward course.   A second winch W₁, with chain K₁, is employed to lift the pile in place; both windlasses are operated by a small steam-engine M, which also serves to move the pile - driver along the railway when a new pile is to be driven.   A jointed pipe Q supplies the steam to the cylinder of the driving apparatus, and follows the latter during its downward movement.

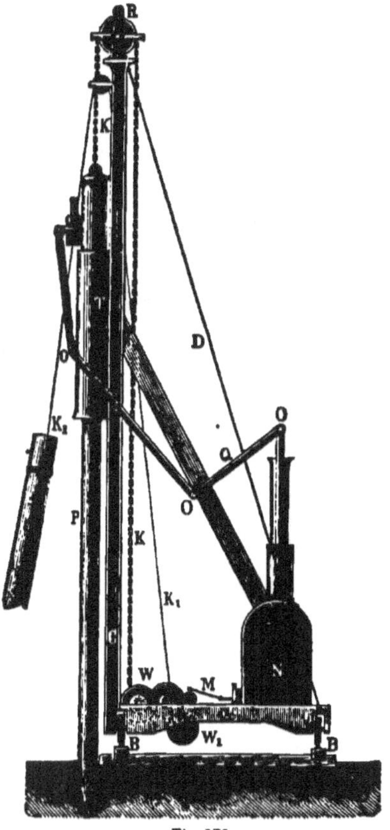

Fig. 172.

The driving apparatus, illustrated in Fig. 173, I and II, consists of the steam cylinder A and the wrought-iron pile-case D, which serves as a guide for the hammer Q, weighing

about 50 cwt., and hung from the end of the piston-rod C.
At F steam enters through the jointed pipe above mentioned

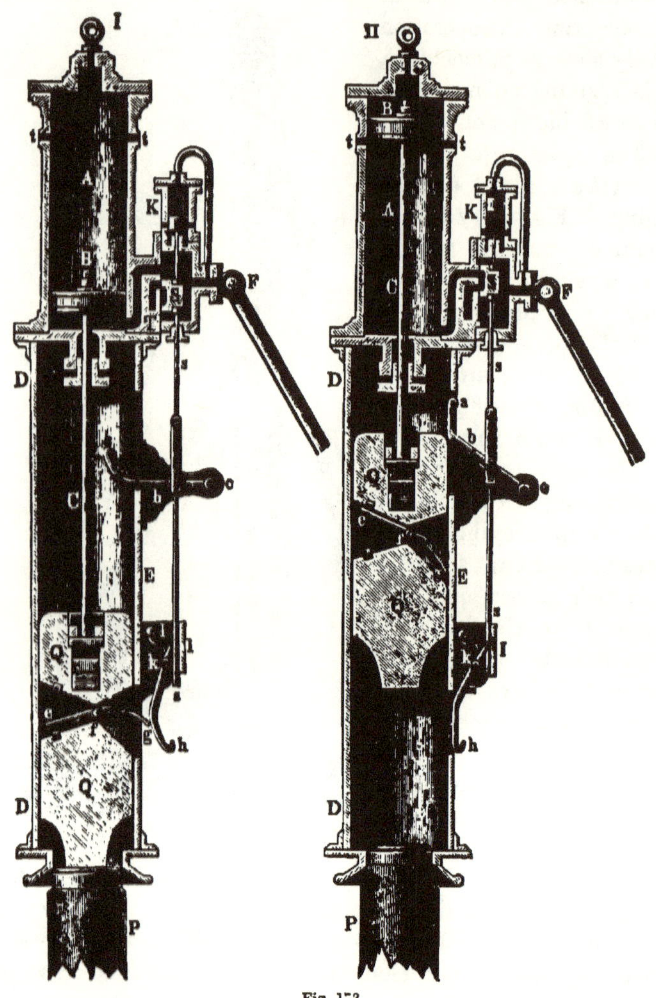

Fig. 173.

to the valve-chamber, in which a slide-valve S effects the
distribution of the steam.   As it is only proposed to *lift* the
piston by the pressure of the steam, the cylinder is made
single-acting.   In position I the steam is admitted from the
valve-chamber to the space below the piston B, thus raising

the hammer Q, which, when approaching the top of its stroke, strikes the valve-lever $a\,b\,c$, and throws it into position II. The lever $a\,c$ passes through a slot in the valve-stem s, by means of which it lifts the valve, the latter being held in its upper position by a trigger k, which springs into a notch at the bottom of the spindle, thus counteracting the tendency of the steam acting on the upper surfaces of the small solid piston K to depress it. It is now evident that in II the steam can escape from the space below the piston B through the interior of the valve into the exhaust pipe at T, and allow the ram to fall upon the head of the pile. In order to follow up the blow immediately, and lift the piston again by allowing fresh steam to enter the cylinder, the latch-lever $e\,f\,g$, pivoted at f, is fitted into a recess in the hammer. While the latter is falling, the lever has position II, but owing to the momentum acquired by the arm $f\,e$ during the fall, the lever is thrown into position I by the force of the blow. The result is that the arm $f\,g$, which protrudes through a mortise in the pile-case E, strikes against the trigger $h\,k$, and releases it from the notch in the valve-spindle s. The valve is now free to obey the downward pressure of the steam on the valve-piston K, and all the parts take position I, whereupon the piston again rises. When the piston is lifted, the upper edge of the mortise in the pile-case causes the lever $e\,f\,g$ to take position II.

To prevent a vacuum being formed above the piston B during the down-stroke, air is allowed to enter through holes made in the cylinder. During the following up-stroke the air can escape through the same holes, until the piston reaches t, when the air above opposes any further motion, and acts as a cushion, thus preventing the piston from striking against the cylinder cover.

As *Nasmyth's* steam pile-driver does not allow the steam to work expansively, at least only in a small degree, another pile-driver has been constructed on the principle of *Daelen's* steam hammer for a better utilisation of the heat-energy of the steam. In the machine referred to the large piston-rod forms the hammer, which is lifted by the pressure of the steam upon the lower surface of the piston. By means of a suitable valve-gear the upper portion of the cylinder freely communicates with the lower portion when the piston is at the end of the up-stroke, so that the steam acts on both sides of the piston during the fall of the hammer. Since the upper surface is expressed by $F = \dfrac{\pi D^2}{4}$, and the lower by $f = \pi \dfrac{D^2 - d^2}{4}$, where D and d represent the diameters of the cylinder and piston-rod, it follows that the excess of pressure on the upper surface above that on the lower surface adds its force to the weight of the falling hammer, and allows the blows to follow each other more rapidly. Moreover, steam is economised, for during the above-mentioned process the steam which was supplied below the piston works expansively in the ratio of $\dfrac{F}{f} = \dfrac{D^2}{D^2 - d^2}$ whenever communication is opened with the upper portion of the cylinder. This is the principle adopted in *Schwartzkopf's* steam pile-driver.[1]

Another construction depends upon the principle of the *Condié* steam hammer. Here the *steam cylinder* is the *hammer*, being guided by the stationary piston-rod suspended from above. The rod is made hollow, and conducts the steam to the upper portion of the cylinder through holes in the piston; when the cylinder has been lifted to a certain height an opening in the hollow rod forms an outlet for the steam. At this instant the cylinder falls, and the air, which was previously compressed in the bottom of it, expands, and increases the force of the blow. Owing to this pneumatic action the guide-frame for the cylinder, from which the piston-rod is suspended, must be *firmly secured* to the head of the pile. This is the arrangement of the steam pile-driver which *Riggenbach* employed in building the railway stations at Biel.[2]

A somewhat different construction, in which, by dispensing with the air-cushion, the necessity of securing the guides to the head of the pile is obviated, was designed by *Lewicki*[3] for the improvement of the Düna near Riga. The details of this machine are here given. Fig. 174 illustrates the driving apparatus, the essential part of which is the heavy hammer or steam cylinder A, which slides on the hollow piston-rod B

[1] *Zeitschr. d. Ver. deutsch. Ing.* 1860, page 224; *Mittheilungen d. Hannov. Gewerbe-Vereins*, 1863, page 243.

[2] *Polytech. Centralblatt*, 1865, page 219.

[3] See *Civil-Ingenieur*, vol. xxi. part i.

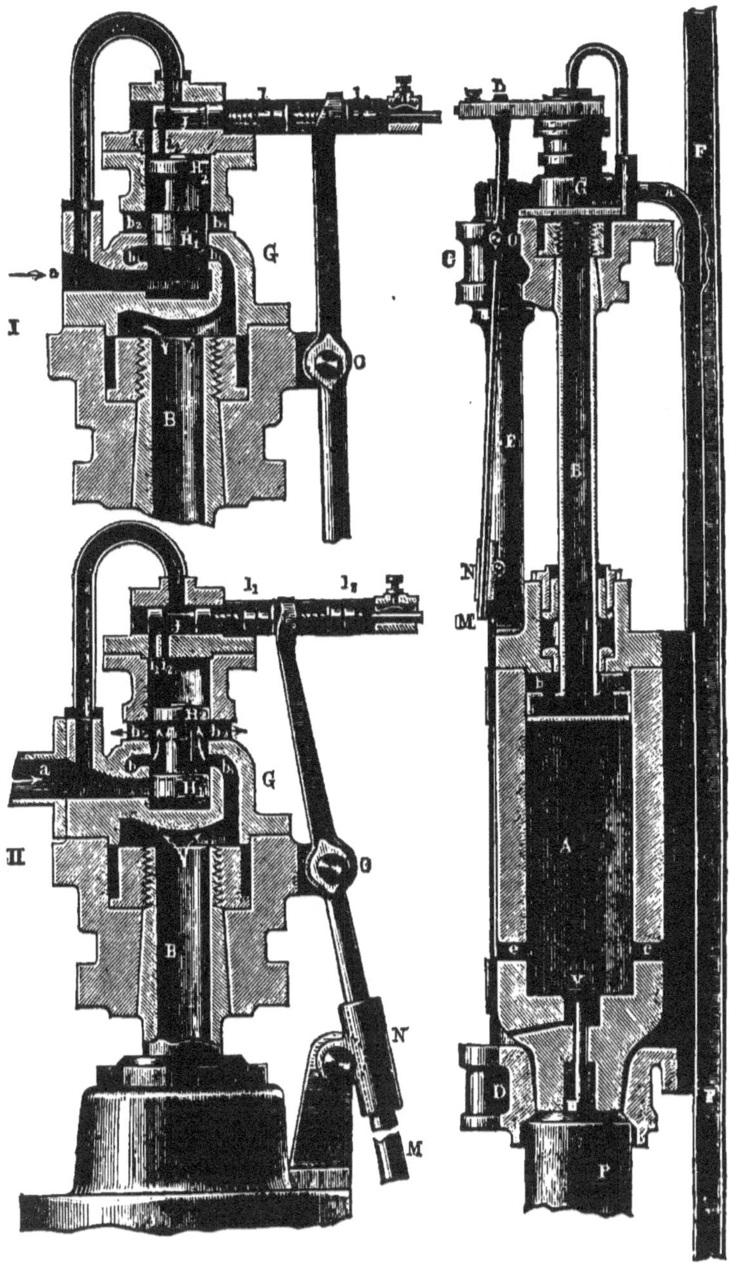

Fig. 175, I and II.

Fig. 174.

suspended from the cross-piece C.　The frame for guiding the cylinder consists of the two cross-pieces C and D bolted together by two wrought-iron rods E.　This frame is guided by the two uprights F belonging to the usual form of pile-driver, and, before the pile-driving begins, is lowered by a windlass, until the lower cross-piece D rests upon the head of the pile P.　By means of a jointed pipe a communicating with a tubular boiler, steam first enters the valve-chamber G, and from here is admitted through the hollow piston-rod B and holes b to the annular space above the piston.　In consequence, the cylinder is lifted, inasmuch as the air below the piston is driven out through the openings e.　If now in the highest position of the cylinder the steam is allowed to escape into the atmosphere through b and the piston-rod, the cylinder falls and drives in the pile P, the driving apparatus following its downward movement.

The peculiar action of the valve-gear is shown in Fig. 175, I and II.　The steam enters, as above mentioned, through the pipe a into the valve-chamber G, in which a double piston-valve $H_1 H_2$ works.　The lower surface of H_1 is always acted upon by the steam-pressure in a, while the upper surface of the somewhat larger piston H_2 is not influenced by the steam in a, unless the small piston-valve J in position II allows the steam to enter through i_2.　Suppose, on the other hand, the valve J to have the position I, the space above H_2 is then in communication with the atmosphere through the opening i_1. From this it is evident that the valve H in position I is lifted by the steam acting on H_1, so that now the steam from the boilers enters at a through the annular passage b_1 and the piston-rod B to the cylinder, thus lifting the hammer.　If, when the latter is in its highest position, the valve J is brought into position II, the valve H will be depressed, owing to the excess of pressure on the larger area H_2.　No steam will then enter from a to B along the above-mentioned route, whereas that in the cylinder can escape into the atmosphere through B, b_1, and b_2, thus liberating the hammer.　To obtain an uninterrupted action of the hammer an automatic device for moving the valve J must be employed, so arranged that when the hammer is at the bottom of its stroke the valve must occupy position I, and when the former is at the top of its stroke the latter must occupy position II. This is accomplished in a neat and simple manner by a lever LOM pivoted at O, whose prongs at L clasp the valve-spindle of J, while the arm OM fits into the sleeve N, which swings on a pivot attached to the cylinder. It is easy to see how, owing to the inclined position of the arm OM, an oscillating motion is given to LM, and also how the adjustable nuts l_1 and l_2 on the valve-spindle of J afford a means of regulating the time at which the motion of the hammer is to be reversed. The function of the small bonnet-valve v in the bottom of the cylinder is to discharge the condensed water, the valve being opened every time a blow is struck on the top of the pile. The air imprisoned between the openings e and the cylinder bottom acts as an elastic cushion.

As to other forms of pile-drivers in which the ram is directly connected with the piston, *compressed air* instead of steam has also been employed for moving the latter. Furthermore, an atmospheric pile-driver,[1] designed by *Clarke* and *Barley*, was employed for building purposes on the Catherine Docks in London. In this pile-driver the rope leading from the hammer was attached to the rod of a piston operating in a cylinder open at the top; in the bottom of the cylinder a vacuum was produced by an air-pump, the hammer being thus lifted by the atmospheric pressure. The difficulty of obtaining air-tight packing in such pneumatic machines is one objection to their extensive use.

In America *Shaw* invented in 1872 a gunpowder pile-driver, in which the raising of the hammer is effected by the explosive force of gunpowder. For this purpose a cast-iron pile-cap is placed on the head of the pile, having a recess in its upper part for the reception of the cartridge. The hammer having been lifted by a windlass, is held in an elevated position by a trigger, which is disengaged by a rope. As the hammer falls a plunger on the under side of it enters the cap, compressing the air therein sufficiently to explode the cartridge. The explosion throws up the hammer, while the reaction of the gases drives the pile to a certain depth. For rapid work a fresh cartridge is thrown into the cap during the ascent of the hammer, which drops again immediately by the action of gravity.

---

[1] *Der Ingenieur*, vol. ii. (first series).

But with rapid firing the cap becomes so hot as to ignite the cartridge before the descent of the hammer. For this reason the hammer is held at its highest point by the above-mentioned trigger until the release of the latter allows the hammer again to fall. A piston at the top of the uprights enters a cavity in the hammer during its upward motion, and acts as a buffer. These pile-drivers, which were exhibited at the Centennial Exhibition at Philadelphia, 1876, are extensively used in the United States. Besides their simplicity and capacity, it is further stated of these pile-drivers that the heads of the piles are not injured by the blows, and need no protection against splitting.

According to *Knight*,[1] the weight of the cartridge is only one-third of an ounce (9·5 g.) for a hammer weighing 675 lbs. (308 kg.) It is stated in a report made to the Franklin Institute by *Prindle* that experiments were made in driving piles with the cartridge and without it, the fall being the same (15 ft.) in both cases, and the penetration of the pile recorded. Results showed that when a cartridge was employed the penetration of the pile was always considerably greater, sometimes four, and even eight times as great as when the hammer worked, as in the ordinary monkey-engine without the cartridge.

§ 43. **Pile-Drivers with Steam-Engines.**—The frequent interruptions to which direct-acting steam pile-drivers are subject when at work are probably the principal reason why of late there has been a return to monkey-engines driven by steam, *i.e.*, drivers in which the hammer or monkey is raised to a great height by hoisting apparatus. In certain cases, especially for very heavy piles, this form of driver is indispensable, because then very powerful blows are needed, which cannot be obtained by the direct-acting steam pile-drivers, on account of their short strokes. The stroke of the hammer in these drivers must always be small, for constructive reasons, and seldom exceeds 1 metre [3·28 ft.] On the other hand, the stroke or fall in the monkey-engine is limited only by the height and strength of the frame. Of course, the blows of the monkey-engine are delivered much less rapidly than those of the steam-drivers, which can deliver up to 120 blows per

---

[1] *American Mechanical Dictionary*, p. 1041.

minute. Under certain circumstances, depending upon the nature of the ground, the rapid succession of blows of the steam pile-driver is very advantageous, and experience has shown that for driving small piles this rapid delivery is better suited than the monkey-engine. On the other hand, for long and heavy piles, the powerful blows of the monkey-engine are better.

Another reason why pile-drivers driven by steam-engines are becoming more popular is that their arrangement is simple, the hammer being lifted by a hoist that is worked by an engine, say a portable one, perhaps already in use for other purposes. For transferring the motion from engine to hoist, belts or chains are used.

Such an arrangement is employed in *Schwartzkopf's* pile-driver, Fig. 176. Here a *Clissold* chain K (see vol. iii. 1, § 65, Weisb. *Mech.*) transmits the motion from a portable engine M to a hoist

Fig. 176.

W, which raises the hammer by means of the rope t. This hoist is composed of two drums B and C, turning loosely on the shaft A; one drum B serves to lift the hammer, and the other C to place the pile in position for driving. The drums are turned by means of two friction-couplings, which

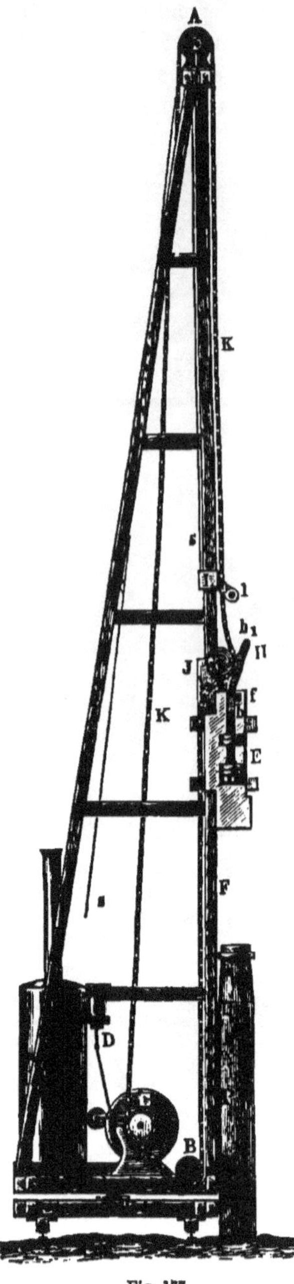

Fig. 177.

receive their motion from a driving-pulley D fastened to the shaft A, and receiving continuous rotation from the portable engine by means of a chain with wedge-shaped links. By shifting the driving-pulley D and its shaft in one or the other direction by means of the screw S and hand-wheel R, sufficient friction arises at the conical surface of contact between D and the drum to wind up the rope that carries the load. When the hammer has been raised to the desired height, a slight reversing of the hand-wheel R will remove the friction between D and B, and the hammer will at once fall, causing the drum B to turn backward. As stated above, this action always weakens the force of the blow to some extent.

To avoid the last-mentioned defect and render the pile-driver automatic, an endless *link-chain* is sometimes employed. This chain is constantly moving upward between the guides, one of the chain-bolts taking hold of a claw on the hammer when the latter is in its lowest position, and then lifting it till the self-acting disengaging device releases the claw and allows the hammer to fall. During the fall the chain is continually moving, and after the blow has been delivered it again lifts the hammer. This pile-driver was applied by *Sisson* and *White*, and was materially improved by *Eassie*.[1] The construction of the apparatus is shown in Fig. 177.

Gall's link-chain K, which lifts the hammer, is guided over the top pulley A, the guide pulley B, and a chain-wheel C, receiving from the latter continuous motion; the wheel in turn receives its motion by means of gearing from the fly-wheel shaft of a portable engine D. The hammer E is guided in the usual manner between the uprights F; a bolt of the chain takes hold of the claw H, which is pressed against the chain K by the spring *f*. In order that no one-sided pull of the chain may cause a cramping action on the hammer, a frame J is employed, which rests on the hammer, embraces the guides F, and carries the guide-roller G, thus guiding the chain so that its pull will be directed through the centre of gravity of the hammer. To release the latter at the desired height, there is a second frame L enclosing the uprights, which frame is kept in position by the rope *s*, its friction-roller L striking against the projection h_1 of the hook H, thus forcing the hook out of the chain, and permitting the hammer to fall. The frame N can be shifted on wheels *r*, and can also be turned about the centre pivot Z. In this pile-driver a hammer, weighing one ton, and falling 4·27 metres (14 ft.), gave very favourable results. In a similar pile-driver,[2] with a hammer weighing one ton, and falling 1·2 to 1·5 m. [4 to 5 ft.], 9 to 10 blows were delivered per minute.

§ 44. **Work of Pile-Drivers.**—The expenditure of work required to drive piles can be calculated from the weight G of the hammer, its fall *h*, and the number of blows *n* needed to drive a pile. This work per blow is given by $A = Gh$, and, therefore, per pile it is $nA = nGh$. The ringing pile-driver is a very imperfect means of performing this work, for experience shows that the useful effect exerted by a labourer pulling down vertically on the rope of the driver is much less than when the labourer is turning a crank. In the first case the exertions of the workmen are so great that for a heat lasting from 40 to 50 seconds, a rest of 2 to 3 minutes is necessary. Another disadvantage is that, as there are a large number of labourers

---

[1] See *Instit. of Mechanical Engineers, Proceedings*, 1867, p. 255; and Rühlmann, *Allgem. Maschinenlehre*, vol. iv. p. 505.

[2] *Zeitschr. d. Hannov. Archit.- u. Ingen.-Vereins*, 1869, p. 279.

(twenty-five to forty) simultaneously pulling on the main rope, the individual hauling lines make a considerable angle a with the vertical direction, thus utilising only the component $P\cos a$ of the pulling force P of a labourer. The latter disadvantage can, in a large measure, be overcome by attaching a horizontal ring to the main rope, and connecting with it the hauling lines, which will then hang vertically.

The hurtful resistances in the ringing engine are comparatively smaller than in the monkey-engine. In the former the resistances are principally due to the stiffness of the main rope and the journal friction of the pulley over which the rope passes. Moreover, when the rope acts on the hammer to one side of the centre of gravity, friction is developed in the guides, the lifting is rendered more difficult, and the effect of the blow is diminished. As the diameter of the main rope of the pile-driver is not usually more than 50 millimetres (2 ins.), these resistances may, according to § 7, be taken as at least equal to 5 per cent of the effective work.

In the monkey-engine we must first deduct the resistances of the hoist or windlass. As the windlass works with a chain, and usually employs a single purchase, about 15 to 16 per cent must be deducted for these resistances. But besides the ram of weight G, the nipper block with appendages must be lifted, the weight of which may be assumed at about 8 to 10 per cent of that of the ram; as this, however, adds nothing to the weight of the fall, no essential error will occur in taking the efficiency of the monkey-engine between 75 and 80.

According to the observations of *Köpke* in building the Custom-house at Harburg, the average work performed by a labourer in ten working hours per day was, with the ringing-engine,

371,950 ft. lbs. = 51,423 metre kilograms ;

and with the monkey-engine,

1,167,970 ft. lbs. = 161,461 metre kilograms.

Lahmeyer gives the network, after deducting the wasteful resistances, at

396,380 ft. lbs. = 54,800 metre kilograms

for the ringing-engine, and

$$792,780 \text{ ft. lbs.} = 109,600 \text{ metre kilograms}$$

for the monkey-engine; so that the work obtained from the latter may be assumed to be from two to two and a half times as much as from the former.

In the direct-acting steam pile-driver the pressure of the steam upon the piston, after deducting the friction of the latter, and in the stuffing-box must exceed the weight of the hammer in order to overcome the friction in the uprights, and impart to the hammer the acceleration necessary to give the desired number of blows per minute. Owing to this acceleration, the hammer, at the instant at which the steam escapes into the atmosphere, has attained a certain velocity v. The energy thus stored in the hammer is expended in lifting it to an additional height before falling. In order to determine the formulas applicable to this case, let F denote the area of the piston of a *Nasmyth* machine, G the weight of the hammer, including the piston and its rod, and h the length of stroke. Further let p represent the effective pressure per unit of area of the steam in the cylinder, and p_0 the atmospheric pressure, and let us reduce all the hurtful resistances, consisting of the friction of the piston, the stuffing-box, and the ram in the uprights, to an equivalent pressure f per square inch, acting on the piston. Let us conceive the piston to have passed upward a distance s_1, then the effective energy exerted by the steam on the piston will be

$$A = F(p - p_0)s_1 \; ;$$

this is expended in lifting the ram to a height s_1, in overcoming the wasteful resistances through the same path s_1, and in accelerating the hammer of weight G. Supposing the hammer to have attained the velocity v, we deduce the equation

$$F(p - p_0 - f)s_1 - Gs_1 = G\frac{v^2}{2g} \quad . \qquad . \qquad . \quad (1),$$

from which we obtain the velocity v of the ram at the moment at which the steam escapes from the cylinder:

$$v = \sqrt{2g}\sqrt{\frac{F(p - p_0 - f) - G}{G}\, s_1}$$

$$= 4 \cdot 429 \sqrt{\frac{F(p - p_0 - f) - G}{G}} s_1 . \qquad . \qquad . \quad (2)$$

($g = 9 \cdot 8087$ in metres and $32 \cdot 187$ in feet).

By virtue of this velocity v the hammer is able to lift itself to an additional height s_2, which follows from

$$G \frac{v^2}{2g} = (G + Ff)s_2 \text{ to}$$

$$s_2 = \frac{G}{G + Ff} \frac{v^2}{2g} \qquad . \qquad . \qquad . \qquad . \quad (3).$$

If in this formula we substitute the value given in (1) we obtain

$$s_2 = \frac{F(p - p_0 - f) - G}{G + Ff} s_1 = ks_1 \qquad . \qquad . \qquad . \quad (4),$$

where

$$\frac{F(p - p_0 - f) - G}{G + Ff} = k.$$

Therefore the total lift of the hammer will be

$$h = s_1 + s_2 = \frac{G + Ff + F(p - p_0 - f) - G}{G + Ff} s_1 = \frac{F(p - p_0)}{G + Ff} s_1 \quad . \quad (5).$$

If now for a given lift h the hammer is to make n strokes per minute, so that the time occupied by each up and down stroke is given by $t = \dfrac{60}{n}$ seconds, we can determine the relations in the following manner. The whole period t for each double stroke is composed of four intervals of time, $t_1 + t_2 + t_3 + t_4$, where t_1 represents the time during which the piston is driven by the steam, and rises through the distance s_1, and t_2 is the time required by the hammer to lift itself to the height s_2 by virtue of its living force ; t_3 is the time required for the fall, and t_4 an interval needed for the blow to exert its effect. The latter interval must be assumed according to circumstances, but the values t_1, t_2, and t_3 can be calculated. The constant pressure $F(p - p_0)$ exerted during the time t_1 gives a uniformly accelerating motion to the piston, the acceleration being expressed by

$$g_1 = \frac{F(p - p_0 - f) - G}{G} g,$$

so that the time required to lift the hammer through a distance s_1 is

$$t_1 = \sqrt{\frac{2s_1}{g_1}} = \sqrt{\frac{2s_1}{g}\frac{G}{F(p-p_0-f)-G}} \qquad . \qquad . \quad (6).$$

The additional motion of the hammer during the time t_2 is uniformly retarded; the retardation, as is obvious from (3), is expressed by

$$g_2 = \frac{G + Ff}{G}g,$$

hence the time is given by

$$t_2 = \sqrt{\frac{2s_2}{g_2}} = \sqrt{\frac{2s_2}{g}\frac{G}{G+Ff}} \qquad . \qquad . \qquad . \quad (7).$$

Finally, the acceleration during the fall through the height $h = s_1 + s_2$ is

$$g_3 = \frac{G - Ff}{G}g,$$

and the time required

$$t_3 = \sqrt{\frac{2h}{g_3}} = \sqrt{\frac{2h}{g}\frac{G}{G-Ff}} \qquad . \qquad . \qquad . \quad (8).$$

If now in any special case the weight of the ram G, the area of the piston F, and the length of stroke h are given, we can, by assuming the effective pressure $p - p_0$, obtain from equation (5) the path s_1 described by the piston. But we have yet to see whether, with the chosen data, the sum of the intervals computed from equations (6), (7), and (8) is less than the assumed time $t = \dfrac{60}{n}$ of a complete operation. If such is the case, the ram must remain in contact with the pile during a certain interval, be it ever so small (0·1 second). If this condition is not fulfilled, we may obtain the necessary requirements, say by altering the assumption as regards the effective pressure $p - p_0$.

In this investigation the expansion of the steam and the compression of the air while being dispelled are not considered. If it is desired to include these factors, we must apply the rules and formulas given in vol. ii., Weisb. *Mech.*, on steam-engines.

We may also follow the method given by *Lewicki*[1] in the article already referred to. Although the application of steam to the direct-acting machine is not very economical on account of the absence of expansion, still this is generally of little importance when compared with the greater expense attending the hand-power pile-driver. Moreover, as all interruptions in the work are attended with drawbacks, it is best in pile-drivers, and in all machines employed for building purposes, to lay more stress upon the simplicity of the construction than upon the efficiency of the machine.

In the preceding remarks attention has only been given to the amount of energy required for lifting the ram, and the corresponding efficiency, of course, applies only so far as the pile-driver is considered a *hoisting machine*. The action of the falling ram upon the pile, the forcing of the latter into the ground, and the compression of the ground, are effects which can be ascertained only by practical experience. We must therefore refer to technical journals for information upon these subjects.

We will note here, however, a fact concerning the comparative action of the different types of pile-drivers and the influence of the height of stroke or fall. When the ram of weight G_1 falls from a height h, this motion represents mechanical work to the amount of $A = G_1 h$, which, leaving frictional resistances out of account, will produce a velocity $v = \sqrt{2gh}$ in the hammer at the moment of impact. Now let G_2 denote the weight of the pile; then, if the bodies be considered inelastic, the amount of energy which disappears after the blow is determined from § 335, vol. i., Weisb. *Mech.*, and is given by

$$A' = \frac{G_1 G_2}{G_1 + G_2} h = A \frac{G_2}{G_1 + G_2} ,$$

while the remaining actual energy imparted to the pile is

$$A'' = A - A' = A \frac{G_1}{G_1 + G_2} ,$$

the latter being utilised in driving; that is, in giving *motion* to the pile. The lost energy A' is essentially employed in doing molecular work, for example, in injuring the pile, which

---

[1] *Civil-Ingenieur*, vol. xxi. part i.

represents work foreign to the purposes of the machine, and we must therefore seek to make this loss as small as possible.   It is obvious that this value

$$A' = A\frac{G_2}{G_1 + G_2} = A\frac{1}{\dfrac{G_1}{G_2} + 1}$$

expresses a smaller fractional part of A, the greater the ratio of the weight of the ram to the weight of the pile.   Were we to assume $G_1 = G_2$, the loss would become $A' = \frac{1}{2}A$, while in ordinary cases where a value from $G_1 = 2G_2$ to $2\frac{1}{2}G_2$ is chosen, the value of A' varies from $0\cdot33\,A$ to $0\cdot268\,A$.   If now a certain amount of work A has to be expended in lifting the hammer G_1 to the height h, it follows that the *loss due to impact will be a larger fraction of this work, the greater the lift h*, i.e. *the smaller* G_1, and consequently, for the same expenditure of work A, the percentage of loss due to impact in the monkey pile-driver is greater than in the steam pile-driver, where the lift h is smaller, and accordingly the weight G_1 greater.   This result agrees with the fact observed in practice that considerable injury is done to the piles when monkey pile-drivers are used.

Now, although the impact of the hammer on the pile is not perfectly inelastic, and consequently the amount A' is not wholly lost, yet the loss due to impact will be greater the smaller the ratio of G_1 to G_2.

Concerning the load which a driven pile will sustain, we must refer to Weisb. *Mech.*, vol. i. § 371, Ger. ed., for more explicit information.   For practical purposes it is advisable, however, to base such estimates upon empirical results rather than upon theoretical deductions.   (See Hagen, *Handbuch der Wasserbaukunst*, vol. i. page 628.)

INDEX

ACCUMULATORS, 125
Action of accumulators, 138
Adriany's hoist, 188
Armstrong's accumulator, 129, 132
   hydraulic crane, 255
Atmospheric pile-driver, 319

BALANCE-CARRIAGE, 199, 268
Balancing the hoisting rope, 197
Bernier's windlass, 100
Borgsmüller's safety catch, 214
Both's dredge, 294
Brake winch, 162
Brakes for lowering, 160
Bridge, 274
Brown's steam crane, 270
Büttenbach's safety catch, 214

CAPSTAN, 94
Cavé's crane, 230
Chinese windlass, 89
Clark's sheers, 255
Clissold chain, 321
Collet and Engelhard's tackle, 68
Compressed-air hoist, 193
   air pile-drivers, 319
Condition of equilibrium of cranes, 238
Conical drums, 199, 206
Cotton rope cranes, 280
Couvreux's dredge, 298, 306
Cranes, 228

DIAMETER of hoisting drums, 79
Differential pulley blocks, 59
   screw-jacks, 35
   windlass, 89
Direct-acting steam pile-driver, 312
Double dredges, 301
   portable crane, 266
Dredging hoist, 112
   machines, 295
Drops, 163

EADE's tackle, 69
Eassie's pile-driver, 322

Edoux lift, 121
Efficiency of differential pulley blocks, 64
   fixed pulleys, 42
   gear shafts, 17
   gear teeth, 14
   hoisting drums, 83
   levers, 11
   mine hoists, 207
   movable pulleys, 46
   pneumatic hoists, 158
   screw gearing, 30
   tackle, 54
Elevator for flour mills, 106
Equilibrium of cranes, 238
Excavators, 286
Eytelwein's formulas, 41

FAIRBAIRN's hydraulic crane, 258
   rotary crane, 234
Floating crane, 229
Fontaine's safety catch, 214
Foundry cranes, 237
French lever-jack, 7
Furnace-lift, 103, 108

GANTRY, 274
Gear wheels, 11
Geared hoisting engine, 196
   mine winch, 173
German lever-jack, 7
Gibbons' pneumatic lift, 154
Gjers' pneumatic lift, 155

HAND- and horse-gins, 174
Hand dredge, 298
   lift, 107
Handle dredge, 286
Hick's experiments, 149
Hohendahl's safety catch, 216
Hoisting machines for mines, 160
   tackle, 46
Holroyd's dredge, 294
Hoppe's safety brake, 218
Horse dredge, 296
Horse gins, 174

Hydraulic cranes, 255
  hoists, 114
  jacks, 116
  lifts, 132
  regulator, 224

INCLINED lifts, 105, 109
  planes, 167

JACKS, 7, 18
Jack screws, 25

KÖPKE'S experiments, 324
Krauss and Kley's safety catch, 217

LAHMEYER'S experiments, 324
Land dredges, 298
Lever, 6
  jacks, 7
Lewicki's steam pile-driver, 316
Life-saving apparatus, 160
Lifts, 102
Locomotive lift, 122
Lohmann's safety catch, 217
Longitudinal dredging, 304
Long's hoist, 92
Lowering crane, 262

MAN-ENGINES, 219
Masting sheers, 252
Mine hoists, 169
  winch, 170, 173
Monkey pile-driver, 310
Movable pulleys, 42

NASMYTH'S pile-driver, 312
Neustadt's crane, 99
Nippers, 312

OTHER forms of hoists, 89
  of tackle, 67

PILE-DRIVERS, 307
  with steam-engines, 320
Plunger friction, 149
Pneumatic hoists, 153
Portable cranes, 264
  hoisting engines, 193
  steam crane, 269, 270
Power cranes, 278
Pressure reservoirs, 120
Prindle's experiments, 320

Pulleys, 38

RACK and pinion jacks, 18
Radial dredging, 302
Railroad crane, 234, 279
Ram, 307
Rammer, 308
Ramsbottom's cranes, 280
Relief-valves, 137, 258
Reversing wheel, 181
Riggenbach's steam pile-driver, 316
Ringing pile-engine, 308
Ritter's hydraulic crane, 262
Roller-bearing for crane-post, 231
Rotary cranes, 230

SAFETY apparatus, 211
  brakes, 212, 218
  catches, 212
Saxon horse-gin, 177
Schwartzkopf's pile-drivers, 316, 321
Scoop dredge, 286
Screw-jack, 25
Shaw's pile-driver, 319
Sheers, 248
Sisson and White's pile-driver, 322
Sparre's safety catch, 217
Spiral drums, 199, 206
Stability of portable cranes, 267
Stauffer's windlass, 101
Steam hoists, 88
  hoisting engines, 191
Swedish lever-jack, 8

TACKLE, 46
Transverse dredging, 302
Travelling cranes, 273
Turbine hoist, 184

VERTICAL drums, 174
  hoisting engines, 195

WAREHOUSE crane, 235
  lifts, 134
Water-power hoists, 180
  pressure hoists, 188
Weighing crane, 236
Weston blocks, 59
White and Grant's safety catch, 214
White's tackle, 55
Windlasses, 12, 74
Work of pile-drivers, 323